This volume brings together authoritative articles summarizing the current state of knowledge of the genetic organization of bacterial populations. Contributions include the evolution of *E. coli* populations, the roles of gene transfer and recombination, the distribution and evolution of plasmids and antibiotic resistance, population genetics of phase-variable systems, population dynamics and epidemiology of bacterial pathogens, and plasmid transfer in the natural environment. Similarities to and differences from the population genetics of higher organisms are emphasized throughout, and the impact of bacterial studies on current evolutionary theories highlighted. The consequences of population structure in clinical practice, environmental situations and commercial exploitation of bacteria are also considered.

POPULATION GENETICS OF BACTERIA

SYMPOSIA OF THE
SOCIETY FOR GENERAL MICROBIOLOGY

Series editor (1991–1996): Dr Martin Collins, Department of Food Microbiology, The Queen's University of Belfast
  Volumes currently available:

CONTENTS

CONTRIBUTORS

BARCUS, V. A. Institute of Cell and Molecular Biology, Darwin Building, King's Buildings, Edinburgh EH9 3JR, UK

BENNETT, P. M. Department of Pathology and Microbiology, University of Bristol, Bristol BS8 1TD, UK

DEVINE, K. M. Department of Genetics, Trinity College, Dublin 2, Ireland

DYKHUIZEN, D. E. Center for Microbial Ecology, Plant and Soil Science Building, Michigan State University, East Lansing, MI 48824-1325, USA

FEAVERS, I. M. Division of Bacteriology, National Institute for Biological Standards and Control, Blanche Lane, South Mimms, Potters Bar, Herts EN6 3QG, UK

FEIL, E. Microbial Genetics Group, School of Biological Sciences, University of Sussex, Falmer, Brighton BN1 9QG, UK

FOSTER, P. L. Department of Environment Health, BU School of Public Health, University of Boston, S107A Boston, USA

JORDING, D. Universität Bielefeld, Biologie VI (Genetik), Postfach 100131, D-33501 Bielefeld, Germany

LENSKI, R. E. Center for Microbial Ecology, Plant and Soil Science Building, Michigan State University, East Lansing, MI 48824-1325, USA

LEVIN, B. R. Emory University, Department of Biology, Atlanta, Georgia GA 30322, USA

MCKANE, M. Department of Biological Sciences, The University of Iowa, Iowa City, Iowa 52242-1324, USA

MAIDEN, M. C. J. Division of Bacteriology, National Institute for Biological Standards and Control, Blanche Lane, South Mimms, Potters Bar, Herts EN6 3QG, UK

MAYNARD SMITH, J. Microbial Genetics Group, School of Biological Sciences, University of Sussex, Falmer, Brighton BN1 9QG, UK

MILKMAN, R. Dept of Biological Sciences, The University of Iowa, 138 Biology Building, Iowa 52242-1324, USA

MURRAY, N. E. University of Edinburgh, Institute of Cell and Molecular Biology, Darwin Building, King's Buildings, Mayfield Road, Edinburgh EH9 3JR, UK

NOLAN, N. C. Department of Genetics, Trinity College, Dublin 2, Ireland

O'ROURKE, M. Microbial Genetics Group, School of Biological Sciences, University of Sussex, Falmer, Brighton BN1 9QG, UK

PICKUP, R. W. Institute of Freshwater Ecology, Windermere Laboratory, The Ferry House, Far Sawrey, Ambleside, Cumbria LA22 0LP, UK

PÜHLER, A. Biologie VI (Genetik), Universität Bielefeld, Postfach 100131, D-33501 Bielefeld, Germany

SAUNDERS, J. R. Department of Genetics and Microbiology, University of Liverpool, PO Box 147, Liverpool L69 3BX, UK

SHARP, P. M. University of Nottingham, Department of Genetics, Queens Medical Centre, Nottingham NG7 2UH, UK

SIMON, R. Universität Bielefeld, Biologie VI (Genetik), Postfach 100131, D-33501 Bielefeld, Germany

SMITH, N. H. Microbial Genetics Group, School of Biological Sciences, University of Sussex, Falmer, Brighton BN1 9QG, UK

SPRATT, B. G. Microbial Genetics Group, School of Biological Sciences, University of Sussex, Falmer, Brighton BN1 9QG, UK

WERNER, S. Universität Bielefeld, Biologie VI (Genetik), Postfach 100131, D-33501 Bielefeld, Germany

WHITTAM, T. Department of Biology, 208 Mueller Laboratory, Pennsylvania State University, University Park, PA 16802, USA

WILKINS, B. M. University of Leicester, Department of Genetics, Adrian Building, University Road, Leicester LE1 7RH, UK

ZHOU, J. Microbial Genetics Group, School of Biological Sciences, University of Sussex, Falmer, Brighton BN1 9QG, UK

EDITORS' PREFACE

Do bacteria have population genetics? Since the subject matter of population genetics is natural genetic diversity, its causes and its consequences, bacteria must, of course, have population genetics of some kind. The pattern of genetic variation in bacteria, as in all organisms, results from the fundamental processes of mutation, recombination, selection, drift and migration. The question is whether conventional population genetics has so far offered an adequate description of bacterial populations. The answer is that it has not. Bacteria are too remote from the 'ideal' sexually reproducing diploids that most of the theory take as a starting point. We need to develop a theoretical framework that takes account of bacterial biology, but a clearer description of the relevant phenomena is needed before that can be done. That is one of the aims of this book.

Several authors tackle recombination, which is a difficult issue for bacterial population genetics. Of course, there are several well-studied mechanisms that permit the transfer of genetic information from one bacterial lineage to another, but it is not clear how we should model their effects at the population level. There is no direct relationship between reproduction and recombination, as in a sexual eukaryote, so empirical evidence is needed on many questions. Are bacterial populations largely clonal or freely recombining? Which transfer mechanisms are actually important in evolution? How large are the segments that recombine? Do plasmid-encoded genes move more freely than those on chromosomes? Do bacteria have discrete species defined by gene exchange? How important is wide-range lateral transfer of genes between distantly related bacteria?

Research into mutation and selection also comes in a distinctive bacterial flavour, partly because the relative simplicity of bacterial genetics allows a high degree of sophistication in the design and interpretation of experiments. Population geneticists normally treat the mutation rate as a constant, and certainly as uninfluenced by selection. However, a lively controversy is now raging over the interpretation of experiments that suggest that some mutations in bacteria are observed much more frequently when a relevant selection pressure is applied. On the other hand, careful studies on bacteria are now providing hard data on the selective effects of differences at the level of individual enzymes. This was, at one time, a controversial topic in eukaryote population genetics. The use of enzyme electrophoresis revolutionized our concepts of natural genetic diversity in bacteria, just as it did in eukaryotes. Its DNA-based successors are now providing a wealth of information on the spatial and temporal structure of bacterial populations, as well as on phylogenetic relationships and rates of evolution.

Bacterial population genetics does not yet have a comprehensive theoretical foundation. However, the last 20 years of observations, experiments and ideas have radically altered and refined our concepts of bacterial populations and their evolution. This seems a good moment to take stock, and an account of our current state of knowledge on all the above questions, and more, can be found in this book.

Understanding bacterial populations is an important intellectual goal, but it is also of real practical importance. The spread of antibiotic resistance genes through many species of medically important bacteria is a population genetics phenomenon that is familiar to everyone. How exactly did it happen? If we had known 40 years ago what we know now, could we have devised a strategy for the deployment of drugs that would have prevented or delayed this spread? Is there still time to do something, or do we have to accept that the golden era of therapeutic antibiotics is drawing to a close? The recent growth of interest in microbial biotechnology has released a host of wider, if more nebulous concerns. If bacteria with novel combinations of genetic traits are to be released into the environment, whether deliberately or accidentally, what will be their fate, what will happen to the genes that they carry, and what will be the effect on the indigenous microbial community? Of course, there are no universal answers to such questions, nor is it ever likely to be possible to predict in detail the future trajectory of any system as complex as a natural biological community. Nevertheless, these questions draw attention to our lamentable ignorance and need to be answered not only by carefully monitored case studies but also with reference to reliable intellectual framework. As the science of bacterial population genetics matures over the next few years, its success will be judged not only by its conceptual elegance but by its ability to offer useful guidance in these practical applications.

This book, then, should be of interest to environmental and medical bacteriologists, as well as to bacterial geneticists and taxonomists. Indeed, if they can be persuaded to read it, it should also be of interest to evolutionary biologists who know nothing about bacteria, because bacteria offer unrivalled opportunities for experimental research into general evolutionary questions. Bacteria certainly have many special features, but they share in the universal characteristics of life, and the combination of small size, rapid reproduction and sophisticated genetics makes it feasible to study aspects of the evolutionary process 'in the test-tube'. Population genetics will continue to contribute to our understanding of bacteria, but bacteria will also provide new insights in population genetics and evolutionary biology. We must thank the authors for providing such an authoritative summary of achievements so far, and hope that this book may help to guide the next phase.

DO BACTERIA HAVE POPULATION GENETICS?

JOHN MAYNARD SMITH

School of Biological Sciences, University of Sussex

WHAT IS POPULATION GENETICS GOOD FOR?

Population geneticists study the variability of natural populations, and formulate theories to account for that variability. The subject first developed in an attempt to reconcile Darwin's theory of evolution by natural selection and the growing science of Mendelian genetics. After the rediscovery of Mendel's laws, the two theories were at first seen as being in conflict: Darwinists stressed selection acting on continuous and graded variation, whereas Mendelians stressed discontinuous and sudden change. It was the great achievement of the early population geneticists to show that this contradiction was only apparent, and to develop a coherent theory of evolution based on modern genetics.

Since the subject first developed to provide a theoretical basis for evolution, and since its formulation required more mathematical expertise than most branches of biology, it is sometimes seen as an esoteric branch of science, of some academic interest but little practical importance. It would, I think, be dangerous if bacteriologists were to take this view, essentially because evolutionary changes in bacteria are of growing practical importance. The short generation times and immense population sizes of bacteria imply that evolutionary changes will be rapid. Even in higher organisms, the spread of insecticide resistance in many insect species has taught us that not all evolutionary changes are too slow to matter. In bacteria, the spread of resistance to antibiotics has now reached a stage at which we can foresee an end to the era in which antibiotics are an effective means of controlling infectious disease. It is perhaps inevitable that such a stage should have been reached, but a better appreciation of population genetics would have ensured that antibiotics continued to be effective for longer. With the release of genetically engineered organisms, and the industrial use of bacteria, for example, in the disposal of polluting byproducts, it is crucial that we know enough population genetics to predict the likely results of our actions. Genes will not stay in the organisms in which they were released, and bacterial populations will not remain unchanged to do the jobs we want them to do. Whenever we wish to apply our knowledge of bacteriology, we will probably need to know some population genetics.

WHY ARE BACTERIA PECULIAR?

Since there already exists a body of population genetics theory, can we not take that theory and apply it directly to bacteria? Indeed, bacteria should be easier to analyse, because they are haploid, so that it is easier to deduce the gentoype from the phenotype, and no complications arise from dominance and over-dominance. Unhappily, this hope of simplicity is unrealized. Much of classical population genetics is based on the assumptions of infinite population size, random mating, and free recombination. Given these assumptions, the alleles present at one locus are independent of those at other loci. Changes in the frequency of an allele at one locus, therefore, are independent of what is happening elsewhere in the genome: each locus can be treated independently. Of course, much effort has gone into studying the effects of finite population size, or non-random mating, and of linkage, but at the back of our minds we have had the comforting feeling that, for most practical purposes, the simple model is a reasonable approximation to the truth.

No such assumption can be made for bacteria. Population sizes may be so large as to be effectively infinite, but recombination is certainly not free: in most populations it is rare enough to ensure that changes produced by selection at one locus will cause changes at others. The genome has to be treated as an inter-related whole, and not as a set of independently changing genes. If recombination were wholly absent, things would not be too bad. We could subject individual cells to phylogenetic analysis, just as taxonomists treat species and higher taxa: bacterial population genetics would become a branch of cladistics. Indeed, many papers on bacterial population genetics proceed in just this manner. After collecting data on individual variation, using electrophoresis or some more recent technique, the authors construct a phylogenetic tree using one of the packages now available, and then rest on their laurels. As we will see in the next section, there may be no more justification for this procedure than there would be for constructing a phylogenetic tree of the scientists attending a conference or the cars passing a particular point on a motorway.

What makes the population genetics of bacteria novel, and difficult, is that there is often too much recombination to justify the pure phylogenetic approach, but rarely enough to justify an assumption of 'random assortment' (by this phrase I mean a population structure in which the alleles at one locus are independent of the alleles at others). Difficulties arise because often we do not know how much, or what kind, of recombination is going on, and because the theory describing a population with a little recombination is a lot harder than for population with no recombination, or a lot. These difficulties are the topic of the next section.

HOW CLONAL ARE BACTERIA?

Studies of bacterial population genetics first became practicable with the introduction of protein electrophoresis. Milkman (1973) was the first to apply these methods to bacteria. His original motivation was to answer the much-debated question of how far electrophoretic variation is maintained selectively, and how far neutral. The logic was as follows. Bacterial population numbers are immense. Hence theory predicts that, if variation is selectively neutral, the amount of variability should also be very large. If, as turned out to be the case, variation, although large compared to eukaryotic species, is limited, this implies selection. Unfortunately, the logic does not apply to populations with restricted recombination, because of the phenomenon of 'recurrent selection' (Atwood, Schneider & Ryan, 1951), or 'hitch-hiking'. If there were *no* recombination, each time a new favourable mutation arose in an individual and spread to fixation by selection, it would make the whole population homogeneous, at all loci, for the genotype of the individual in which the mutant first occurred. As Levin (1981) pointed out, in a population with limited recombination, selection of favourable mutants reduces the effective population size. Thus Milkman's work did not settle the controversy over neutrality, but it did open up the field of bacterial population genetics.

An early conclusion from electrophoretic studies of *E. coli* and other bacteria (Whittam, Ochman & Selander, 1983; Selander, Caugant & Whittam, 1987) was that populations of bacteria consist of a number of independently evolving clones; a clone being defined as a set of genetically similar cells, derived from a common ancestor, without chromosomal recombination. Essentially, there are three kinds of evidence that can be used to demonstrate clonal structure:

(i) Selander & Levin (1980) showed that the same electrophoretic type (ET) at 20 loci could be found in *E. coli* from unassociated hosts. Similar findings in other bacteria demonstrated that recombination could not be a frequent event. There are two reasons why this argument from the widespread occurrence of a single ET should be treated with caution. First, one must show that the widespread ET is more frequent than would be expected by random assortment. For example, 35 of 227 isolates of *Neisseria gonorrhoeae* were identical at the nine loci analysed (O'Rourke & Stevens, 1993), but this is not evidence for clonality: the number expected from random assortment was 32.2. Secondly, there is a possibility, even in a population in which recombination is reasonably common, of the explosive increase of a particular ET, which then exists at high frequency for a time, before being lost by recombination. The data of Caugant *et al.* (1987) for *N. meningitidis* suggest such an 'epidemic structure', despite frequent recombination (Maynard Smith *et al.*, 1993).

(ii) Stronger evidence for restricted recombination comes from measure-
      ments of linkage disequilibrium: that is, of the tendency for particular
      alleles at different loci to co-occur. Whittam *et al.* (1983) demonstrated
      highly significant linkage disequilibrium in *E. coli*: their method is
      described below, when discussing evidence for recombination.
(iii) Linkage disequilibrium (and the too-frequent occurrence of particular
      ETs, which is a manifestation of such disequilibrium) shows that
      recombination is restricted, but not that it is absent. A third line of
      evidence can, in principle, provide stronger evidence for clonality. If
      there is no recombination, the ancestral history of all chromosomal
      genes is identical, and phylogenetic trees derived from sequences of
      different genes should be similar. Nelson, Whittam & Selander (1991)
      sequenced the *gapA* gene from 16 strains of *Salmonella*, and found that
      the phylogeny derived from these sequences was very similar to that
      based on multiple locus electrophoresis, providing strong evidence for
      the clonality of *Salmonella*: unfortunately, the *gapA* genes from 13 *E.
      coli* strains were too similar to be informative. I return to evidence
      derived from comparing the phylogenies of different genes from the
      same species below, when discussing the nature of species.

   I turn now to evidence pointing in the opposite direction. Evidence for the
importance of recombination in bacterial populations is also of three kinds:

 (i) From the mosaic structure of nucleotide sequences.
(ii) From a statistical analysis of the association, or lack of it, between
      alleles at different loci.
(iii) From a comparison of gene trees.

   The first clear evidence for the importance of recombination in natural
populations came from comparing gene sequences from different strains of
E. coli (Stoltzfus, Leslie & Milkman, 1988, for the trp operon; Dubose,
Dykhuizen & Hartl, 1988, for the phoA gene; Dykhuizen & Green, 1986,
for the gnd gene). In all these cases, gene sequences from different strains
were found to be rather similar for much of the region sequenced, but very
different in particular regions, suggesting that short pieces of DNA had been
inserted from an unknown source by recombination. The same phenomenon, in a more dramatic and easily analysed form, was described in the
penicillin-binding protein (PBPs) of *Streptococcus* (Dowson *et al.*, 1989) and
Neisseria (Spratt *et al.*, 1989 and this volume). There are two reasons why
the evidence for recombination is particularly clear in these two genera.
First, unlike *E. coli* and *Salmonella*, they are competent for transformation.
There are good reasons to think that recombination is frequent, and that it
can occur between species differing by up to 23% of nucleotides. Secondly,
the PBP gene has been under particularly strong selection for change since
the clinical use of penicillin: the observed changes in the gene are therefore

recent, and there has not yet been time for the mosaic structure to be obscured by later mutation.

Because of strong selection for penicillin resistance, the frequency with which distinct transformation events have been established in the pathogenic *Neisseria* is higher at the PBP locus than at other loci. Such events, however, have occurred in a housekeeping gene, *argF* (Zhou & Spratt, 1992). Horizontal transfer has not been confined to species competent for transformation. Examples from *E. coli* were mentioned above: in *Salmonella*, there is evidence of horizontal transfer between strains at a flagellin-determining locus, *fliC* (Smith, Beltran & Selander, 1990).

Before turning to the statistical evidence for recombination, it is important to emphasize that there is no contradiction between the electrophoretic data suggesting limited recombination, and sequence data indicating a mosaic structure arising from horizontal transfer. This is because of the local nature of recombination (Maynard Smith, Dowson & Spratt, 1991). In eukaryotes, whole groups of genes are exchanged simultaneously, but in bacteria the main mechanisms of exchange, transduction and transformation, transfer only short pieces of DNA: the sequence data suggest that the transferred pieces are often in the range 100–1000 bases only. Thus a transfer event will not alter the ET of the recipient unless it happens to affect one of the electrophoretic loci. Local transformation events have to be common, relative to mutations, before they will generate linkage equilibrium (Maynard Smith, 1994).

What deductions can be drawn from the frequencies of alleles at different loci? Whittam *et al.* (1983) used the following method, first proposed by Brown, Feldman & Nevo (1980) in a different context, to test for associations between loci. Suppose that n strains have been characterized at m polymorphic loci. The difference, d_{ab}, between a pair of strains, a and b, is the number of loci at which they differ: d_{ab} can therefore vary from 0 to m, although extreme values may not exist in a particular data set. There are $n(n-1)/2$ pairs of strains, and hence values of d. It is easy to calculate $\bar{d}$ and V_{obs}, the mean and variance of the $n(n-1)/2$ values of d. To test for association between loci, we need to know V_{exp}, the expected value of the variance of d if the alleles present at different loci are independent. In fact,

$$V_{exp} = \sum_{j=1}^{m} h_j(1 - h_j),$$

where h_j is the probability that two individuals are different at the jth locus: this probability can easily be calculated from the data. By comparing V_{obs} and V_{exp}, one can test for the association between loci.

Maynard Smith *et al.* (1993) compared bacterial species, using the statistic

$$I_A = V_{obs}/V_{exp} - 1.$$

Table 1.

Species	Number of isolates	Number of loci	Mean distance per locus	V_{obs}	V_{exp}	I_A
Neisseria gonorrhoeae[a]	227	9	0.309	1.77	1.68	0.04 ± 0.09
Salmonella[b]	1495	24	0.355	16.04	3.90	3.11 ± 0.04
S. panama	99	24	0.038	1.85	0.79	1.34 ± 0.16
S. paratyphiB	118	24	0.074	3.49	1.30	2.68 ± 0.13
S. typhimurium	340	24	0.035	1.44	0.71	1.08 ± 0.10
S. paratyphiC	100	24	0.054	5.77	1.02	4.66 ± 0.15
S. choleraesius	174	24	0.036	2.64	0.74	2.57 ± 0.12
Rhizobium meliloti[c]	232	14	0.238	17.17	2.34	6.34 ± 0.09
Division A	208	14	0.101	1.59	1.08	0.47 ± 0.10
Division B	23	14	0.219	2.30	1.85	0.24 ± 0.29
Bacillus[d]	95	10	0.697	8.11	1.86	3.37 ± 0.14
B. licheniformis	35	10	0.628	10.79	2.17	3.97 ± 0.23
B. subtilis	60	13	0.526	5.68	2.30	1.47 ± 0.18
Division B	28	13	0.356	2.52	2.10	0.20 ± 0.26
Division D	27	13	0.377	2.89	1.98	0.46 ± 0.26

Sources of data:
[a]O'Rourke & Stevens (1993).
[b]Caugant *et al.* (1987).
[c]Eardly *et al.* (1990).
[d]Istock *et al.* (1992) and personal communication.

Table 1 gives values of I_A for several species. Note first the contrast between *Neisseria gonorrhoeae* (collected worldwide over 25 years) and *Salmonella* (also a worldwide sample). In the former, alleles at different loci assort independently, a fact that sheds some light on the sexual habits both of the bacterium and its host. In contrast, *Salmonella* is highly clonal. The 'species' are now regarded as serovars of a single species, *S. enterica* (Ewing, 1986). The great reduction of genetic distance within 'species' indicates that these serovars are genuine genetic entities, but the genetic structure within serovars is still clonal.

The data on *Rhizobium meliloti* indicate that this 'species' consists of two populations, within which recombination is frequent, but between which it is rare. The data on *Bacillus* are of particular interest, because they were collected from a single patch of desert soil, in three successive annual samples: they therefore more nearly represent a 'population' as the term is used by students of higher organisms. *Bacillus* is competent for transformation, and in the laboratory genes are transferred both within species and, at a reduced rate, between species. The distinction between *B. licheniformis* and *B. subtilis* was made biochemically: the values of genetic distance suggest that the entities so recognized may not have much reality. The distinction between groups B and D, which together included 55 of the 60 isolates, was based on the electrophoretic data. The I_A values suggest that recombination within these groups is frequent.

The point of this analysis is to show how data on allele frequencies can be used to identify sets of genetically similar organisms, and to indicate whether recombination is rare or absent (*Salmonella*, within and between serovars; *Rhizobium* and *Bacillus*, between subgroups), or common enough to generate random assortment of alleles (*N. gonorrhoeae*; within subgroups of *Rhizobium meliloti* and *Bacillus subtilis*).

It is probably not accidental that the two species, *N. gonorrhoeae* and *B. subtilis*, that are known to be competent for transformation, also approach random assortment. However, it is not the case that all species competent for transformation show random assortment. *Haemophilus* (Musser *et al.*, 1990) shows evidence of clonal structure. *N. meningitidis* also shows significant values of I_A, but this is at least in part due to the explosive spread of particular ETs.

ARE THERE BACTERIAL SPECIES?

It is conventional to use the Linnaean binomial system to give names to bacteria. Do the named species correspond to real genetic entities? If not, is there some other naming system which would more correctly reflect the real pattern of variation?

Before attempting the answer these questions, it will be convenient to discuss what use is made of the species concept in higher organisms: there is no point in arguing about the term 'species' unles we have some idea what use we want to make of it. To make things easier, consider the naming of birds, which are rather easily classified into species, for several reasons. Their reproduction is exclusively sexual: they are large, conspicuous and diurnal; we recognize them by the use of the same senses, sight and sound, by which they recognize one another. Starting from the practical and progressing to the theoretical, the species concept is useful in the following ways;

(i) A bird-watcher expects to be able to identify every bird seen, and this expectation rarely fails. Recognition depends partly on phenotype (plumage, song), partly on habitat (one does not expect a reed warbler in a cornfield or a corn bunting in a marsh), and partly on geographical location (a phalarope in Canada is Wilson's phalarope). In other words, species can be defined phenotypically, and members of a given species have a known ecology and a known geographical distribution.

(ii) It is generally thought that the properties just described are true because of sexual reproduction. They arise because mating and gene flow is common within species and rare or absent between them.

(iii) It follows that a favourable mutation arising in one species can spread to fixation in that species, but is unlikely to spread to a second species. Because of recombination, the fixation of a favourable mutation does not subsequently reduce variation at other loci.

(iv) Species are monophyletic: that is, they include all the descendants, and only the descendants, of a single ancestral population.

 (v) In evolution, the splitting of a single species into two typically requires a period of geographical isolation.

None of the above statements is universally true, but they are true often enough for the species concept to be useful, and for it to be worthwhile to ensure that species names are given only to populations of which the statements are, at least approximately, true. In general, the same objectives hold when classifying other taxa, although they may be harder to achieve. In flowering plants, for example, hybridization between species may be rather common, and it may be sufficient for the origin of new species that populations be isolated by habitat and not by geography.

In bacteria, no species concept could be devised that would carry this heavy load of meanings and theoretical implications. One clear distinction is that, in at least some bacteria, the geographical differentiation characteristic of higher organisms is absent. To anyone familiar with higher organisms, the most surprising feature of bacterial population genetics is that most species have a worldwide distribution, but a local population may contain the full range of variation that exists worldwide: for example, O'Rourke & Stevens (1993) found this to be true for *N. gonorrhoeae*. Oddly enough, this is a characteristic we humans share with bacteria, but with few other animals. In general, the pattern of local races, or subspecies, replacing one another geographically is unusual in bacteria, and there is no reason to think that the origin of genetic differences between species depended on geographical isolation.

More important, phenotypically recognized species often do not correspond to evolving populations, sharing a common gene pool, isolated from the gene pool of other species. In *Neisseria*, the evolving unit is wider than the named species: the evolution of penicillin resistance has involved gene exchange across the whole genus. In contrast, in *Salmonella*, there is little gene exchange, and hence no population that shares a gene pool, or that can be identified as an evolving unit. Hence there is no way of defining a species that will correspond both to a phenotypically recognized entity and an evolving unit, and that will have, even approximately, the same meaning for all bacterial taxa.

Nevertheless, there are real discontinuities in the pattern of variation. For example, there seems to be no intermediates between *E. coli* and *Salmonella*, and no clear evidence of gene exchange between them (although it is worth noting that *E. coli* is genetically more similar to *Salmonella* than it is to some bacteria that have been placed in the genus *Escherichia*). Despite gene exchange, *N. meningitidis* and *N. gonorrhoeae* are not only phenotypic entities, but are genetically distinct. How, then, should we proceed in naming bacteria?

Two proposals have been made aimed at regularizing the naming procedure. One is that bacteria should be placed in different species, genera, etc. according to the degree of genetic distance between them, judged, for example, by DNA hybridization. This proposal seems to me arbitrary and without merit. The frequency of gene exchange seems to vary more or less continuously with genetic distance (e.g. Roberts & Cohan, 1993). The object of naming should be to recognize real distinctions, not to impose arbitrary divisions upon a continuum. A more rational proposal is that of Dykhuizen & Green (1991), who suggest that the phylogenetic trees of different gene loci from members of a single species should be different (because members of a species exchange genes), but the phylogenies of different genes from members of different species should be the same. This suggestion has the merit of corresponding to the 'biological species concept' as used in higher organisms, but it would lead us to recognize very small differences as being of specific rank in *Salmonella*, and to group together very different bacteria in a single species in *Neisseria*.

All the same, it is interesting to ask how the proposal of Dykhuizen & Green would work out in practice. They report that the phylogenies for the genes *gnd*, *trp* and *phoA* in eight strains of *E. coli* are different, indicating that there has been recombination, and implying that *E. coli* is, by this definition, properly regarded as a species.

Figure 1 shows gene trees for the *argF* and *PBP*2 genes for seven named species of *Neisseria*. The species were named for phenotypic traits, before sequences were available. The similarity of the two trees suggests that the species are real entities. However, the agreement depends on the fact, that only 'frame DNA' (in the sense of Milkman & Bridges, 1990) was included in the analysis, and inserted sequences were ignored. If this had not been done, the trees would have depended on the particular isolates chosen to represent each species: for example, there are strains of *N. meningitidis* whose *PBP*2 gene is identical to that of *N. mucosa*.

It seems then that Dykhuisen & Green's proposal, although it has real theoretical merits, has two practical disadvantages. It would result in entities of very different kinds being recognized as species in different bacterial taxa, and it would require gene sequences to be available for several genes from several representatives of each potential species.

In practice, I think that species will continue to be named on phenotypic criteria: if a group of strains do not share properties of interest to the investigator, there is little point in naming them. However, increased genetic knowledge should constrain our behaviour in two ways:

(i) It is highly desirable that named taxa (species, genera, etc.) should be

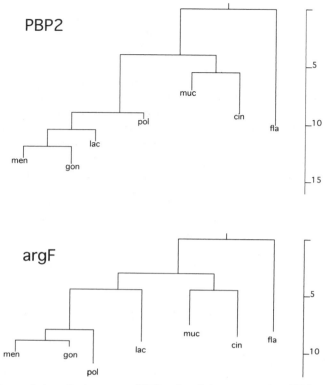

Fig. 1. Phylogenetic trees for two genes, *PBP2* and *argF*, in seven species of *Neisseria*: men, *N. meningitidis*; gon, *N. gonorrhoeae*; pol, *N. polysaccharea*; lac, *N. lactamica*; muc, *N. mucosa*; cin, *N. cinerea*; fla, *N. flavescens*. The scale represents percentage nucleotide divergence. The trees are based only on 'frame' DNA, ignoring inserted sequences. Since only statistically significant insertions have been omitted, some real insertions may have been included in the analysis: this may have led to a misplacing of *N. lactimica* in the *argF* tree.

monophyletic, as defined above, In the case of genera like *Neisseria*, species can be only approximately monophyletic: that is, most of the DNA of the members of a given species has been derived from a common ancestor, but there are many inserted pieces that have not. But there can be little justification for wholly polyphyletic 'species' like *Shigella*, which consists of two or more unrelated groups of strains, all of which are phylogenetically included within *E. coli* (Selander *et al.*, 1987).

(ii) Named species in different bacterial taxa will prove to have very different amounts of recombination within them. This is a fact of life we will have to learn to live with.

CONCLUDING REMARKS

In this paper, I have concentrated on the significance of recombination for bacterial population genetics. This is by no means all that there is to discuss. Among important topics I have not mentioned are the role of plasmids; the significance of codon bias (Sharp & Li, 1987); the use of population data to decide whether variation is neutral or selective, and the experimental study of laboratory populations, pioneered by Hartl & Dykhuizen (1984). I have also not felt competent to review the practical implications of bacterial population genetics, which were admirably reviewed by Istock (1990).

REFERENCES

Atwood, K. C., Schneider, L. K. & Ryan, F. J. (1951). Periodic selection in *Escherichia coli*. *Proceedings of the National Academy of Sciences, USA*, **37**, 146–55.

Brown, A. H. D., Feldman, L. W. & Nevo, E. (1980). Multilocus structure of natural populations of *Hordeum spontaneum*. *Genetics*, **96**, 523–6.

Caugant, D. A., Mocca, L. F., Frashc, C. E., Froholm, L. O., Zollinger, W. D. & Selander, R. K. (1987). Genetic structure of *Neisseria meningitidis* populations in relation to serogroup, serotype, and outer membrane protein pattern. *Journal of Bacteriology*, **169**, 2781–92.

Dowson, C. G., Hutchison, A., Brannigan, J. A., George, R. C., Hansman, D., Linares, J., Tomasj, A., Maynard Smith, J. & Spratt, B. G. (1989). Horizontal transfer of penicillin-binding genes in penicillin-resistant clinical isolates of *Streptococcus pneumoniae*. *Proceedings of the National Academy of Sciences, USA*, **86**, 8842–6.

Dubose, R. F., Dykhuizen, D. E. & Hartl, D. L. (1988). Genetic exchange among natural isolates of bacteria: recombination within the *phoA* gene of *Escherichia coli*. *Proceedings of the National Academy of Sciences, USA*, **85**, 7036–40.

Dykhuizen, D. E. & Green, L. (1986). DNA sequence variation, DNA phylogeny and recombination. *Genetics*, **113**, s71.

Dykhuizen, D. E. & Green, L. (1991). Recombination in *Escherichia coli* and the definition of biological species. *Journal of Bacteriology*, **173**, 7257–68.

Eardly, B. D., Materou, L. A., Smith, N. H., Johnson, D. A., Rumbaugh, M. D. & Selander, R. K. (1990). Genetic structure of natural populations of the nitrogen-fixing bacterium *Rhizobium meliloti*. *Applied Environmental Microbiology*, **56**, 187–94.

Ewing, W. H. (1986). *Edwards and Ewing's Identification of Enterobacteriaceae*. Elsevier, New York.

Hartl, D. L. & Dykhuizen, D. E. (1984). The population genetics of *Escherichia coli*. *Annual Review of Genetics*, **18**, 31–68.

Istock, C. A. (1990). Genetic exchange and genetic stability in bacterial populations. In *Assessing the Risk of Biotechnology*. (Ginzberg, L. R. ed.). Butterworth.

Istock, C. A., Duncan, K. E., Ferguson, N. & Zhou, X. (1992). Sexuality in a natural population of bacteria – *Bacillus subtilis* challenges the clonal paradigm. *Molecular Ecology*, **1**, 95–103.

Levin, B. R. (1981). Periodic selection, infectious gene exchange and the genetic structure of *E. coli* populations. *Genetics*, **99**, 1–23.

12   JOHN MAYNARD SMITH

Maynard Smith, J. (1994). Estimating the minimum rate of genetic transformation in bacteria. *Journal of Evolutionary Biology* (in press).

Maynard Smith, J., Dowson, C. G. & Spratt, B. G. (1991). Localized sex in bacteria. *Nature, London*, **349**, 29–31.

Maynard Smith, J., Smith, N. H., O'Rourke, M. & Spratt, B. G. (1993). How clonal are bacteria? *Proceedings of the National Academy of Sciences, USA*, **90**, 4384–8.

Milkman, R. (1973). Electrophoretic variation in *Escherichia coli* from natural sources. *Science*, **182**, 1024–6.

Milkman, R. & Bridges, M. M. (1990). Molecular evolution of the *Escherichia coli* chromosome. III. Clonal frames. *Genetics*, **126**, 505–17.

Musser, J. M., Kroll, J. S., Branoff, D. M., Moxon, D. R., Brodeur, D. R., Campos, J. *et al.* (1990). Global genetic structure and molecular epidemiology of encapsulated *Hemophilus influenzae*. *Review of Infectious Diseases*, **12**, 75–111.

Nelson, K., Whittam, T. S. & Selander, R. K. (1991). Nucleotide polymorphism and evolution in the glyceraldehyde-3-phosphate dehydrogenase gene (*gapA*) in natural populations of *Salmonella* and *Escherichia coli*. *Proceedings of the National Academy of Sciences, USA*, **88**, 6667–71.

O'Rourke, M. & Stevens, E. (1973). Genetic structure of *Neisseria gonorrhoeae* populations: a non-clonal pathogen. *Journal of General Microbiology*, **139**, 2603–11.

Roberts, M. S. & Cohan, F. M. (1993). The effect of DNA sequence divergence on sexual isolation in *Bacillus*. *Genetics*, **134**, 402–8.

Selander, R. K. & Levin, B. R. (1980). Genetic diversity and structure in *Escherichia coli* populations. *Science*, **210**, 545–7.

Selander, R. K., Caugant, D. A. & Whittam, T. S. (1987). Genetic structure and variation in natural populations of *Escherichia coli*.

Sharp, P. M. & Li, T. S. (1987). The rate of synonymous substitution in enterobacterial genes is inversely related to codon usage bias. *Molecular Biology and Evolution*, **4**, 222–30.

Smith, N. H., Beltran, P. & Selander, R. K. (1990). Recombination of *Salmonella* Phase 1 flagellin genes generates new serovars. *Journal of Bacteriology*, **172**, 2209–16.

Spratt, B., Zhang, Q.-Y., Jones, D. M., Hutchison, A., Brannigan, J.A. & Dowson, C.B. (1989). Recruitment of a penicillin-binding protein gene from *Neisseria flavescens* during the emergence of penicillin resistance in *Neisseria meningitidis*. *Proceedings of the National Academy of Sciences, USA*, **86**, 8988–92.

Spratt, B. G., Smith, N. H., Zhou, J., O'Rourke, M. & Feil, E. (1994). The population genetics of the pathogenic *Neisseria*. (this volume).

Stoltzfus, A., Leslie, J. F. & Milkman, R. (1988). Molecular evolution of the *Escherichia coli* chromosome. I. Analysis of structure and natural variation in a previously uncharacterized region between *trp* and *tonB*. *Genetics*, **120**, 345–58.

Whittam, T. S., Ochman, H. & Selander, R. K. (1983). Multilocus genetic structure in natural populations of *Escherichia coli*. *Proceedings of the National Academy of Sciences, USA*, 80, 1751–5.

Zhou, J. & Spratt, B. G. (1992). Sequence diversity within the *argF*, *fbp* and *recA* genes of natural isolates of *Neisseria meningitidis*: interspecies recombination within the *argF* gene. *Molecular Microbiology*, **6**, 2135–46.

ADAPTIVE MUTATION

PATRICIA L. FOSTER

*Department of Environmental Health, Boston University School of
Public Health, Boston University School of Medicine,
80 E. Concord St., Boston, MA 02118, USA*

Spontaneously arising mutations are the source of variation upon which
natural selection works, producing evolutionary change. We all grew up
with the belief that these mutations arise at random during exponential
growth of the cell, probably as the result of unavoidable replication errors.
However, six years ago, John Cairns, Julie Overbaugh, and Stephan Miller
(1988) published a paper presenting evidence that mutations also arise
among static populations of bacteria exposed to non-lethal selections.
Cairns, Overbaugh & Miller (1988) further suggested that the process giving
rise to these mutations had the unusual property of producing only useful
mutations, and not useless or deleterious ones. This phenomenon has been
called 'directed', 'selection-induced' and 'Cairnsian' mutation. I will refer to
it as adaptive or post-selection mutation.

This paper focuses on three aspects of adaptive mutation. First, what is
the evidence that mutations do arise in cells that are apparently not dividing
or replicating their genomes? Secondly, what is the evidence that the
mutations that arise are adaptive? Separating the evidence into these two
categories allows the phenomenon to be thought of as two processes: a
generating process that produces genetic variation in non-growing cells, and
an editing process that prevents useless mutations from occurring, or
eliminates them after they have occurred. Finally, possible mechanisms for
adaptive mutations are considered.

MUTATIONS OCCUR AFTER SELECTION IS APPLIED

What is the evidence that mutations are occurring among non-growing cells
after selection is applied? Our belief that mutations occur at random during
non-selective growth is based on classical genetic experiments, particularly
the Luria–Delbrück fluctuation test (Luria & Delbrück, 1943). That mutations can also arise after selection is applied, provided that the selection is
non-lethal, is also based, in part, on the Luria–Delbrück fluctuation test, but
in this case on significant deviations from the predicted distributions of
mutants. However, it must be emphasized that such deviations are just half
of the evidence; the other half is that mutants continue to appear with time
after selection is applied.

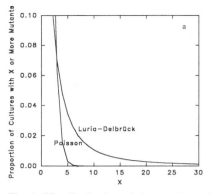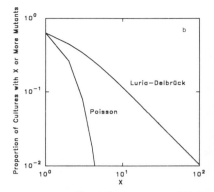

Fig. 1. The distribution of the numbers of mutants among replica cultures expected if the mutations occur during growth (Luria–Delbrück) or after plating on a selective medium (Poisson). The curves are as in Cairns *et al.*, 1988 with m, the mean number of mutations occurring during growth, and μ, the mean number of mutations occurring after plating, both equal to 1. a: linear scale; b: log scale.

In a fluctuation test a large number of identical cultures are inoculated with a few cells such that no culture contains a pre-existing mutant. These are allowed to grow up under non-selective conditions and are then plated on a selective medium where mutants can form colonies. Any mutant that arises during non-selective growth will produce a clone of identical descendants, each of which will give rise to a colony after plating. Rare mutants arising early during growth of a culture will result in a large number of mutant progeny, and a large number of colonies after plating (a jackpot). The resultant distribution of mutant numbers among the replica cultures, known as the Luria–Delbrück distribution, is characterized by a large variance. If, however, a mutant arises after the cultures are plated on a selective medium, it will also produce a clone of descendants, but these will result in only one colony and will be scored as one. So, if mutations occur after selection, the distribution of the number of mutants among replica cultures will be Poisson with a variance equal to the mean (Luria & Delbrück, 1943).

The two distributions are shown in Fig. 1. Such data are usually presented as a log–log plot (Fig. 1b) because the Luria–Delbrück distribution results in a straight line with slope-1 (from which the mutation rate can be obtained by extrapolation to the origin (Luria, 1951)). In contrast, the Poisson distribution has a conspicuous bulge at low numbers and then a steep drop-off. When Luria and Delbrück did the experiment they, of course, got the jackpot distribution, thus demonstrating that mutations occur before selection (Luria & Delbrück, 1943). However, it has been largely overlooked (except by Delbrück (1946)) that the experiment did not demonstrate the converse, that mutations might not also occur after selection. Luria and

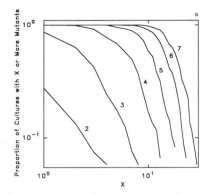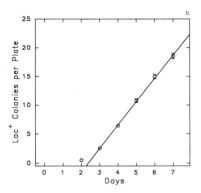

Fig. 2. Results of a fluctuation test with FC40, which carries the *lacZ* frameshift allele. a. The distribution, among 120 replica cultures, of Lac$^+$ colonies that appeared during a week that the cells were on lactose minimal plates (modified from Cairns & Foster, 1991); b: the accumulation of Lac$^+$ colonies with time on lactose plates. The Figure shows the mean and the standard error of the mean (SEM) for 118 cultures; two cultures were not included because they contained jackpots.

Delbrück could not have shown this because the selection they used, to phage resistance, is immediately lethal, and dead cells cannot mutate.

So, what happens when a non-lethal selection is used? In most of the experiments done in my and John Cairns' laboratories, mutant strains of *Escherichia coli* that cannot use lactose as a carbon or energy source are plated on lactose minimum plates to select for reversion of the Lac$^-$ phenotype. Figure 2a shows the distribution of the numbers of Lac$^+$ mutants among 120 cultures in a fluctuation test using FC40, a strain that is Lac$^-$ because of a frameshift mutation that is polar on the *lacZ* gene (Cairns & Foster, 1991). On day 2, Lac$^+$ colonies that result from mutations occurring during non-selective growth make their appearance, and the distribution is a reasonable approximation to the Luria–Delbrück. However, with time, more Lac$^+$ colonies appear, and their distribution is Poisson. Thus, the total distribution becomes increasingly Poisson with time. Furthermore, as shown in Fig. 2b, the rate at which new Lac$^+$ colonies appear is nearly constant over the course of a week or so. From these data, we estimated that 90 to 95% of the Lac$^+$ mutants that arose during the experiment were due to mutations that occurred after the cells were placed under selection for lactose utilization (Cairns & Foster, 1991).

There are artefactual reasons why the distribution of mutants in a fluctuation test might be skewed away from the Luria–Delbrück toward the Poisson. Several of these have been modelled by Stewart, Gordon & Levin (1990): poor growth of mutant cells during non-selective growth, poor plating efficiency of mutants, and a generation-independent component of the mutation rate. None of these can account for the results shown in Fig. 2. There is no evidence that significant numbers of Lac$^+$ revertants are disadvantaged during non-selective growth; when tested, Lac$^+$ and Lac$^-$ cells have the same growth rate on glycerol (Cairns & Foster, 1991). The effects of even very low plating efficiencies modelled by Stewart *et al.* (1990) are far too weak to account for 90% of the mutants falling into the Poisson component. In addition, neither of these factors would account for Lac$^+$ revertants appearing at a constant rate, or for *late* appearing mutants having a Poisson distribution. The last factor modelled by Stewart *et al.* (1990), a time-dependent but generation-independent component of mutation, is, of course, another way of saying that mutations arise in non-dividing cells.

CELLS DO NOT GROW OR DIE DURING SELECTION

There is one artefactual explanation for the appearance of mutants during selection that can account for the results presented above. Obviously, if the cells are actually growing or replicating their DNA under the selective conditions, mutations might occur by normal replication-dependent mutational processes. The cells might be growing because the allele is 'leaky', the selection is not perfect, or because of contaminants in the medium. Or, possibly, some cells might be growing on nutrients released by pre-existing mutants (cross-feeding) or by dying cells (turnover).

It is instructive to consider how much growth or turnover would be required to account for the mutations that arise during selection if they arise at the normal replication-dependent mutation rates. Table 1 presents several experiments where there were good data on the mutation rates both during non-selected growth and during selection. These rates can be used to calculate the number of cell divisions or genome replications that would have had to have occurred during the course of the experiments to account for the mutations at the normal pre-selection rates (labelled 'replication equivalents' in Table 1). For example, in the experiment shown in Fig. 2 with the *lacZ* frameshift allele, nearly 100 cell divisions or genome replications would be needed *per cell*. Thus, when 10^9 Lac$^-$ cells are plated on a minimum lactose plate and 200 or so mutants arise over the next week, if the mutants arose due to normal replication-dependent mutations, then 10^{11} Lac$^-$ cells must have existed on that plate at some time during the week. That number of cells is about as many Lac$^+$ as a lactose plate will support.

It is easy to show that no such gross population increase is taking place among the Lac$^-$ cells. By simply taking plugs off a plate and titering the cells, we found less than a 2-fold increase in the number of Lac$^-$ cells over 5 days on lactose plates (Fig. 3, (Cairns & Foster, 1991)). It might be argued that early-arising Lac$^+$ colonies crossfeed the Lac$^-$ cells, allowing them to grow, and we would not have detected this because the growth would be localized around the Lac$^+$ colonies. Typically, we plate an excess of non-revertible Lac$^-$ cells to scavenge contaminants. When the Lac$^-$ frameshift strain is plated at various dilutions with the same density of scavengers, the

Table 1. *Comparison of spontaneous mutation rates during exponential growth and during non-lethal selection*

Reference	Mutational target	Mutation rates		Replication equivalents	
		Mutations per cell per generation during non-selected growth[a]	Mutations per cell per hour during selection[b]	Total for experiment[c]	Per day[d]
Escherichia coli					
Cairns & Foster (1991)	*lacI::lacZ* frameshift	1.4×10^{-9}	1.1×10^{-9}	96	19
Cairns *et al.* (1988)	*lacZ* amber	1.7×10^{-10}	1.1×10^{-10}	75	25
Cairns & Foster (1991) & unpublished	*trpE* frameshift	3.0×10^{-10}	3.1×10^{-10}	50	25
Hall (1991)	*lacZ* missense	3.1×10^{-10}	7.9×10^{-11}	43	6.1
Hall (1990)	*trpB* missense	1.5×10^{-10}	1.4×10^{-11}	26	2.2
	trpA missense	9.4×10^{-11}	4.5×10^{-12}	14	1.2
Ryan (1955)	*his⁻*	3.0×10^{-8}	4.1×10^{-10}	2	0.33
Saccharomyces cerevisiae					
Steele & Jinks-Robertson (1992)	*lys2* frameshift	4.9×10^{-9}	3.0×10^{-10}	7	1.5

Reprinted from Foster (1993).
[a]Calculated as $m/(2N)$ where m = average number of mutations per culture, N = cells per culture. These rates are slightly different from published rates calculated as $[m \ln(2)]/N$ (Hall, 1990, 1991).
[b]Calculated as $M/(N_0 h)$ where M = total number of mutants appearing under selection. N_0 = cells originally plated, and h = hours the experiment lasted after pre-existing mutations appeared. These rates are different from published rates that took into account cell death (Hall, 1990; Ryan, 1955).
[c]Calculated as M/N_0 divided by the exponential phase mutation rate.
[d]Total turnover divided by the days the experiment lasted.

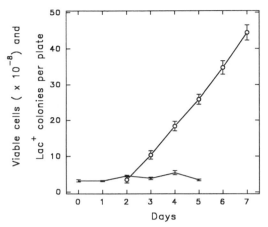

Fig. 3. The accumulation of Lac⁺ colonies (o) and the number of viable cells among FC40 cultures plated on lactose plates. The Figure shows the mean and SEM for 6 cultures, except for viable counts, where $n = 5$ for days 3, 4, and 5 (modified from Cairns & Foster, 1991).

number of mutants that arise *per revertible cell* is a constant (Cairns & Foster, 1991). This means that the same number of Lac⁺ mutants arise whether they do so in the presence of 1, 10, or 100 pre-existing Lac⁺ colonies. Therefore, crossfeeding cannot be important in these experiments.

It is also possible that cryptic growth, or turnover, is occurring. That is, the number of viable cells might only appear constant because some of the cells divide while others die. Again, it is instructive to consider the amount of turnover that would be required to account for the mutations that arise if the normal replication-dependent mutation rate pertains (Table 1). Considering that a generation in minimum medium is about an hour, the rate of appearance of new Lac⁺ revertants of the frameshift allele *per unit time* is about the same both before and during selection. It follows, therefore, that the cells would have to be turning over on lactose plates at the same rate as they grow exponentially in non-selective minimal medium to account for the mutants. This is inherently improbable. Furthermore, the amount of cell death on lactose plates is less than 10% (Foster, 1994).

The growth of cells may be irrelevant if the amount of DNA synthesis that occurs is uncoupled from cell division. However, estimates of the amount of DNA synthesis taking place in static populations ranges from 0.5 to 1% of the genome per cell per day, Nakada & Schneider, 1961; Grivell & Hanawalt, 1975; Boe, 1990). If these measurements are correct, then either the DNA synthesis must be severely restricted to only certain parts of the genome, or the error rate of this synthesis must be two or three orders of magnitude higher than it is during exponential growth, or both.

It has also been suggested that the post-selection mutations arise from

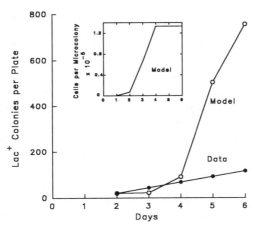

Fig. 4. Comparison between the accumulation of Lac$^+$ revertants of a strain bearing a *lacZ* amber allele and the number of revertants predicted by the intermediate model. Data are from Cairns *et al.*, 1988, Fig. 3, but calculated per plate (3×10^9 cells). The model is as given in Lenski *et al.*, 1989. The mutation rate from the intermediate to the final phenotype is assumed to be 10^{-5} per generation per cell, and 10^3 intermediates are assumed to be present. The growth rate of the intermediate was adjusted to make p, the probability that at least one mutation has occurred per microcolony, equal to 0.07 on day 2, as given in Lenski *et al.*, 1989. To keep the number of cells per microcolony fewer than 10^5, this growth rate was decreased with time, giving the curve shown in the inset. The value for day 2, which is predominantly due to full Lac$^+$ revertants that arose during non-selected growth, was set at 20.

among a subpopulation that is growing imperceptibly on the plate. For example Lenski, Slatkin & Ayala (1989) proposed that a mutation occurs during exponential growth that confers a partial phenotype, then these weakly Lac$^+$ cells grow into microcolonies out of which the final Lac$^+$ mutants arise. One can think of the first mutants as partial revertants, as tRNA suppressors, or as cells that have duplicated or further amplified the unreverted allele. However, in order to account for mutants appearing at frequencies of about 10^{-9} per cell, the two mutation rates, that from the original to the intermediate phenotype and that from the intermediate to the final full Lac$^+$ phenotype, have to be set extremely high. To model the post-plating reversion of a strain carrying a *lacZ* amber allele (Cairns *et al.*, 1988), Lenski *et al.* (Lenski *et al.*, 1989) imagined about 1000 initial intermediates per plate (requiring a mutation rate of about 10^{-8} per cell per generation during non-selective growth), that these would grow into 'microcolonies' of about 10^5 (which is, in fact, a readily visible colony), and sustain a mutation rate of 10^{-5} per cell per generation to the final Lac$^+$ phenotype. Leaving aside the questions of whether this last rate is plausible, of why the 'microcolonies' are not observed, and of whether the final distribution of Lac$^+$ mutants would be Poisson, the Lenski hypothesis fails to model the time course of appearance of Lac$^+$ mutants. As shown in Fig. 4 the actual data are linear, whereas with the parameters used by Lenski *et al.* (1989), the number of mutants increases exponentially. This is inevitable. If the number of cells at risk for mutation is increasing, then the rate at which mutations arise must increase in parallel unless, for some reason, mutation rates are inversely proportional to population size.

A linear increase in the number of mutants with time during selection, indicating that the mutations occur at a constant rate, is typical of a number of point mutations, including the *lacZ* amber and frameshift mutations shown here, and reversion of a histidine auxotrophy studied by Ryan *et al.* (1961). Hall (1990) found that revertants of point mutations in the *trpA* and *trpB* genes, scored as papillae on aging colonies, also accumulated at a constant rate per colony, but, because the number of viable cells detected within the colonies was declining, the mutation rate per viable cell increased with time. Mutations due to movement of insertional elements appear to increase sharply with time in starving populations (Shapiro, 1984; Cairns *et al.*, 1988; Hall, 1988; Tormo, Almiron & Kolter, 1990). As at least some of these latter mutations can occur in the absence of apparent selection for them (Mittler & Lenski, 1990, 1992; Naas *et al.*, 1994, Foster & Cairns, 1994; Maenhaut-Michel & Shapiro, 1994; but see Cairns, 1990; Cairns, 1993; Hall, 1994), it is unclear at this point if the movement of insertional elements can generate adaptive mutations.

ADAPTIVE MUTATIONS OCCUR DURING SELECTION

Two related criteria can be applied to determine whether the process that produces mutations under selection is adaptive. First, mutants should not accumulate when the cells are subjected to a non-specific stress, such as simple starvation; and, secondly, when cells are under selection for one phenotype, mutants with a non-selected phenotype should not accumulate. Both of these ask whether there is some generalized mutagenic state induced by starvation, stress, or a specific selective condition, that results in mutations appearing without regard to their utility. Note that the issue here is not comparative mutation rates, as supposed by Lenski and Mittler (1993); what is important is when mutations arise and what phenotypes they confer.

It is easy to show with the *lacZ* alleles that starvation per se is not mutagenic by using the delayed overlay experiment devised by Cairns *et al.* (1988). Lac$^-$ cells are plated in top agar on minimal medium without a carbon source and at various times thereafter the plates are overlaid with top agar containing lactose. If mutations are induced by starvation, the number of mutants appearing after a delay should be greater than the number of mutants that appeared on plates that received lactose immediately after plating. For several *lacZ* alleles that have been tested (Cairns *et al.*, 1988; Cairns & Foster, 1991; Foster & Cairns, 1992) this is clearly not the case. In fact, the time-course of appearance of Lac$^+$ mutants among cells that have spent several days without a carbon source exactly parallels the time-course of appearance of Lac$^+$ mutants among cells given lactose immediately. This result cannot be explained by cell death during the period of starvation, as less than 10% of the Lac$^-$ cells die or turn over during the course of such experiments (Foster, 1994).

It is more difficult to show that a population under selection for one phenotype is not also accumulating mutations giving a second, non-selected phenotype. The difficulty arises because of constraints on the second phenotype. The mutation conferring the non-selected phenotype has to normally occur at a rate high enough to make the results meaningful, and the phenotype has to be immediately expressed in order to be observed in non-growing cells. There is also an unanswerable question of whether the mutational events being scored in the two cases are comparable. Valine resistance (Val^r) has been used as the second phenotype by several workers (Cairns *et al.*, 1988; Hall, 1988, 1990; Benson, DeCloux & Munro, 1991). High concentrations of valine are bacteriostatic to *E. coli* K12 strains because valine inhibits the biosynthesis of isoleucine; different kinds of mutations in a number of genes can relieve this inhibition (De Felice *et al.*, 1979). All such experiments have given the same result: cells under selection for one phenotype do not at the same time accumulate Val^r mutants. This significance of this result has been called into question by MacPhee (1993), who argued that glucose, the carbon source that was used to allow the Val^r cells to grow, suppresses some (or all) mutational processes. However, MacPhee's argument is not valid. For example, in the experiments reported by Cairns *et al.* (1988), the Val^r mutants, had they appeared, would have been due to mutations occurring when the Lac$^-$ cells were on plates with lactose as the sole carbon source, exactly the condition which MacPhee predicts should result in the highest mutation rates (MacPhee, 1993).

In another test of the occurrence of non-selected mutations, Hall (1990) used a strain with auxotrophies for both tryptophan and cysteine. Within colonies incubating on filters placed on plates with an excess of one amino acid but a limiting amount of the other amino acid, mutants appeared that could grow and produce papillae. After several days, the colonies were shifted to the opposite conditions, and no new papillae appeared within 40 hours. Thus, populations that were accumulating Cys$^+$ mutants were not at the same time accumulating Trp$^+$ mutants, and vice versa. This experiment, although perfect in concept, suffers from the lingering doubt of whether 40 hours was sufficient to allow pre-existing mutants to make their appearance. A similar and more definitive experiment was done by Steele and Jinks-Robertson (1992). They showed that a strain of *Saccharomyces cerevisiae* auxotrophic for tryptophan, leucine, and lysine accumulated Lys$^+$ revertants only when starved for lysine. No Lys$^+$ revertants accumulated when cells were starved for typtophan, leucine, or lysine and tryptophan. What makes this experiment particularly persuasive is that when the experiment was repeated with a strain that could become Lys$^+$ by mitotic recombination between heteroalleles, instead of by mutations, a different result was obtained. Although the number of Lys$^+$ recombinants increased during starvation, they did so no matter which amino acid was missing (Steele & Jinks-Robertson, 1993). This latter result eliminates artifactual explanations for the adaptive appearance of Lys$^+$ mutants.

A number of more problematic second phenotypes have also been used to test the hypothesis that only useful mutations occur during selection. These include mutations to *lacI*$^-$ (Hall, 1990) and bacteriophage and drug resistances (Benson *et al.*, 1991) in *E. coli*, and resistance to inositol-less death in *S. cerevisiae* (Hall, 1992). Recently I have shown that cells under selection to revert the *lacZ* frameshift allele do not accumulate mutations giving rifampicin resistance (Foster, 1994). This was accomplished by allowing the cells a period of non-selected growth to express the rifampicin resistant phenotype. Although none of these experiments are conclusive, they add weight to the argument that cells under selection have access to a mechanism that insures that the only mutations that accumulate are those that allow the cell to grow.

MECHANISMS FOR ADAPTIVE MUTATION

As suggested above, the process of adaptive mutation can be considered as two steps: the generation of variation and the editing of variants. Then, the first question about mechanism is how might genetic variation be generated in cells that are not dividing and, apparently, engaging in only limited amounts of DNA metabolism? A number of possibilities have been suggested.

Even stationary cells transcribe genes, and transcription might be inherently mutagenic. For example, Davis (1989) postulated that the regions of single stranded DNA produced by transcription would be especially sensitive to DNA damaging agents. In addition, transcribed genes are preferentially repaired by excision repair (Mellon & Hanawait, 1989), and this might generate errors during repair synthesis. However, it is also true that cells that are defective in excision repair show post-plating mutations perfectly well (Cairns *et al.*, 1988; Bridges, 1994; Foster, 1993). There are also various types of DNA base damage that might occur spontaneously in stationary cells, for example, oxidation (Ames, 1983), depurination, (Lindahl & Nyberg, 1972), deamination (Coulondre *et al.*, 1978), and alkylation by endogenous alkylating agents (Rebeck & Samson, 1991). These lesions might be directly mutagenic, or they might stimulate localized repair synthesis. Miscoding lesions occurring on the transcribed strand could give rise to mutant transcripts without DNA synthesis and then, if a useful protein resulted, cause the same sequence change in the DNA during subsequent replication. Localized DNA synthesis could be initiated by recombination (Foster & Cairns, 1992; Harris, Longerich & Rosenberg,

1994; J. Roth and F. Stahl, pers. comm.) or transcription (Stahl, 1992). Cairns *et al.* (1988) suggested that transcriptional errors could be immortalized as mutations by reverse transcriptase, although, to date, reverse transcriptase has not been found in *E. coli* K12 strains.

If the limited amount of DNA synthesis occurring in non-growing cells is particularly error-prone, then it might be sufficient to account for the number of mutations that appear. For example, as discussed below, error-correcting pathways may be deficient or slow to act in nutritionally deprived cells (Stahl, 1988; Boe, 1990). It is also possible that transcriptional and translational errors could lead to transient states of hypermutation by creating error-prone polymerases or defects in error-correcting pathways (Ninio, 1991; Boe, 1992). However, Ninio (1991) calculated that such mutators would only be significant in the case of rare double mutations.

And so, there is no lack of ways that genetic variation might be produced in stationary cells. But why do adaptive mutations arise, i.e. what is the editing mechanism that can prevent such processes from producing mutations willy-nilly? The most radical hypothesis was proposed by Cairns *et al.* (1988). They suggested that there might be a reverse information flow from a good (mutant) protein back to its DNA, which could be mediated by reverse transcription. A highly efficient adaptive process would be achieved if the cell had some mechanism of associating a given transcript with its protein, and then reverse transcribing only the transcript that coded for the successful protein. However, revertants of a *lacZ* amber allele include tRNA suppressors as well as true revertants (Foster & Cairns, 1992). In the case of the suppressors, the mutation that gives rise to the successful protein is not in the DNA that encodes the protein, but in the DNA that encodes the tRNA, making a direct reverse information flow unlikely (unless, of course, the cell also has a mechanism to associate the entire translational machinery with the protein, and reverse transcribe the tRNAs as well, a charming but improbable idea).

Davis (1989) proposed that the sole role of the selective agent was to induce transcription of the relevant gene and, as mentioned above, that transcription might be inherently mutagenic. This hypothesis predicts that constitutively expressed genes should be indifferent to the presence of the selective agent. However, this is not the case. The *lacZ* frameshift allele is constitutively expressed, and transcription of the *lacZ* amber allele can be induced with a non-metabolizable inducer. In both cases, Lac^+ mutations do not arise unless lactose is present (Cairns & Foster, 1991; Foster & Cairns, 1992). This result does not mean that transcription might not be mutagenic, nor that transcription and/or the repair synthesis associated with it might not contribute to the production of genetic variants in stationary cells. But some other process must be responsible for insuring that only adaptive mutations arise.

The two hypotheses discussed above suppose that the selective conditions result in a biased mutation rate in the genes that are under selection. An alternative hypothesis is that variants arise at random, but they are transient and only immortalized as mutations if they allow the cell to grow (Cairns *et al.*, 1988). This can be thought of as evolutionary 'trial and error' which would allow a cell (or a population of cells) to increase its genetic variability without suffering deleterious mutations. There are three classes of such hypotheses: the 'slow repair' model, the 'hypermutable state' model, and the 'extra DNA (or RNA)' model.

Stahl (1988) and Boe (1990) proposed that error-correcting enzymes, specifically the methyl-directed mismatch repair system, might be slow to act in nutritionally deprived cells. An error arising due to some limited DNA synthesis would persist for a while, but eventually be corrected back to the original sequence. If a good protein was made before the error was corrected, then the cell would replicate its DNA, resulting in a mutation. However, this model predicts that in the absence of mismatch repair, mutations should accumulate in the absence of selection, and this is not the case (Foster & Cairns, 1992). So mismatch repair is not the editing mechanism. None the less, recent evidence has suggested that the first part of this idea, that starving cells may be deficient in mismatch repair, may be correct (see below).

Hall (1990) shifted the idea of trial and error from individual cells to the entire population. He proposed that at any given time a subset of cells enter into an hypermutable state in which they accumulate mutations at a high rate throughout their genome. If such a cell achieved a useful mutation, it would exit the hypermutable state and begin to grow. However, if no useful mutation occurred after a time, the cell would die. On the level of the population, this process would appear to be adaptive because the majority of cells would have no mutations at all. Thus deleterious mutations would not accumulate in the population, but would accumulate in the cells that achieved success. (In fact, all the trial and error models predict that some level of non-selected mutations would occur in cells that achieve success, because any variants present at the time the cells began to grow would be immortalized).

The final class of trial and error models begins with the supposition that the cells under selection have extra copies of their genetic material which can be variant, and which are discarded if no useful variant arises. This could be at the mRNA level, and is the inefficient, random version of the reverse transcriptase model proposed by Cairns *et al.* (1988). The same result could be achieved if there were extra DNA copies (Cairns *et al.*, 1988). Prompted by our finding that adaptive reversion of the *lacZ* frameshift allele requires *E. coli*'s major recombinational enzyme, RecA (Cairns & Foster, 1991), we (Foster & Cairns, 1992) proposed that cells under selection are amplifying at random various parts of their genomes, and these extra copies accumulate mutations during the amplification process. If a useful mutation occurred in one copy, a good protein would be made, the cell would begin to grow, and the amplified array would be resolved by recombination. If, however, a useful mutation did not occur, the amplified array would still be resolved, but since the majority of the copies would not be mutant, the original sequence would be retained. Stahl (1992) has pointed out, however, that this model requires a large amplification array to arise in the absence of selection for it. He proposed an alternative source of extra DNA copies. In certain situations, RNA transcription, by producing R-loops, can initiate a form of DNA replication that is RecA-dependent (Kogoma, 1986). This synthesis might not proceed very far in a starving cell, and the new DNA would eventually be degraded. But, if this DNA contained a useful mutation and a good protein were made, normal replication would take over and the mutation would be immortalized (Stahl, 1992).

Recently Harris *et al.* (1994) showed that, in addition to RecA, the appearance of late revertants of the *lacZ* frameshift allele required RecBC and was enhanced in cells deficient for RecD. RecBCD is exonuclease V, which loads onto double strand breaks and degrades the DNA in both directions. After encountering a Chi site, RecD apparently leaves the complex, and RecBC(D⁻) helicase activity generates the single-stranded ends that initiate RecA-dependent recombination (Myers & Stahl, 1994). Harris *et al.* (1994) have proposed that double strand breaks accumulate in cells under stress, and these provide the molecular mechanism for the hypermutable state. In some cells the double strand breaks can be repaired by recombination with another copy of the gene, or with a partially homologous sequence. If during such recombination a useful mutation is produced, the cell will begin to grow and cease accumulating double strand breaks. If not, the cell will eventually die.

There is an alternative to this hypothesis that also accounts for all of the genetic results but that does not require cell death to eliminate useless variants (Foster & Trimarchi, 1994). Cells have a second form of RecA-dependent DNA replication that is initiated at special origins after DNA damage or other blocks to DNA replication (Magee *et al.*, 1992). This replication requires RecA and RecBC, and is enhanced if RecD is deficient, suggesting that it is initiated by D-loops (Asai *et al.*, 1993). Thus, the Stahl hypothesis (Stahl, 1992) can be simply modified by substituting D-loops for R-loops. By this model, D-loops would initiate DNA synthesis, but this synthesis would be limited in extent, and only become a productive replication fork if the new DNA contained a useful mutation and the cell received the benefit. If not, the new DNA would be degraded, and any useless mutations eliminated. It would be a satisfying simplification if the same D-loop origins are active in starving cells as are active after DNA replication is blocked.

Of course, both this hypothesis and the Harris hypothesis (Harris *et al.*,

1994) requires that at least two copies of the relevant region of the DNA be present. Thus, these events would be restricted to cells that have more than one copy of the episome (in the case of the *lacZ* frameshift allele) or of the chromosome, or that have duplicated or further amplified the relevant region of the DNA.

The molecular basis of adaptive reversion of the *lacZ* frameshift allele has recently been clarified by identifying the mutations responsible for early- and late-arising Lac$^+$ revertants (Foster & Trimarchi, 1994; Rosenberg *et al.*, 1994). One result was that multiple sequence changes, expected if revertants arise by recombination with only partially homologous sequences (Harris *et al.*, 1994; Hastings & Rosenberg, 1992; Higgins, 1992), were not found. More surprising, the mutations that gave rise to late-arising revertants were a single subset of all the possible mutations that can revert the allele. Although the mutations that occurred during growth included deletions, duplications, and possibly other complex changes, nearly all of the late-arising mutations were simple one base pair deletions, and 90% of them occurred at runs of 3 or more identical bases. Mutations at iterated bases are enhanced in cells that are deficient in mismatch repair (Cupples *et al.*, 1990; Schaaper & Dunn, 1987; Strand *et al.*, 1993; Fishel *et al.*, 1993; Leach *et al.*, 1993; Parsons *et al.*, 1993). Thus, this result supports the hypothesis (Stahl, 1988; Boe, 1990) that the DNA synthesis that takes place in starving cells, however it is initiated, may be highly error prone because mismatch repair activities are low, or, possibly, because the newly synthesized DNA is inaccessible to the mismatch repair enzymes (Foster & Trimarchi, 1994).

SOME CONCLUDING QUESTIONS

One major question is, how general is the phenomenon of adaptive mutation? To date, the phenomenon has been documented in bacteria and yeast, but not in any other organisms. Because adaptive mutation requires interactions between the organism's genome and the environment, it seems unlikely that it would occur in organisms that have evolved to isolate their germ lines from the environment. However, an adaptive mutational process occurring in somatic cells might lead to the success (for them) that we call cancer (Cairns, 1988; Foster 1991; Strauss, 1992).

A second question is how general is the phenomenon even among bacteria? It seems clear that point mutations, such as base substitutions and frameshifts, can give rise to adaptive changes, but the evidence about the movement of insertional elements is contradictory. This may not be surprising if we think of insertional elements as parasites under different selective pressures than their hosts. The bacteriophage lambda provides a useful model. Lambda utilizes the host's SOS response, which normally allows an uninfected bacterium to survive DNA damage, to escape the damaged cell, thereby killing its host.

A final question is, how general is the recombinational mechanism that underlies the adaptive reversion of the *lacZ* frameshift allele? In a number of other cases of adaptive, or late-arising, mutations, the process is clearly not dependent on RecA (Foster, 1993). Thus, there must be more than one mechanism for adaptive mutation. But, as discussed above, the RecA-dependent part of the process may be the initiation of DNA synthesis, and this initiation may be restricted to only certain regions of the genome. The underlying mechanism by which mutations are generated, eg. limited but error-prone DNA synthesis, may be general. And the mechanism by which non-adaptive variants are eliminated, eg. by destruction of the DNA or death of the cell, may be general. What may differ is the method by which DNA synthesis is initiated in different regions of the genome in cells under stress.

ACKNOWLEDGEMENTS

I thank my colleagues for stimulating discussions, especially J. Cairns, J.W. Drake, E. Eisenstadt, M.S. Fox and F.W. Stahl. Work in my laboratory was supported by USA National Science Foundation grant MCB-9213137.

REFERENCES

Ames, B. N. (1983). Dietary carcinogens and anticarcinogens. Oxygen radicals and degenerative diseases. *Science*, **221**, 1256–64.

Asai, T., Sommer, S., Bailone, A. & Kogoma, T. (1993). Homologous recombination-dependent initiation of DNA replication from DNA damage-inducible origins in *Escherichia coli*. *EMBO Journal*, **12**, 3287–95.

Benson, S. A., DeCloux, A. M. & Munro, J. (1991). Mutant bias in nonlethal selections results from selective recovery of mutants. *Genetics*, **129**, 647–58.

Boe, L. (1990). Mechanism for induction of adaptive mutations in *Escherichia coli*. *Molecular Microbiology*, **4**, 597–601.

Boe, L. (1992). Translational errors as the cause of mutations in *Escherichia coli*. *Molecular and General Genetics*, **231**, 469–71.

Bridges, B. A. (1994). Spontaneous mutation in stationary phase *Escherichia coli* WP2 carrying various DNA repair alleles. *Mutation Research*, **302**, 173–6.

Cairns, J. (1988). Origin of mutants disputed. *Nature, London*, **336**, 527–8.

Cairns, J., Overbaugh, J. & Miller, S. (1988). The origin of mutants. *Nature, London*, **335**, 142–5.

Cairns, J. (1990). Causes of mutation and Mu excision. *Nature, London*, **345**, 213.

Cairns, J. (1993). Directed Mutation, *Science*, **260**, 1221–2.

Cairns, J. & Foster, P. L. (1991). Adaptive reversion of a frameshift mutation in *Escherichia coli*. *Genetics*, **128**, 695–701.

Coulondre, C. Miller, J. H., Farabough, P. J. & Gilbert, W. (1978). Molecular basis of base substitution hotspots in *Escherichia coli*. *Nature, London*, **274**, 775–80.

Cupples, C. G., Cabrera, M., Cruz, C. & Miller, J. H. (1990). A set of *lacZ* mutations in *Escherichia coli* that allow rapid detection of specific frameshift mutations. *Genetics*, **125**, 275–80.

Davis, B. D. (1989). Transciptional bias: a non-Lamarckian mechanism for substrate-induced mutations. *Proceedings of the National Academy of Sciences, USA*, **86**, 5005–9.

De Felice, M., Levinthal, M., Laccarino, M. & Guardiola, J. (1979). Growth inhibition as a consequence of antagonism between related amino acids: effect of valine in *Escherichia coli* K-12. *Microbiology Review*, **43**, 42–58.

Delbrück, M. (1946). Heredity and variations in microorganisms. *Cold Spring Harbor Symposia in Quantitative Biology*, **11**, 154.

Fishel, R., Lescoe, M.K., Rao, M. R. S., Copeland, N. B., Jenkins, N. A., Garber, J., Kane, M. & Kolodner, R. (1993). The human mutator gene homolog MSH2 and its association with hereditary nonpolyposis colon cancer. *Cell*, **75**, 1027–38.

Foster, P. L. (1991) Directed mutation in *Escherichia coli*: theory and mechanisms. In *Organism and the Origin of Self*. (Tauber, A. I, ed.) pp. 213–234. Kluwer Academic Publishers, The Netherlands.

Foster, P. L. (1993). Adaptive mutation: the uses of adversity. *Annual Review Microbiology*, **47**, 467–504.

Foster, P. L. (1994). Population dynamics of a Lac⁻ strain of *Escherichia coli* during selection for lactose utilization. *Genetics*, in press.

Foster, P. L. & Cairns, J. (1992). Mechanisms of directed mutation. *Genetics*, **131**, 783–9.

Foster, P. L. & Cairns, J. (1994). The occurrence of heritable *Mu* excisions in starving cells of *Escherichia coli*. *EMBO Journal*, in press.

Foster, P. L. & Trimarchi, J. M. (1994). Adaptive reversion of a frameshift mutation in *E. coli* occurs by simple base deletions at homopolymeric runs. *Science*, **265**, 407–9.

Grivell, A. R., Grivell, M. B. & Hanawait, P. C. (1975). Turnover in bacterial DNA containing thymine or 5-bromouracil. *Journal of Molecular Biology*, **98**, 219–33.

Hall, B. G. (1988). Adaptive evolution that requires multiple spontaneous mutations. I. Mutations involving an insertion sequence. *Genetics*, **120**, 887–97.

Hall, B. G. (1990). Spontaneous point mutations that occur more often when they are advantageous than when they are neutral. *Genetics*, **126**, 5–16.

Hall, B. G. (1991). Spectrum of mutations that occur under selective and non-selective conditions in *E. coli*. *Genetica*, **84**, 73–6.

Hall, B. G. (1992). Selection-induced mutations occur in yeast. *Proceedings of National Academy of Sciences, USA*, **89**, 4300–3.

Hall, B. G. (1994). On alternatives to selection-induced mutation in the BgI operon of *Escherichia coli*. *Molecular Biology of Evolution*, **11**, 159–68.

Harris, R. S., Longerich, S. & Rosenberg, S. M. (1994). Recombination in adaptive mutation. *Science*, **264**, 258–60.

Hastings, P. J. & Rosenberg, S. M. (1992). Gene conversion. In *Encyclopedia of Immunology*. (Roitt, I. M. & Delves, P. J. eds.) pp. 602–605. Saunders Scientific Publications, London.

Higgins, N. P. (1992). Death and transfiguration among bacteria. *TIBS*, **17**, 207–11.

Kogoma, T. (1986). RNase H-defective mutants of *Escherichia coli*. *Journal of Bacteriology*, **166**, 361–3.

Leach, F. S., Nicolaides, N. C., Papadopoulos, N., Liu, B., Jen, J., Parsons, R., Peltomäki, P., Sistonen, P., Aaltonen, L. A., Nyström-Lahti, M., Guan, X.-Y., Zhang, J., Meltzer, P. S., Yu, J.-W., Kao, F.-T., Chen, D. J., Cerosaletti, K. M., Fournier, R. E. K., Tood, S., Lewis, T., Leach, R. J., Naylor, S. L., Weissenbach, J., Mecklin, J.-P., Järvinen, H., Petersen, G. M., Hamilton, S. R., Green, J., Jass, J., Watson, P., Lench, H. T., Trent, J. M., de la Chapelle, A., Kinzler, K. W. & Vogelstein, B. (1993). Mutations of a mutS homolog in hereditary nonpolyposis colorectal cancer. *Cell*, **75**, 1215–25.

Lenski, R. E., Slatkin, M. & Ayala, F. J. (1989). Mutation and selection in bacterial populations: alternatives to the hypothesis of directed mutation. *Proceedings of the National Academy of Sciences, USA*, **86**, 2775–8.

Lenski, R. E. & Mittler, J. E. (1993). The directed mutation controversy and neo-Darwinism. *Science*, **259**, 188–94.

Lindahl, T. & Nyberg, B. (1972). Rate of depurination of native deoxyribonucleic acid. *Biochemistry*, **11**, 3610–18.

Luria, S. E. (1951). The frequency distribution of spontaneous bacteriophage mutants as evidence for the exponential rate of phage reproduction. *Cold Spring Harbor Symposia in Quantitative Biology*, **16**, 463–70.

Luria, S. E. & Delbrück, M. (1943). Mutations of bacteria from virus sensitivity to virus resistance. *Genetics*, **28**, 491–511.

MacPhee, D. G. (1993). Directed evolution reconsidered. *American Scientist*, **81**, 554–61.

Maenhaut-Michel, G. & Shapiro, J. A. (1994). The roles of starvation and selective substrates in the emergence of *araB–lacZ* fusion clones. *EMBO Journal*, in press.

Magee, T. R., Asai, T., Malka, D. & Kogoma, T. (1992). DNA damage-inducible origins of DNA replication in *Escherichia coli*. *EMBO Journal*, **11**, 4219–25.

Mellon, I. & Hanawalt, P. C. (1989). Induction of the *Escherichia coli* lactose operon selectively increases repair of its transcribed DNA strand. *Nature, London*, **342**, 95–8.

Mittler, J. E. & Lenski, R. E. (1990). New data on excisions of Mu from *E. coli* MCS2 cast doubt on directed mutation hypothesis. *Nature, London*, **344**, 173–5.

Mittler, J. E. & Lenski, R. E. (1992). Experimental evidence for an alternative to directed mutation in the *bgl* operon. *Nature, London*, **356**, 446–8.

Myers, R. K. & Stahl, F. W. (1994). Chi and the RecBCD enzyme of *Escherichia coli*. *Annual Review in Genetics*.

Naas, T., Blot, M., Fitch, W.M. & Archer, W. (1994). Insertion sequence-related genetic variation in resting *Escherichia coli* K-12. *Genetics*, **136**, 721–30.

Ninio, J. (1991) Transient mutators: a semiquantitative analysis of the influence of translation and transcription errors on mutation rates. *Genetics*, **129**, 957–62.

Parsons, R., Li, G.-M., Longley, M. J., Fang, W.-H., Papadopoulos, N., Jen, J., de la Chapella, A., Kinzler, K. W., Vogelstein, B. & Modrich, P. (1993). Hyper-mutability and mismatch repair deficiency in RER[+] tumor cells. *Cell*, **75**, 1227–36.

Rebeck, G. W. & Samson, L. (1991). Increased spontaneous mutation and alkylation sensitivity of *Escherichia coli* strains lacking the *ogt* O[6]-methylguanine DNA repair methyltransferase. *Journal of Bacteriology*, **173**, 2068–76.

Rosenberg, S. M., Longerich, S., Gee, P. & Harris, R. S. (1994). Adaptive mutation by deletions in small mononucleotide repeats. *Science*, **265**, 405–7.

Ryan, F. J. (1955). Spontaneous mutation in non-dividing bacteria. *Genetics*, **40**, 726–38.

Ryan, F. J., Nakada, D. & Schneider, M. J. (1961). Is DNA replication a necessary condition for spontaneous mutation? *Z. Vererbungsl*, **92**, 38–41.

Schaaper, R. M. & Dunn, R. L. (1987). Spectra of spontaneous mutations in *Escherichia coli* strains defective in mismatch correction: the nature of *in vivo* DNA replication errors. *Proceedings of National Academy of Sciences, USA*, **84**, 6220–4.

Shapiro, J. A. (1984). Observations on the formation of clones containing *araB–LacZ* cistron fusions. *Molecular and General Genetics*, **194**, 79–90.

Stahl, F. W. (1988). A unicorn in the garden. *Nature, London*, **335**, 112–13.

Stahl, F. W. (1992). Unicorns revisited. *Genetics*, **132**, 865–7.

Steele, D. F. & Jinks-Robertson, S. (1992). An examination of adaptive reversion in *Saccharomyces cerevisiae*. *Genetics*, **132**, 9–21.

Steele, D. F. & Jinks-Robertson, S. (1993). Time-dependent mitotic recombination in *Saccharomyces cerevisiae*. *Current Genetics*, **23**, 423–9.

Stewart, F. M., Gordon, D. M. & Levin, B. R. (1990). Fluctuation analysis: the probability distribution of the number of mutants under different conditions. *Genetics*, **124**, 175–85.

Strand, M., Prolla, T. A., Liskay, R. M. & Petes, T. D. (1993). Destabilization of tracts of simple repetitive DNA in yeast by mutations affecting DNA mismatch repair. *Nature, London*, **365**, 274–6.

Strauss, B. S. (1992). The origin of point mutations in human tumor cells. *Cancer Research*, **52**, 249–53.

Tormo, A., Almiron, M. & Kolter, R. (1990). *surA*, an *Escherichia coli* gene essential for survival in stationary phase. *Journal of Bacteriology*, **172**, 4339–47.

BARRIERS TO RECOMBINATION: RESTRICTION

VICTORIA A. BARCUS AND NOREEN E. MURRAY

Institute of Cell and Molecular Biology, Darwin Building, King's Buildings, Edinburgh, EH9 3JR, UK

INTRODUCTION

It is now more than 40 years since it was recognized that the ability of a bacteriophage to infect a particular host could be influenced by the strain in which it was last propagated (Luria & Human, 1952; Bertani & Weigle, 1953). The molecular basis, but not necessarily the biological significance, of this observation is now well understood. It has been shown that bacteria impose distinctive patterns of modification not only on their own DNA, but similarly on the DNA of phages replicating within them. As a consequence, DNA lacking the resident modification pattern, classically the methylation of specific nucleotide sequences (targets), can be distinguished as foreign (Dussoix & Arber, 1962) and attacked by a target-specific endonuclease against which the specific pattern of modification protects the resident DNA (Kühnlein, Linn & Arber, 1969). The phenomenon whereby phages grown in one host were prevented from infecting another was termed restriction, the relevant endonucleases became known as restriction enzymes and we now commonly talk rather loosely of 'restricting' DNA *in vitro* by the addition of an endonuclease. A number of recent reviews (Bickle, 1987; Wilson & Murray, 1991; Bickle & Krüger, 1993; Halford *et al.*, 1993; Heitman, 1993) provide more detailed accounts of published material.

The biological barrier provided by restriction is far from complete, and it can be quantified by determining the titre of unmodified phages on the restricting strain relative to the titre on a non-restricting strain. For phage genomes with small numbers of targets, the barrier increases with target number, as witnessed by a decrease in the efficiency of plating (eop) (Murray, Manduca de Ritis & Foster, 1973a). Foreign DNA can be recognized and attacked by restriction endonucleases irrespective of its mode of entry into the cell and nature has many tricks for combating restriction. Nevertheless, a recipient strain with a restriction system is endowed with a potential barrier to the intrusion of large DNA molecules should these molecules include the relevant, but unmodified, target sequences. Whether or not the cutting of large DNA molecules into smaller ones is necessarily a barrier to genetic recombination is of critical concern, since DNA breaks can be substrates for recombination.

The restriction of foreign DNA is dependent on the donor bacterium lacking a modification system that protects its DNA against attack by a restriction system present within a recipient cell. Any evaluation of the importance of restriction to the transfer of genetic information requires an awareness of the diversity of restriction specificities within a bacterial species. This review considers what is known about the diversity of restriction systems within the *Enterobacteriaceae*, but most particularly within the best studied species, *Escherichia coli*. In the absence of any truly systematic analyses, the focus will be primarily on screens for chromosomally encoded restriction and modification (R–M) systems.

TYPES OF RESTRICTION SYSTEMS

The classical view of a restriction system is one in which a restriction enzyme (endonuclease) is necessarily accompanied by a modification enzyme (methyltransferase). Both recognize the same target sequence but the endonuclease only attacks DNA if the target sequence is unmethylated. Restriction systems that meet this definition are commonly divided into three types, although one enzyme has been designated type IV (Janulaitis *et al.*, 1992).

The type I enzymes, the first to be studied in any detail, are the most complex. *E. coli* K-12 possesses such an enzyme, designated *Eco*KI. This R–M system was detected when phage λ grown on *E. coli* strain C was found to form plaques with poor efficiency (an eop of 2×10^{-4}) on *E. coli* K-12 (Fig. 1). All type I R–M systems are made up of three different subunits encoded by three genes, *hsdR*, M and S (where *hsd* is the acronym for *host specificity of DNA*). They require, in addition to Mg^{2+}, ATP and S-adenosylmethionine as cofactors, the S-adenosylmethionine being required for both restriction and modification. Type I R–M systems confounded experimenters by making cuts at non-specific sequences quite remote from the unmethylated target sequence (Horiuchi & Zinder, 1972; Adler & Nathans, 1973; Murray, Batten & Murray, 1973b; Bickle, Brack & Yuan, 1978; Rosamond, Endlich & Linn, 1979). For this reason type I systems are not of significant use to molecular biologists and consequently have not been the subject of extensive searches by biotechnologists. The genes that encode *Eco*KI map at 98.5 min on the bacterial chromosome, though not all determinants of type I R–M systems, even in *E. coli*, are chromosomally located. It is absolutely characteristic of type I enzymes that a single complex of three different subunits (HsdR, M and S) functions as both an endonuclease and a methyltransferase, the activity being determined by the methylation state of the target sequence (see Bickle, 1987). All target sequences so far identified are bipartite with the two components separated by a non-specific spacer of six to eight nucleotides. One component is a trinucleotide, the other a tetra- or penta-nucleotide; *Eco*KI, for example, recognizes the sequence AAC (N_6) GTGC (Kan *et al.*, 1979).

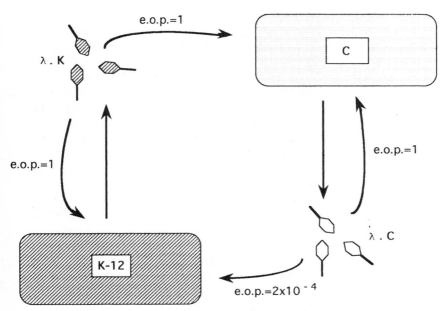

Fig. 1. Host-controlled restriction of bacteriophage λ. *E. coli* K-12 possesses, while *E. coli* C lacks, a type I R–M system. Phage λ propagated in *E. coli* C (λ.C) is not protected from restriction by *Eco*KI and thus plates with reduced efficiency on *E. coli* K-12 as compared to *E. coli* C. Phages escaping restriction are modified by the *Eco*KI methyltransferase (λ.K), and consequently plate equally well on *E. coli* K-12 and C. Modified DNA is indicated by hatch marks. (Adapted from Arber, 1971).

Type II restriction enzymes include those used as laboratory reagents, for example *Eco*RI and *Eco*RV. These are the simplest R–M systems. A dimeric endonuclease usually recognizes and cuts within, or close to, a rotationally symmetrical target sequence and a separate, usually monomeric, modification enzyme is responsible for the methylation of the same target sequence. While some type II enzymes do not meet all aspects of this definition, they are always relatively simple and require only Mg^{2+} as cofactor for their endonucleolytic activity (see reviews cited earlier).

Type III systems are intermediate in complexity between type I and type II. The archetypal type III R–M enzyme is encoded by coliphage P1. This enzyme is a complex of two different subunits, requires ATP and, like type I systems, has the potential for cutting or modifying DNA. The DNA is cut at a fixed, but short, distance from a target sequence.

Other restriction endonucleases fail to meet the classical expectations. These are active only when certain cytosine or adenine residues are methylated and therefore such restriction enzymes are not associated with cognate modification enzymes. Methylation dependent restriction (MDR) systems were the cause of the very first restriction barriers detected (Luria &

Human, 1952). These systems, now termed McrA and McrBC, are activated by methylated cytosines, though they were first identified by their activity on T-even phages, but only when the phage DNA lacked its protective glycosyl residues. In the absence of glycosylation but in the context of the correct nucleotide sequences, hydroxymethylcytosine sensitizes the phage DNA to attack by McrA and McrBC. The latter are also referred to as Rg1A and Rg1B on the basis of their ability to *restrict glucoseless* T-even phage DNA (Raleigh, Trimarchi & Revel, 1989).

All known types of R–M systems are found in *E. coli*. Both type I and the MDR systems are encoded by the genome of *E. coli* K-12. The genes specifying *Eco*KI are immediately flanked by genes encoding two MDR systems (*mrr* and *mcrBC*). Mrr is noteworthy in that its substrates include some sequences in which an adenosyl residue is methylated, and others which contain 5-methylcytosine (Waite-Rees *et al.*, 1991; Kelleher & Raleigh, 1991). The segment of the *E. coli* K-12 genome encoding Mrr, *Eco*KI and McrBC has been referred to as the 'immigration control region' (see Raleigh, 1992). All the relevant genes are readily removed by a single deletion and growth of the resulting strain under laboratory conditions appears to be normal. Additional R–M systems, types I, II and III, are readily provided by natural plasmids.

Salmonella typhimurium LT2 (renamed *S. enterica* serovar typhimurium LT2), the second most commonly used enteric bacterium in the laboratory, has three chromosomally encoded R–M systems each of which has been removed by mutations (Bullas & Ryu, 1983). *Sty*LTI is a type III system (Dartois, De Backer & Colson, 1993). *Sty*LTIII (previously called *Sty*SB) is type I (Van Pel & Colson, 1974), and *Sty*LTII (previously called *Sty*SA) remains to be characterized. The genetic determinants of *Sty*LTII and III are closely linked, the latter being alleles of the *hsd* genes of *E. coli* K-12.

THE FAMILY CONCEPT

Some indication of the variety of restriction systems in *E. coli* was available even in the late 1960s when four commonly used laboratory strains of *E. coli*, namely K-12, B, C and 15T⁻, were known to differ from each other with respect to R–M systems that recognize the provenance of bacteriophage λ. *E. coli* C remains the one well-documented *E. coli* strain that naturally lacks such a system, while K-12, B and 15T⁻ have allelic, chromosomal, *hsd* genes which confer diagnostic specificities to their respective strains. In addition, *E. coli* 15T⁻ harbours a plasmid encoding a type III R–M system very similar to that encoded by coliphage P1 (Iida *et al.*, 1983).

Of fundamental value to our understanding of type I R–M systems has been the finding that they exist as closely related members of a family. This was first indicated by the demonstration that mutants with defects in the genes encoding *Eco*KI and *Eco*BI could complement each other. It was inferred from such tests that each enzyme comprised three subunits, that the subunits were interchangeable and that the subunit encoded by one gene, *hsdS*, confers sequence *S*pecificity to the multimeric complex (Boyer & Roulland-Dussoix, 1969; Glover & Colson, 1969; Hubácek & Glover, 1970).

A first hint that allelic genes might also encode sufficiently dissimilar type I R–M systems to warrant their separation into a different family came from hybridization screens of bacterial DNAs and serological screens of cell extracts (Murray *et al.*, 1982). It was found, as expected, that the nucleotide sequences of the *hsd* genes for *Eco*KI and *Eco*BI would hybridize to each other and antibodies raised against *Eco*KI reacted with *Eco*BI. In contrast, DNA probes comprising the *Eco*KI genes failed to hybridize with those of *E. coli* 15T⁻ which encoded *Eco*AI; similarly antibodies against *Eco*KI did not cross-react with *Eco*AI. The *hsd* genes in these two strains behave as alleles in genetic tests, but are of very different nucleotide sequence (Daniel *et al.*, 1988).

At least 13 chromosomally encoded type I R–M systems have been identified in enteric bacteria (*E. coli*, *Salmonella* and *Citrobacter*), with *Eco*KI having six relatives and *Eco*AI five (Table 1). The former are referred to as type IA and the latter as type IB (Bickle, 1987). A third family, type IC, headed by *Eco*R124I has plasmid-encoded members, although genes for a chromosomally encoded relative of *Eco*R124I have been identified in a region of the genome distinct from the 'immigration control region' (Tyndall, Meister & Bickle, 1994). There is now evidence for a fourth family of type R–M enzymes (see Table 1). The first representative (*Sty*SBLI) is the R–M system of *Salmonella enterica* serovar blegdam initially identified on the basis of biological tests (Bullas, Colson & Neufeld, 1980). The genes behave as alleles of those encoding *Eco*KI or *Eco*AI, but they fail to cross-hybridize to the *hsd* genes encoding type IA, IB or IC enzymes (A.J.B. Campbell & N.E. Murray, unpublished observations).

In summary, type I R–M systems are readily divided into discrete families with members within each family differing in the subunit that determines specificity. It seems relevant that a change in the specificity subunit will concomitantly affect both modification and restriction. Type I R–M systems of new specificity can be made quite easily, and currently these are the only examples where R–M systems of new specificities have emerged in the laboratory by well-documented, sometimes unsolicited, natural processes.

DIVERSIFICATION OF TYPE I SPECIFICITY SYSTEMS

The sequences of a variety of specificity (*hsdS*) genes of type IA systems were first compared to identify differences that would correlate with the recognition of different target sequences (Gough & Murray, 1983). Two long (~450 bp) regions were identified within each *S* gene, so-called

Table 1. Type I R–M systems identified in enteric bacteria

Family	Enzyme	Target	Reference[a]
IA	EcoKI	AAC (N$_6$) GTGC	Kan et al., 1979
	EcoBI	TGA (N$_8$) TGCT	(Lautenberger et al., 1978
			(Ravetch, Horiuchi & Zinder, 1978
			(Somer & Schaller, 1979
	EcoDI	TTA (N$_7$) GTCY	Nagaraja et al., 1985a
	StyLTIII[b]	GAG (N$_6$) RTAYG	Nagaraja et al., 1985b
	StySPI	AAC (N$_6$) GTRC	Nagaraja et al., 1985b
	EcoR5I	unknown[c]	V.A. Barcus, unpublished observations
	EcoR23I	unknown[c]	V.A. Barcus, unpublished observations
IB	EcoAI	GAG (N$_7$) GTCA	(Kröger & Hobom, 1984
			(Siri, Shepherd & Bickle, 1984
	EcoEI	GAG (N$_7$) ATGC	Cowan et al., 1989
	EcoR42I	unknown[c]	V.A. Barcus, unpublished observations
	CfrAI	GCA (N$_8$) GTGG	Kannan et al., 1989
	StySTI	unknown[c]	A.J.B. Campbell & N.E. Murray, unpublished observations
	StySKI	unknown[c]	D. Ternent & N. E. Murray, unpublished observations
IC	EcoR124I	GAA (N$_6$) RTCG	Price, Shepherd & Bickle, 1987
	EcoDXXI	TCA (N$_7$) RTTC	Gubler et al., 1992
	EcoprrI	CCA (N$_7$) RTGC	Tyndall et al., 1994
ID?	StySBLI	unknown[c]	A.J.B. Campbell & N.E. Murray, unpublished observations
	EcoR9I	unknown[c]	A.J.B. Campbell & N.E. Murray, unpublished observations

[a]Where the target sequence has been determined, the reference for this work is given.
[b]StyLTIII previously known as StySB.
[c]Target sequences unknown but each enzyme shown to differ from other members of the same family.

'variable' regions since they were different in S genes conferring different specificities. The variable regions were shown to correlate with target recognition domains (TRDs) in the specificity polypeptide. The amino-variable region specifies the trinucleotide component of the target sequence (Cowan, Gann & Murray, 1989) and the carboxy-variable region is believed to equate with the TRD for the tetra- (or penta-) nucleotide component (Fuller-Pace et al., 1984; Nagaraja, Shepherd & Bickle, 1985c). The hsdS genes of members of the same family, whether type IA, B or C, have a central region of nearly identical sequence flanked by TRDs which are very different in genes that confer different specificities (Gough & Murray, 1983; Kannan et al., 1989; Meister et al., 1993) (Fig. 2). When two enzymes share a common component in their bipartite recognition sequence, they have similar TRDs (Fuller-Pace & Murray, 1986; Cowan et al., 1989). EcoAI and EcoEI, for example, both recognize the trinucleotide 5'GAG and their amino TRDs share 80% identity (Cowan et al., 1989). This organization of two variable regions separated by a sequence common to all members of the same family immediately suggests that homologous recombination between

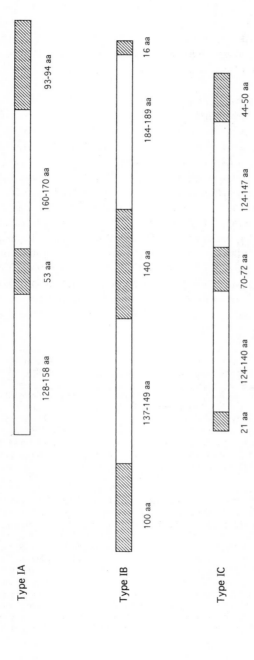

Fig. 2. A schematic diagram of the specificity polypeptides of three families of type I R–M systems. The hatched regions are conserved between family members. A central conserved region separates two variable regions (unhatched). These correspond to the two target recognition domains (TRDs). The ranges of sizes for the different conserved and variable regions are given in amino acids (aa).

different *hsdS* genes could generate new combinations of TRDs. A new R–M system encountered as the by-product of P1 mediated transduction (Bullas, Colson & Van Pel, 1976) arose in this way (Fuller-Pace *et al.*, 1984) and was shown to recognize the predicted target sequence (Nagaraja *et al.*, 1985c). More recent experiments have relied on either recombination *in vivo* (Gann *et al.*, 1987) or the joining of DNA fragments *in vitro* (Gubler *et al.*, 1992) to produce new combinations of TRDs by design rather than chance (see Fig. 3).

A second serendipitous change in DNA specificity arose in the plasmid encoded type IC system *Eco*R124I (Hughes, 1977). This reversible change can also be attributed to recombination. The recombination event would require unequal crossing-over between a misaligned 12 bp duplication located in the central conserved region of the *hsdS* gene. As a consequence, the new gene has a triplication of the 12 bp sequence and the two components of the new target sequence are separated by seven, rather than six, base pairs (Price *et al.*, 1989; Gubler & Bickle, 1991). A third spontaneous variant in the type IC family arose by transposition of Tn5 into the distal part of the *hsdS* gene of the *Eco*DXXI system, resulting in a truncated coding sequence retaining the TRD for the trinucleotide, but not the tetranucleotide, component of the *Eco*DXXI target sequence. The target of this novel specificity system is still a hyphenated sequence, but the trinucleotide flanks the spacer in a symmetrical arrangement (Meister *et al.*, 1993). A similar result has been achieved for *Eco*R124I, but this time the deletions

---

Fig. 3. Evolution of type I R–M systems with new specificities. (a) Recombination between *hsdS* genes produces hybrid genes and chimaeric S polypeptides. *Sty*SPI and *Sty*LTIII are naturally occurring type I R–M systems (see Table 1). *Sty*SQ and SJ have hybrid *hsdS* genes (Fuller-Pace *et al.*, 1984; Gann *et al.*, 1987). The regions originating from *Sty*SPI are hatched, those from *Sty*LTIII are stippled. Reassortment of the target recognition domains (TRDs) accordingly gave rise to recombinant recognition sequences (Nagaraja *et al.*, 1985c; Gann *et al.*, 1987). Site-directed mutagenesis of the central conserved region of the *Sty*SQ *hsdS* gene produced *Sty*SQ*, comprising only the amino variable region from *Sty*SPI and the remainder from *Sty*LTIII. The *Sty*SQ* target sequence confirms that the amino variable region is in fact a TRD responsible for recognition of the trinucleotide component of the sequence (Cowan *et al.*, 1989). (b) Sequence specificity may also be altered by changing the length of the non-specific spacer of the target sequence. The S polypeptides of *Eco*R124I and *Eco*R124II differ only in the number of times a short amino acid motif (X = TAEL) is repeated within their central conserved regions (Price *et al.*, 1989), resulting in the extension of the spacer in the target sequence from N_6 in *Eco*R124I to N_7 in *Eco*R124II. The recognition sequence of *Eco*DXXI also contains a non-specific spacer of N_7, corresponding to three TAEL repeats in its S polypeptide (Gubler *et al.*, 1992). Chimaeric S polypeptides recognize the predicted target sequences (Gubler *et al.*, 1992). (Adapted from Heitman, 1993.) (c) Mutations in the *hsdS* genes resulting in truncated S polypeptides may produce R–M systems with new specificities. The specificity polypeptides, lacking carboxyl TRDs, recognize sequences which are interrupted palindromes in which the trinucleotide component of the wild-type site is repeated in inverted orientation (Abadjieva *et al.*, 1993; Meister *et al.*, 1993). In the case of *Eco*DXXsI, the spacer has an additional base pair than the wild-type sequence, maintaining the proper spacing between the methylatable adenines (Meister *et al.*, 1993). R, either purine; Y, either pyrimidine; N, any base.

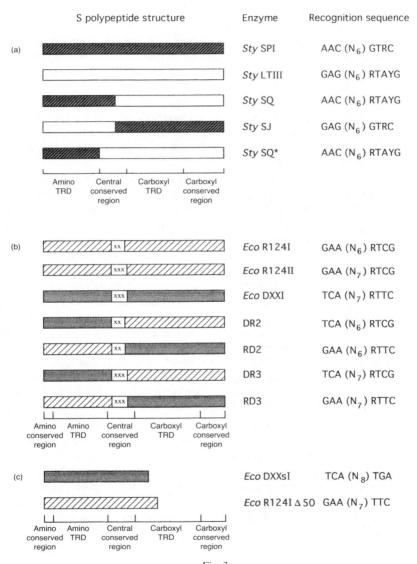

Fig. 3.

were the products of a mutational analysis (Abadjieva *et al.*, 1993). The new variants, their modes of origin and target sequences are summarized in Fig. 3.

The HsdM (M) and HsdS (S) subunits are necessary and sufficient for an active methyltransferase (see Bickle, 1987). For *Eco*R124I (Taylor *et al.*, 1992) and *Eco*KI (Dryden, Cooper & Murray, 1993) the stoichiometry of the active enzyme is M_2S_1, a ratio consistent with one M subunit per TRD. The composition of *Eco*KI, the complex active in both restriction and modification, is $R_2M_2S_1$ (Weiserova *et al.*, 1993; D.T.F. Dryden, unpublished observations).

The hyphenated, symmetrical, target sequences detected for derivative systems with truncated S genes, imply that two 'half' specificity subunits can substitute for the single wild-type specificity subunit with two TRDs. This would be consistent with a complex of $R_2M_2S^*_2$ rather than $R_2M_2S_1$, where S^* is a truncated S subunit.

The comparative sequence analyses that identified the TRDs also provided evidence for gene duplication in the origin of the present *hsdS* genes (Gough & Murray, 1983; Kannan *et al.*, 1989). Ancestral enzymes may have had the composition $R_2M_2S_2$, and perhaps originally even $R_1M_1S_1$. While the evidence for gene duplication is relatively weak for the type IA family (Argos, 1985), because the first part of the gene appears to have been deleted, the evidence is very obvious in the IB family. In this case a sequence of 192 nucleotides in the N-terminal conserved region is repeated in the central conserved region, with 55 of the 64 encoded amino acids being identical (Kannan *et al.*, 1989).

All the above lines of evidence indicate routes for the evolution of new type I R–M systems. While it is easy to chart changes involving either new combinations or different spatial arrangements of TRDs, the generation of new TRDs has not been witnessed, despite experiments designed to select for a relaxation in the specificity of *Eco*KI (5' AAC(N_6) GTGC) to *Sty*SPI (5' AAC(N_6) GTG/AC) (N.E. Murray, unpublished observations). Since two recognition domains that confer the same specificity share many amino acids in common, between 80 and 100% when the enzymes are members of the same family, but less than 50% when they are in different families (Cowan *et al.*, 1989), many amino acid changes may be permissible without loss of specificity. It seems probable that the accumulation of a substantial number of changes may occur, and may well be necessary, before a particular amino acid substitution will define a different nucleotide in the target sequence.

RELATIONSHIPS BETWEEN FAMILIES OF TYPE I R–M SYSTEMS

The *nucleotide* sequences of the genes for different families of type I R–M systems are so dissimilar that with one exception they fail to provide evidence of relatedness even when the 'allelic' *hsd* genes of type IA and IB systems are compared. The coding sequences for TRDs specifying 5'GAG are the exception. In this instance a comparison of nucleotide sequences quite clearly identified interfamily similarities that reflect a common origin rather than convergent evolution (Cowan, 1988).

The M polypeptides of all three families of type I systems share motifs common to adenine methylases but the alignment of their sequences indicated only between 23 and 33% identity, with the higher figures for comparisons between the systems encoded by allelic genes (IA/IB) and the lower figures for comparisons between chromosomal and plasmid-borne genes (e.g. IA/IC). The higher figures are above the 'twilight zone' in which it is difficult to differentiate between divergence and convergence (Doolittle *et al.*, 1986), and Sharp *et al.* (1992) have argued that the *hsdM* genes have a common origin that predates recent branches of the eubacterial phylogenetic tree. The identities shared by R polypeptides are at best only 26%, but there seems no reason to propose that type I R–M systems have evolved more than once, rather that there has been selection for diversity, probably aided by horizontal transfer of R–M genes (Murray *et al.*, 1993). These views are further supported by the following screens of natural isolates of *E. coli*.

SCREENING WILD-TYPE POPULATIONS OF *E. COLI* FOR RESTRICTION AND MODIFICATION SYSTEMS

The utility of type II restriction enzymes to modern molecular biology has resulted in extensive searches for these R–M systems in almost every genus of bacteria. Type II R–M systems may be detected by incubating cell extracts with suitable DNA substrates, then looking for discrete fragments separated by agarose gel electrophoresis. Such methods have allowed enzymes with over 200 different restriction specificities to be identified (Kessler & Manta, 1990; Roberts & Macelis, 1993).

In spite of the identification of so many restriction endonucleases, little has been reported about the relative frequency, distribution and diversity of R–M systems in natural populations of bacteria. Janulaitis and coworkers have screened a number of species of the *Enterobacteriaceae* for sequence-specific endonucleases using the method described above (Janulaitis *et al.*, 1988; A. Janulaitis, pers. comm.). Specific endonuclease activity was detected in 25% of nearly one thousand *E. coli* strains tested. A second screening experiment looked for restriction activity encoded by transmissible antibiotic resistance plasmids in *E. coli* (Glatman *et al.*, 1980). The plasmids were transferred to *E. coli* K-12 and the exconjugants tested for restriction of λ phage *in vivo*. Approximately 10% of the transmissible antibiotic resistance plasmids were correlated with restriction of λ, and the endonucleases responsible were shown to be type II.

To date, many type II and type III systems in *E. coli* have been shown to be plasmid borne. Because of the transmissible nature of many plasmids, the frequency with which they transfer R–M systems between strains could be quite high and their maintenance subject to a variety of selection pressures not associated with the *hsd* genes. While some type I R–M systems are plasmid encoded, the majority of *hsd* genes have been found to be chromosomal. A more informative indicator of the general presence of R–M systems in *E. coli* may emerge from the analysis of chromosomally encoded systems.

Detection of type I R–M systems has relied on restriction of bacteriophages *in vivo* (Bertani & Weigle, 1953; Colson & Colson, 1971; Bullas & Colson, 1975; Bullas *et al.*, 1980), or molecular screens using probes derived from *hsd* genes of known families (Daniel *et al.*, 1988). Unfortunately, a biological screen for R–M systems in wild-type bacterial isolates using bacteriophages is limited to a minority of strains, because most strains are not sensitive to phage infection. Bullas and coworkers (1980) screened *Salmonella* strains representing 85 serotypes for evidence of restriction of four bacteriophages. Of these, only 28 were found to be sensitive to at least one of the tester phages. The restriction phenotypes of these 28 strains were then assayed by determining the eop of phages carrying DNA modified by one or more of the systems of *S. enterica* serovar typhimurium LT2. In this way, restriction systems were detected in 12 *Salmonella* strains. Even resistance to one of the four phage types meant that restriction systems could be overlooked; phage L, for example, was the only indicator of *Sty*LTII restriction (Bullas *et al.*, 1980). The genes for eight R–M systems were transferred to *E. coli/Salmonella* hybrids by P1 cotransduction with *serB*. Four systems proved to be allelic to *Sty*LTIII (then known as *Sty*SB); the resident *hsd* genes were replaced by genes for novel systems. These results were interpreted as an indication that these *serB*-linked systems were type I. Further assays demonstrated that each of the eight systems had a novel specificity. It may be that other isolates of the 85 serovars tested also encode type I R–M systems but were not detectable owing to resistance to the phages used in the screen.

The majority of wild-type *E. coli* strains are resistant to some aspect of productive infection by most common laboratory bacteriophages including P1 (Barcus, 1993). For *Streptomyces albus* G, though not so far for *E. coli*, restriction can cause complete resistance to phage attack (Chater & Wilde, 1980). Moreover, the absence of progeny phages precludes a check for modification. Thus, although a screen using phages as indicators of restriction may detect some systems, it will not give a clear indication of their prevalence in nature.

Screening by DNA hybridization, though not confined to phage-sensitive bacteria, is restricted by the availability of probes. Genes encoding members of known families of type I systems may be detected by hybridization to DNA probes constructed from representative *hsd* genes. Using probes made from type IA and type IB *hsd* genes, Daniel *et al.* (1988) screened a number of wild-type *E. coli* isolates for evidence of related type I R–M systems. Five out of sixteen strains tested (31%) hybridized to either the IA-specific probe or the IB-specific probe. Their results suggested that the majority of the wild-type *E. coli* strains screened did not possess *hsd* genes of either the IA or IB family of type I systems. Results of hybridization studies carried out by Ryu, Rajadas & Bullas (1988) may suggest that other families of type I R–M systems exist in *Salmonella*. They tested the eight *serB*-linked systems of Bullas *et al.* (1980) for sequences that hybridize to either a IA or IB-specific probe. One of the eight appeared to hybridize weakly to the IA-specific probe, while the rest were negative with either probe. These results, coupled with those from complementation analyses, implied that the other systems were not closely related to the IA or IB families. However, negative results can be misleading as shown by the finding that two of the systems, those from serovars kaduna and thompson, have since been shown to be members of the IB family (see Table 1).

Screening members of the ECOR collection for type IA and type IB related sequences

The hybridization survey of Daniel *et al.* (1988) has recently been extended to selected members of the ECOR collection of wild-type *E. coli* strains (Barcus *et al.*, ms in prep). The ECOR collection is composed of 72 *E. coli* strains isolated from a number of sources and geographical locations (Ochman & Selander, 1984). The collection is considered to be representative of the genotypic diversity of the *E. coli* species, and particularly the major subspecific groups, on the basis of multilocus enzyme electrophoresis (MLEE) (Ochman *et al.*, 1983; Whittam, Ochman & Selander, 1983). MLEE detects mobility variation of proteins in starch gel electrophoresis owing to differences in electrostatic charge (Selander *et al.*, 1986). As these differences are due to amino acid substitutions, they may be equated with different alleles at a given gene locus. A number of polymorphisms, or electromorphs, may exist for a particular enzyme. Isolates may be characterized by their electromorph profiles for a number of enzymes, and the genetic relatedness determined by comparing these profiles. Estimates of nucleotide diversity as determined by Southern hybridization, restriction analysis and sequence analysis generally agree with the classification by MLEE (Ochman *et al.*, 1983; Milkman & Bridges, 1990; 1993).

A phenogram displaying the relatedness of the ECOR strains on the basis of MLEE (Selander, Caugant & Whittam, 1987; Herzer *et al.*, 1990) provides a useful framework within which the distribution of other biological traits may be determined and analysed in evolutionary terms (Fig. 4). Screening members of the ECOR collection with *hsd* probes specific for type

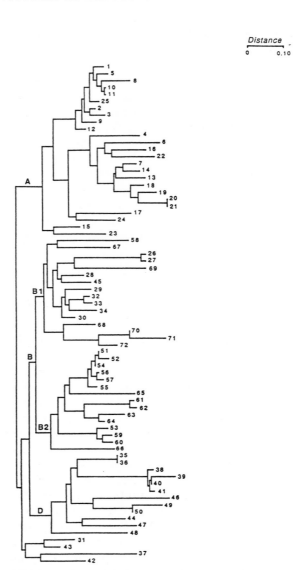

Fig. 4. Genetic relationships of the 72 strains of the ECOR collection. The phenogram is based on comparisons, using the neighbour-joining method (Saitou & Nei, 1987), of polymorphisms of enzymes encoded by 38 loci. The letters correspond to the major subspecific groups of the *E. coli* species. A similar phenogram, based on 35 enzyme loci and using an average-linkage cluster analysis, may be found in Selander *et al.* (1987). The major features of the two phenograms are essentially the same, and the differences do not affect the analysis of the results reported in the text. (Reprinted from Herzer *et al.*, 1990).

IA and IB genes will give information on the frequency and distribution of these families of enzymes in *E. coli*.

A total of 36 strains was probed, comprising the 25 strains of Group A, plus others distributed throughout the other groups. These include eight which have been used in other surveys of the genetic diversity of *E. coli* (Milkman & Crawford, 1983; DuBose, Dykhuizen & Hartl, 1988; Dykhuizen & Green, 1991).

Eleven of the 25 strains (44%) of Group A had DNA that hybridized to either one of the two probes, nine to the type IA-specific probe and two to the IB-specific probe. Outside of Group A, two out of eleven (19%) were positive, one with the type IA-specific probe and the other with the type IB-specific probe.

Though not members of the ECOR collection, *E. coli* K-12 and B would be classified as Group A on the basis of their electromorph profiles (Herzer *et al.*, 1990). Therefore, the higher frequency of type IA-like sequences than IB-like, among strains of Group A, might not be surprising. While both members of some pairs of strains which were shown to be closely related on the basis of MLEE encoded type IA-like systems (ECOR10 and 11; ECOR 13 and 14; see Fig. 4), members of other pairs of strains encoded type I R–M systems of different families (ECOR17 and 24; ECOR15 and 23).

Preliminary analyses of some of the IA and IB-like *hsd* genes have shown whether the specificities of the type I systems they express are the same as or different from ones previously identified (see Table 1). Currently, details of the specificities of six of the systems detected in the ECOR collection are known. ECOR12 and 24 have been shown to encode type IA R–M systems with the same specificity as *Eco*KI, ECOR70, another type IA-like strain, possesses a system with the specificity of *Eco*BI. Two strains in Group A, ECOR5 and ECOR23, encode systems with specificities novel to the type IA and IB families. ECOR42 possesses a type IB R–M system with a specificity that differs from the other two known *E. coli* type IB systems, *Eco*AI and *Eco*EI.

A member of the putative type ID family in the ECOR collection

A probe, made from cloned *hsd* genes of *Salmonella enterica* serovar blegdam, and hence specific for the putative type ID family (see Table 1) was used to screen members of ECOR Group A. One strain contained DNA sequences that hybridized to the probe. These sequences were cloned and shown to encode a functional R–M system. Complementation experiments confirmed the relatedness of these two systems, and demonstrated that each has a unique specificity (A.J.B. Campbell & N.E. Murray, unpublished observations). The new R–M systems are certainly complex, but they are not yet proven to have all the characteristics of type I R–M enzymes.

Evolutionary implications of the present analysis of the distribution of hsd *genes in* E. coli

Sharp *et al.* (1992) reported that the intraspecific divergence between members of the IA and IB families is so great that horizontal transfer from distantly related species was likely to have been responsible for the presence of both families of enzymes in *E. coli.* Therefore if, as is generally believed, MLEE provides a reliable classification of strains, then the presence of different families of enzymes in strains that are closely related is difficult to reconcile without invoking horizontal transfer from distantly related strains or species. The appearance of a representative of yet another family of R–M systems in Group A of the ECOR collection and in a *Salmonella* spp. adds further support for horizontal transfer.

Determination of the *hsd* genotypes of all members of a significant sample of the ECOR strains must precede a definitive analysis of the type I R–M systems of *E. coli.* However, the pattern already evident, involving the occurrence of type I systems of alternative specificities and even families among strains deemed to be closely related on the basis of MLEE, suggests evolutionary pressure for variation of specificity. This is consistent with the theory of Levin and his coworkers (Levin, 1986, 1988; Korona & Levin, 1993; Korona, Korona & Levin, 1993) that bacteriophages will exert frequency-dependent selection on their hosts, so that bacteria with rare specificities will be favoured. Bacteriophages sharing a habitat with restriction and modification proficient bacteria are most likely to be modified with the specificity of the commonest R–M system. Although experimental studies suggest that bacteria sharing a habitat with phage rapidly acquire resistance to infection, these studies also suggest that restriction proficient bacteria encoding a system with novel specificity will be at an advantage when colonizing a new habitat in which phage are already present (Levin, 1988; Korona & Levin, 1993; Korona *et al.*, 1993). As a result, rare specificities will be maintained, and there will be pressure for the evolution of new specificities.

RESTRICTION ALLEVIATION

The evolution of diverse defence systems that protect bacteria against phage attack has been accompanied by the evolution of many strategies for the avoidance of these defences (for a recent review, see Bickle & Krüger, 1993). The majority of antirestriction functions known at present are phage encoded but some conjugative plasmids also encode them (see Wilkins this volume, pp. 59–88 and references).

If restriction is relevant to recombination then so too are the restriction alleviation functions associated with vectors of gene transfer. These functions may be encoded by either conjugative plasmids or transducing phages.

At present, no antirestriction function has been associated with the F factor, which provides one of the two most common ways of mediating genetic transfer in the laboratory. In contrast, transduction by bacteriophage P1 appears to be unhindered by type I R–M systems. The antirestriction proteins of bacteriophage P1 reside within the virion and apparently prevent restriction of the injected DNA regardless of whether it is a phage genome or a segment of bacterial DNA (Iida *et al.*, 1987). It is, nevertheless, possible that this protection against restriction by type I systems is incomplete (see Milkman, this volume, pp. 127–141).

Restriction alleviation phenomena of uncertain importance have been detected in response to UV light; the DNA damage not only stimulates recombination, but it also alleviates restriction by type I and MDR systems. This damage-induced response may be achieved by more than one route, but none is understood. It has been suggested that the restriction alleviation of type I systems provides time for the modification of newly synthesized DNA associated with repair (Thoms & Wackernagel, 1982). An alternative explanation invokes titration of the restriction complex, possibly by unmethylated target sequences (Kelleher & Raleigh, 1994). *Eco*KI, for example, is believed to be present in low amounts and, once bound to an unmodified target sequence, does not 'turn-over'.

A quite different source of restriction alleviation may facilitate the transfer of R–M genes from one bacterial strain to another, an event essential for horizontal transfer of *hsd* genes. Following infection by bacteriophage P1, the phage encoded modification system is expressed long before restriction (Arber & Dussoix, 1962). A transient depression of restriction activity in the recipient cell has also been noted during conjugation, though this effect is not limited to the transfer of genes controlling restriction and modification. It was suggested that this could reflect either the physiological state of exconjugant cells or saturation of the restriction system (Glover & Colson, 1965).

The *hsd* genes of type I systems have two promoters, one that provides transcription of the two genes that encode the subunits essential for modification activity (*hsdM* and *S*), and a second necessary for transcription of *hsdR* and hence the formation of a complex with endonuclease activity (Loenen *et al.*, 1987). This organization offers the opportunity for transcriptional regulation, but experiments in which the *hsd* promoters are fused to reporter genes fail to provide evidence of transcriptional regulation (Loenen *et al.*, 1987; Prakash, Valinluck & Ryu, 1991). Nevertheless, recent experiments support a lag in the expression of restriction activity when an F′ including *hsd* genes enters an *hsd⁻* cell (Prakash-Cheng, Chung & Ryu, 1993). An understanding of this control now seems likely with the identification of a gene, *hsdC*, whose product permits the efficient acquisition by an *hsd⁻* recipient of an F′ that includes the *hsd* genes encoding *Eco*KI. The *hsdC* gene product is required only when the F′ encodes a functional restriction system; a mutation in *hsdR* relieves the barrier present in an *hsdC⁻* recipient. These experiments imply that there is normally a host function that delays restriction activity until the recipient chromosome has been modified, thereby preventing destruction of the recipient chromosome. A post-transcriptional mechanism has been invoked (Prakash-Cheng & Ryu, 1993). Whatever the mechanism, the product of the *hsdC* gene apparently permits efficient transfer of genes encoding *Eco*KI into a cell in which the DNA is not protected by modification. The reported experiments examined only the *hsd* genes of this one type IA member, and it is not known whether the *hsdC* gene product is of relevance to other families of type I R–M systems. The antirestriction functions of phage P1 (Dar) protect against type I R–M systems (Iida *et al.*, 1987), while the Ral polypeptide of phage λ is specific for members of the type IA family (Loenen & Murray, 1986). Type I R–M systems appear more vulnerable to antirestriction systems than type II enzymes, but this bias may reflect the usual choice of *E. coli* K-12 as host and the ease with which phage λ can be used to monitor type I R–M systems.

DISCUSSION

An *E. coli* strain lacking a restriction system does not appear to be disadvantaged under laboratory conditions, but R–M systems are both widespread and diverse in nature and are therefore presumed to confer an advantage. It has been argued that phage provide the selective force for the maintenance of diversity (Levin, 1986) and it is evident that phage have evolved a variety of defences against restriction systems (see, for example, Bickle & Krüger, 1993). A key role for R–M systems as a barrier to phage infection is consistent with the apparent resistance of restriction-proficient *S. albus* G to a variety of phages (Chater & Wilde, 1980). Even so, it cannot be proven that protection is the only key role of R–M systems in natural populations. It has been suggested, on the one hand, that restriction systems provide a general barrier to the transfer of genetic information, creating clonal lines of bacteria and, on the other hand, that the double-stranded DNA ends that result from cleavage are highly recombinogenic (Price & Bickle, 1986).

Only one natural isolate of *E. coli* (*E. coli* C) has been proven to lack R–M systems and it remains quite likely that most wild-type strains possess them, and if not they can be acquired readily on plasmids. Current estimates for the occurrence of R–M systems inevitably give minimal values. A large survey for type II restriction enzymes indicated their presence in 25% of isolates (Janulaitis *et al.*, 1988). A survey for type I systems using probes for the known alleles of the *hsd* locus of *E. coli* K-12 identified *hsd* genes in 48% (12/25) of members of Group A of the ECOR collection and 26% (7/27) of another sample of strains screened with probes for two of the three families of chromosomally encoded type I systems. The real figure for the frequency with which *E. coli* strains have an *hsd* allele could be as low as these estimates or it could approach 100%. *S. typhimurium* LT2, it should be remembered, has three chromosomally encoded R–M systems, two of which may be type I systems, but only one (*Sty*LTIII) will be detected by the probes used in the survey.

The diversity of R–M systems is obvious. The type I systems seem particularly distinguished for their plasticity as emphasized both by their subdivision into families and their readily witnessed ability to evolve new specificities. Their bipartite, non-symmetrical, target sequence in which the length of the spacer is somewhat flexible, greatly enhances the options for target sequence when compared with the symmetrical target sequence of type II R–M systems within which the sequence of the first half dictates that of the second half. Type I systems, despite their cumbersome complexity, have tremendous potential for diversification of sequence specificity (Wilson & Murray, 1991). It should also be borne in mind that they differ from other systems in that they do not cut DNA to give defined fragments. Given a minimum recognition sequence of seven bases, restriction fragments, on average, should approach 20 kb in length. It has been argued (Endlich & Linn, 1985; Price & Bickle, 1986) that, because these enzymes cut at sites remote from their recognition sequences, perhaps even randomly, any gene has the chance to be close to the end of the large DNA fragments generated by a type I restriction enzyme. The precise nature of the ends of the fragments remains to be documented, but many experiments suggest the absence of 5′ ends that are susceptible to polynucleotide kinase, even after treatment with phosphatase (Eskin & Linn, 1972; Murray, Batten & Murray, 1973b). For *Eco*BI there is evidence for the presence of single-stranded 3′ tails which, it was suggested, could initiate recombination (Endlich & Linn, 1985).

In apparent conflict with the hypothetical role of restriction enzymes as stimulators of recombination is the demonstration that the DNA fragments generated by most, if not all, restriction enzymes are degraded by the RecBCD nuclease, alias ExoV (Simmon & Lederberg, 1972). The products of restriction, therefore, are substrates for degradation by the very enzyme that is an essential component of the major recombination pathway in *E. coli* (see Radding, 1973). In this early review, Radding cites experiments of S. Lederberg indicating recombination between the products of restriction in the *absence* of the RecBCD nuclease. There are now well-documented demonstrations (for example, Thaler, Stahl & Stahl, 1987; Eddy & Gold, 1992) that DNA ends generated by cutting with *Eco*RI can serve to stimulate recombination. These experiments relied on the recombination pathways of either phage λ or the RecE system of the cryptic prophage *rac* and, moreover, they were done under conditions in which DNA breakdown by the RecBCD nuclease was inhibited. Other experiments have taken advantage of mutations (*recD*) that inactivate the degradative activity of the RecBCD enzyme.

Of more significance are experiments in which it was shown that a vulnerable type II restriction target can stimulate RecBCD-dependent recombination, even in the presence of the nucleolytic activity of the wild-type RecBCD enzyme (Stahl *et al.*, 1983). The recombination activity of RecBCD is stimulated by a specific 8 base 'recombinator' sequence designated Chi (see, for example, Smith, 1991). To activate RecBCD-dependent recombination in λ, Chi must be oriented in the appropriate direction with respect to a double-strand break, normally generated by cutting at the termini of the phage genome. In an experiment in which the orientation of Chi is reversed, restriction was shown to activate RecBCD-mediated recombination. This experiment shows that the RecBCD enzyme can promote recombination when it enters a DNA molecule in response to the cutting or nicking of an *Eco*RI target and then encounters a Chi sequence in the appropriate orientation. More generally, the disposition of Chi sequences within the products of restriction could influence the opportunity for recombination as an alternative to DNA breakdown. The current view of Chi is not simply as a sequence that stimulates cutting by the RecBCD enzyme, but as a unique regulatory element that acts by attenuating the degradative function of RecBCD, thereby enhancing its recombinative function (Stahl *et al.*, 1990; Dixon & Kowalczykowski, 1993). Two Chi sequences inversely and appropriately oriented with respect to both ends of a DNA fragment would impede degradation and promote recombination.

Earlier experiments in which recombination was monitored when unmodified donor DNA from an Hfr strain entered a restriction proficient, *recBCD*[+] recipient showed the acquisition of early markers to be inefficient and linkage much reduced by either type I or type III R–M systems (Boyer, 1964; Pittard, 1964; Arber & Morse, 1965). In the case of P1 transduction, transfer of bacterial genes is normally unaffected by type I R–M systems since restriction by them is blocked by the antirestriction function (Dar) of the phage. Even the prophage genome of phage λ is immune to attack by *Eco*KI when it enters the cell via P1, despite the presence of five unmodified targets (Iida *et al.*, 1987). When a P1 mutant defective in the antirestriction function was used, a very substantial reduction in transduction frequency was seen (Iida *et al.*, 1987). The conjugation and transduction experiments provide evidence that restriction by types I and III systems severely depresses general recombination. At the present time, we are unaware of experiments that monitor the effects of either type II or MDR systems on either conjugation or transduction, although type II systems have been shown to reduce transformation frequency in *Bacillus subtilis* strains differing in their R–M systems (Cohan, Roberts & King, 1991).

Following conjugation, most recombination is believed to occur by the incorporation of long 'chunks' of donor DNA (Smith, 1991), with one exchange occurring within the DNA that is close to the origin of transfer. This recombination is mediated by the RecBCD pathway. It seems inevitable that fragmentation of DNA by restriction would greatly reduce the opportunity for the incorporation of long stretches of DNA. Some recombination achieved by the incorporation of smaller 'chunks' of DNA may occur by the RecBCD pathway, some by alternative pathways (Smith, 1991). It has been pointed out, on the basis of sequence comparisons, that members of the ECOR collection differ by short lengths of nucleotide sequence (Milkman & Bridges, 1993), one implication being that restriction could reduce the size of the DNA fragments incorporated by recombination.

It has been suggested (Rayssiguier, Thaler & Radman, 1989) that R–M systems are not required as *inter*specific barriers to recombination, since DNA sequence differences in themselves are sufficient to hinder recombination between *E. coli* and *Salmonella*. Also, it is evident that selection maintains *intra*specific diversity and consequently restriction barriers are presumed to be of significance at this level. While the few experiments in which interspecific recombination has been monitored in the presence of R–M systems indicate that restriction reduces the transfer of genetic information and loosens linkage, it remains likely that the cutting of DNA by restriction enzymes will sometimes promote recombination. Perhaps, also, the random fragmentation characteristic of type I R–M systems might occasionally expose short segments of DNA sequences that are normally inaccessible to recombination because they are embedded in sequences sufficiently dissimilar that they hinder recombination. Given a thorough documentation of R–M systems in *E. coli* K-12, and a knowledge of recombination pathways, the effects of restriction on recombination warrant further molecular and genetic investigation.

This chapter has focused on the R–M systems of the *Enterobacteriaceae*, the bacteria most commonly used in early genetic analyses. There is no known reason to believe that the conclusions will not be generally applicable to other bacteria. Screens of bacterial extracts for sequence-specific endonucleases have indicated their presence in roughly one-quarter of the 10 000 isolates tested (see Wilson & Murray, 1991). This figure is the same as that obtained for 1000 isolates of *E. coli* (A. Janulaitis, pers. comm.). Until recently, type I systems, however, have been associated with only the enteric bacteria. The genetic analyses of *Mycoplasma pulmonis* now identify an R–M system which, on the basis of sequence comparisons, is shown to be related to the type IC family (Dybvig & Yu, 1994). It seems likely that information on type I systems has been limited by the nature and aims of past investigations.

ACKNOWLEDGEMENTS

We thank many colleagues for their helpful criticisms of the manuscript, particularly David Leach, Frank Stahl and Keith Chater, and Isobel Black for her patient help in the preparation of the manuscript. We are indebted to Annette Campbell and Diane Ternent for permitting us to cite their unpublished results. V.A.B. acknowledges support from the ORS awards scheme. The research from this laboratory was funded by grants from the Medical Research Council.

REFERENCES

Abadjieva, A., Patel, J., Webb, M., Zinkevich, V. & Firman, K. (1993). A deletion mutant of the type IC restriction endonuclease *Eco*R124I expressing a novel DNA specificity. *Nucleic Acids Research*, **21**, 4435–43.

Adler, S. P. & Nathans, D. (1973). Conversion of circular to linear SV40 DNA by restriction endonuclease from *Escherichia coli* B. *Biochemica et Biophysica Acta*, **299**, 177–88.

Arber, W. (1971). Host-controlled variation. In *The Bacteriophage Lambda*. (Hershey, A. D. ed.), pp. 83–96. Cold Spring Harbor Laboratory, New York.

Arber, W. & Dussoix, D. (1962). Host specificity of DNA produced by *Escherichia coli*. I. Host controlled modification of bacteriophage lambda. *Journal of Molecular Biology*, **5**, 18–36.

Arber, W. & Morse, M. L. (1965). Host specificity of DNA produced by *Escherichia coli* VI. Effects on bacterial conjugation. *Genetics*, **51**, 137–48.

Argos, P. (1985). Evidence for a repeating domain in type I restriction enzymes. *EMBO Journal*, **4**, 1351–5.

Barcus, V. A. (1993). The type I restriction systems of *Escherichia coli*. PhD thesis.

Bertani, G. & Weigle, J. J. (1953). Host controlled variation in bacterial viruses. *Journal of Bacteriology*, **65**, 113–21.

Bickle, T. A. (1987). DNA restriction and modification systems. In Escherichia coli *and* Salmonella typhimurium: *Cellular and Molecular Biology*. (Neidhart, F. E., Ingraham, J. L., Low, K. B., Magasanik, B., Schaechter, M. and Umbarger, H. E. eds.) pp. 692–6. American Society for Microbiology, Washington, DC.

Bickle, T. A., Brack, C. & Yuan, R. (1978). ATP-induced conformational changes in the restriction endonuclease from *Escherichia coli* K-12. *Proceedings of the National Academy of Sciences, USA*, **75**, 3099–103.

Bickle, T. A. & Krüger, D. H. (1993). Biology of DNA restriction. *Microbiological Reviews*, **57**: 434–50.

Boyer, H. W. (1964). Genetic control of restriction and modification in *Escherichia coli*. *Journal of Bacteriology*, **88**, 1652–60.

Boyer, H. W. & Roulland-Dussoix, D. (1969). A complementation analysis of the restriction and modification of DNA in *Escherichia coli*. *Journal of Molecular Biology*, **41**, 459–72.

Bullas, L. R. & Colson, C. (1975). DNA restriction and modification systems in Salmonella III. SP, a *Salmonella potsdam* system allelic to the SB system in *Salmonella typhimurium*. *Molecular and General Genetics*, **139**, 177–88.

Bullas, L. R., Colson, C. & Neufeld, B. (1980). Deoxyribonucleic acid restriction and modification systems in *Salmonella*: chromosomally located systems of different serotypes. *Journal of Bacteriology*, **141**, 275–92.

Bullas, L. R., Colson, C. & Van Pel, A. (1976). DNA restriction and modification systems in Salmonella. SQ, a new system derived by recombination between the SB system of *Salmonella typhimurium* and the SP system of *Salmonella potsdam*. *Journal of General Microbiology*, **95**, 166–72.

Bullas, L. R. & Ryu, J. (1983). *Salmonella typhimurium* LT2 strains which are r–m$^+$

for all three chromosomally located systems of DNA restriction and modification. *Journal of Bacteriology*, **156**, 471–4.

Chater, K. F. & Wilde, L. C. (1980). *Streptomyces albus* G mutants defective in the *SalGI restriction-modification system. Journal of General Microbiology*, **116**, 323–34.

Cohan, F. M., Roberts, M. S. & King, E. C. (1991). The potential for genetic exchange by transformation within a natural population of *Bacillus subtilis*. *Evolution*, **45**, 1393–421.

Colson, C. & Colson, A. M. (1971). A new *Salmonella typhimurium* DNA host specificity. *Journal of General Microbiology*, **69**, 345–51.

Cowan, G. M. (1988). A new family of type I restriction and modification systems. PhD thesis.

Cowan, G. M., Gann, A. A. F. & Murray, N. E. (1989). Conservation of complex DNA recognition domains between families of restriction enzymes. *Cell*, **56**, 103–9.

Daniel, A. S., Fuller-Pace, F. V., Legge, D. M. & Murray, N. E. (1988). Distribution and diversity of *hsd* genes in *Escherichia coli* and other enteric bacteria. *Journal of Bacteriology*, **170**, 1775–82.

Dartois, V., De Backer, O. & Colson, C. (1993). Sequence of the *Salmonella typhimurium StyLT1* restriction-modification genes: homologies with *Eco*P1 and *Eco*P15 type-III R–M systems and presence of helicase domains. *Gene*, **127**, 105–10.

Dixon, D. A. & Kowalczykowski, S. C. (1993). The recombination hotspot X is a regulatory sequence that acts by attenuating the nuclease activity of the *E. coli* RecBCD enzyme. *Cell*, **73**, 87–96.

Doolittle, R. F., Feng, D. F., Johnson, M. S. & McClure, M. A. (1986). Relationships of human protein sequences to those of other organisms. *Cold Spring Harbor Symposia on Quantitative Biology*, **51**, 447–55.

Dryden, D. T. F., Cooper, L. P. & Murray, N. E. (1993). Purification and characterization of the methyltransferase from the type I restriction and modification system of *Escherichia coli* K12. *Journal of Biological Chemistry*, **268**, 13228–36.

DuBose, R. F., Dykhuizen, D. E. & Hartl, D. L. (1988). Genetic exchange among natural isolates of bacteria: Recombination within the *phoA* gene of *Escherichia coli*. *Proceedings of the National Academy of Sciences, USA*, **85**, 7036–40.

Dussoix, D. & Arber, W. (1962). Host specificity of DNA produced by *Escherichia coli*. II. Control over acceptance of DNA from infecting phage lambda. *Journal of Molecular Biology*, **5**, 37–49.

Dybvig, K. & Yu, H. (1994). Regulation of a restriction modification system via DNA inversion in *Mycoplasma pulmonis*. *Molecular Microbiology*, **12**, 547–60.

Dykhuizen, D. E. & Green, L. (1991). Recombination in *Escherichia coli* and the definition of biological species. *Journal of Bacteriology*, **173**: 7257–68.

Eddy, S. R. & Gold, L. (1992). Artificial mobile DNA element constructed from the *Eco*RI endonuclease gene. *Proceedings of the National Academy of Sciences, USA*, **89**, 1544–7.

Endlich, B. & Linn, S. (1985). The DNA restriction endonuclease of *Escherichia coli* B. *Journal of Biological Chemistry*, **260**, 5729–38.

Eskin, B. & Linn, S. (1972). The deoxyribonucleic acid modification and restriction enzymes of *Escherichia coli* B. II. Purification, subunit structure, and catalytic properties of the restriction endonuclease. *Journal of Biological Chemistry*, **247**, 6183–91.

Fuller-Pace, F. V., Bullas, L. R., Delius, H. & Murray, N. E. (1984). Genetic recombination can generate altered restriction specificity. *Proceedings of the National Academy of Sciences, USA*, **81**, 6095–9.

Fuller-Pace, F. V., Cowan, G. M. & Murray, N. E. (1985). *Eco*A and *Eco*E: alternatives to the *Eco*K family of type I restriction and modification systems of *Escherichia coli*. *Journal of Molecular Biology*, **186**, 65–75.

Fuller-Pace, F. V. & Murray, N. E. (1986). Two DNA recognition domains of the specificity polypeptides of a family of type I restriction enzymes. *Proceedings of the National Academy of Sciences, USA*, **83**, 9368–72.

Gann, A. A. F., Campbell, A. J. B., Collins, J. F., Coulson, A. F. W. & Murray, N. E. (1987). Reassortment of DNA recognition domains and the evolution of new specificities. *Molecular Microbiology*, **1**, 13–22.

Glatman, L. I., Moroz, A. F., Iablokova, M. B., Rebentish, B. A. & Kholmina, G. V. (1980). New plasmid-mediated restriction-modification systems of DNA in clinical strains of *Escherichia coli*. *Doklady Akademii Nauk USSR*, **252**, 993–5.

Glover, S. W. & Colson, C. (1965). The breakdown of the restriction mechanism in zygotes of *Escherichia coli*. *Genetical Research*, **6**, 153–5.

Glover, S. W. & Colson, C. (1969). Genetics of host-controlled restriction and modification in *Escherichia coli*. *Genetical Research*, **13**, 227–40.

Gough, J. A. & Murray, N. E. (1983). Sequence diversity among related genes for recognition of specific targets in DNA molecules. *Journal of Molecular Biology*, **166**, 1–19.

Gubler, M. & Bickle, T. A. (1991). Increased protein flexibility leads to promiscuous protein-DNA interactions in type IC restriction-modification systems. *EMBO Journal*, **10**, 951–7.

Gubler, M., Braguglia, D., Meyer, J., Piekarowicz, A. & Bickle, T. A. (1992). Recombination of constant and variable modules alters DNA sequence recognition by type IC restriction-modification enzymes. *EMBO Journal*, **11**, 233–40.

Halford, S. E., Taylor, J. D., Vermote, C. L. M. & Vipond, I. B. (1993). In *Nucleic Acids and Molecular Biology*, vol. 7. (Eckstein, F. and Lilley, D. M. J. eds.), pp. 47–69. Springer-Verlag, Berlin.

Heitman, J. (1993). On the origins, structures and functions of restriction-modification enzymes. In *Genetic Engineering*, vol. 15. (Setlow, J. K. ed.) pp. 57–108. Plenum Press, New York.

Herzer, P. J., Inouye, S., Inouye, M. & Whittam, T. S. (1990). Phylogenetic distribution of branched RNA-linked multicopy single-stranded DNA among natural isolates of *Escherichia coli*. *Journal of Bacteriology*, **172**, 6175–81.

Horiuchi, K. & Zinder, N. D. (1972). Cleavage of bacteriophage f1 DNA by the restriction enzyme of *Escherichia coli* B. *Proceedings of the National Academy of Sciences, USA*, **69**, 3220–4.

Hubáček, J. & Glover, S. W. (1970). Complementation analysis of temperature-sensitive host specificity mutations in *Escherichia coli*. *Journal of Molecular Biology*, **50**, 111–27.

Hughes, S. G. (1977). Studies of plasmid encoded restriction and modification systems. PhD thesis.

Iida, S., Meyer, J., Bachi, B., Stalhammar-Carlemalm, M., Schrickel, S., Bickle, T. A. & Arber, W. (1983). DNA restriction-modification genes of phage P1 and plasmid p15B. *Journal of Molecular Biology*, **165**, 1–18.

Iida, S., Streiff, M. B., Bickle, T. A. & Arber, W. (1987). Two DNA antirestriction systems of bacteriophage P1, *darA*, and *darB*: characterization of *darA⁻* phages. *Virology*, **157**, 156–66.

Janulaitis, A., Kazlauskiene, R., Lazareviciute, L., Gilvonauskaite, R., Steponaviciene, D., Jagelavicius, M., Petrušyte, M., Bitinaite, J., Vezeviciute, Z., Kiuduliene, E. & Butkus, V. (1988). Taxonomic specificity of restriction-modification enzymes. *Gene*, **74**, 229–32.

Janulaitis, A., Petrušyte, M., Maneliene, Z., Klimasauskas, S. & Butkus, V. (1992). Purification and properties of the *Eco*571 restriction endonuclease and methylase - prototypes of a new class (type IV). *Nucleic Acids Research*, **20**, 6043–9.

Kan, N. C., Lautenberger, J. A., Edgell, M. H. & Hutchison, III, C. A. (1979). The nucleotide sequence recognized by the *Escherichia coli* K12 restriction and modification enzymes. *Journal of Molecular Biology*, **130**, 191–209.

Kannan, P., Cowan, G. M., Daniel, A. S., Gann, A. A. F. & Murray, N. E. (1989). Conservation of organization in the specificity polypeptides of two families of type I restriction enzymes. *Journal of Molecular Biology*, **209**, 335–44.

Kelleher, J. E. & Raleigh, E. A. (1991). A novel activity in *Escherichia coli* K12 that directs restriction of DNA modified at CG dinucleotides. *Journal of Bacteriology*, **173**, 5220–3.

Kelleher, J. E. & Raleigh, E. A. (1994). Response to UV damage by four *E. coli* K-12 restriction systems. *Journal of Bacteriology*, in press.

Kessler, C. & Manta, V. (1990). Specificity of restriction endonucleases and DNA modification methyltransferases – a review. *Gene*, **92**, 1–248.

Korona, R., Korona, B. & Levin, B. R. (1993). Sensitivity of naturally occurring coliphages to type I and type II restriction and modification. *Journal of General Microbiology*, **139**, 1283–90.

Korona, R. & Levin, B. R. (1993). Phage-mediated selection and the evolution and maintenance of restriction-modification. *Evolution*, **47**, 556–75.

Kröger, M. & Hobom, G. (1984). The nucleotide sequence recognized by the *Escherichia coli*: a restriction and modification enzyme. *Nucleic Acids Research*, **12**, 887–99.

Kühnlein, U., Linn, S. & Arber, W. (1969). Host specificity of DNA produced by *Escherichia coli*. XI. *In vitro* modification of phage fd replicative form. *Proceedings of the National Academy of Sciences, USA*, **63**, 556–62.

Lautenberger, J. A., Kan, N. C., Lackey, D., Linn, S., Edgell, M. H. & Hutchinson III, C. A. (1978). Recognition site of *Escherichia coli* B restriction enzyme on ϕXsB1 and simian virus 40 DNAs: an interrupted sequence. *Proceedings of the National Academy of Sciences, USA*, **75**, 2271–5.

Levin, B. R. (1986). Restriction-modification immunity and the maintenance of genetic diversity in bacterial populations. In *Evolutionary Processes and Theory*. (Karlin, S. & Nero, E. eds.), pp. 669–88. Academic Press, Inc., New York.

Levin, B. R. (1988). Frequency-dependent selection in bacterial populations. *Philosophical Transactions of the Royal Society, London*, Ser. B, **319**, 459–72.

Loenen, W. A. M., Daniel, A. S., Braymer, H. D. & Murray, N. E. (1987). Organization and sequence of the *hsd* genes of *Escherichia coli* K-12. *Journal of Molecular Biology*, **198**, 159–70.

Loenen, W. A. M. & Murray, N. E. (1986). Modification enhancement by the restriction alleviation protein (Ral) of bacteriophage λ. *Journal of Molecular Biology*, **190**, 11–22.

Luria, S. E. & Human, M. L. (1952). A nonhereditary, host-induced variation of bacterial viruses. *Journal of Bacteriology*, **64**, 557–69.

Meister, J., MacWilliams, M., Hübner, P., Jütte, H., Skrzypek, E., Piekarowicz, A. & Bickle, T. A. (1993). Macroevolution by transposition: drastic modification of DNA recognition by a type I restriction enzyme following Tn5 transposition. *EMBO Journal*, **12**, 4585–91.

Milkman, R. & Bridges, M. M. (1990). Molecular evolution of the *Escherichia coli* chromosome. III. Clonal frames. *Genetics*, **126**, 505–17.

Milkman, R. & Bridges, M. M. (1993). Molecular evolution of the *Escherichia coli* chromosome. IV. Sequence comparisons. *Genetics*, **133**, 455–68.

Milkman, R. & Crawford, I. P. (1983). Clustered third-base substitution among wild strains of *Escherichia coli*. *Science*, **221**, 378–80.

Murray, N. E., Batten, P. L. & Murray, K. (1973b). Restriction of bacteriophage lambda by *Escherichia coli* K. *Journal of Molecular Biology*, **81**, 395–407.

Murray, N. E., Daniel, A. S., Cowan, G. M. & Sharp, P. M. (1993). Conservation of motifs within the unusually variable polypeptide sequences of type I restriction and modification enzymes. *Molecular microbiology*, **9**, 133–43.

Murray, N. E., Gough, J. A., Suri, B. & Bickle, T. A. (1982). Structural homologies among type I restriction-modification systems. *EMBO Journal*, **1**, 535–9.

Murray, N. E., Manduca de Ritis, P. & Foster, L. A. (1973a). DNA targets for the *Escherichia coli* K restriction system analysed genetically in recombinants between phages phi80 and lambda. *Molecular and General Genetics*, **120**, 261–81.

Nagaraja, V., Shepherd, J. C. W. & Bickle, T. A. (1985c). A hybrid recognition sequence in a recombinant restriction enzyme and the evolution of DNA sequence specificity. *Nature, London*, **316**, 371–2.

Nagaraja, V., Shepherd, J. C. W., Pripfl, T. & Bickle, T. A. (1985b). Two type I restriction enzymes from *Salmonella* species: purification and DNA recognition sequences. *Journal of Molecular Biology*, **182**, 579–87.

Nagaraja, V., Steiger, M., Nager, C., Hadi, S. M. & Bickle, T. A. (1985a). The nucleotide sequence recognised by the *Escherichia coli* D type I restriction and modification system. *Nucleic Acids Research*, **13**, 389–99.

Ochman, H. & Selander, R. K. (1984). Standard reference strains of *E. coli* from natural populations. *Journal of Bacteriology*, **157**, 690–3.

Ochman, H., Whittam, T. S., Caugant, D. A. & Selander, R. K. (1983). Enzyme polymorphism and genetic population structure in *Escherichia coli* and *Shigella*. *Journal of General Microbiology*, **129**, 2715–26.

Pittard, J. (1964). Effect of phage-controlled restriction on genetic linkage in bacterial crosses. *Journal of Bacteriology*, **87**, 1256–7.

Prakash, A., Valinluck, B. & Ryu, J. (1991). Genomic *hsd*-Mu(*lac*) operon fusion mutants of *Escherichia coli* K-12. *Gene*, **99**, 9–14.

Prakash-Cheng, A., Chung, S. S. & Ryu, J. (1993). The expression and regulation of hsd_K genes after conjugative transfer. *Molecular and General Genetics*, **241**, 491–6.

Prakash-Cheng, A. & Ryu, J. (1993). Delayed expression of in vivo restriction activity following conjugal transfer of *Escherichia coli* hsd_K (restriction-modification) genes. *Journal of Bacteriology*, **175**, 4905–6.

Price, C. & Bickle, T. A. (1986). A possible role for DNA restriction in bacterial evolution. *Microbiological Sciences*, **3**, 296–9.

Price, C., Lingner, J., Bickle, T. A., Firman, K. & Glover, S. W. (1989). Basis for changes in DNA recognition by the *Eco*R124 and *Eco*R124/3 Type I DNA restriction and modification enzymes. *Journal of Molecular Biology*, **205**, 115–25.

Price, C., Shepherd, J. C. W. & Bickle, T. A. (1987). DNA recognition by a new family of type I restriction enzymes: a unique relationship between two different DNA specificities. *EMBO Journal*, **6**, 1493–7.

Radding, C. M. (1973). Molecular mechanisms in genetic recombination. *Annual Review of Genetics*, **7**, 87–111.

Raleigh, E. A. (1992). Organization and function of the *mcrBC* genes of *Escherichia coli* K-12. *Molecular Microbiology*, **6**, 1079–86.

Raleigh, E. A., Trimarchi, R. & Revel, H. (1989). Genetic and physical mapping of the *mcrA* (*rglA*) and *mcrB* (*rglB*) loci of *Escherichia coli* K-12. *Genetics*, **122**, 279–96.

Ravetch, J. V., Horiuchi, K. & Zinder, N. D. (1978). Nucleotide sequence of the recognition site for the restriction-modification enzyme of *Escherichia coli* B. *Proceedings of the National Academy of Sciences, USA*, **75**, 2266–70.

Rayssiguier, C., Thaler, D. S. & Radman, M. (1989). The barrier to recombination between *Escherichia coli* and *Salmonella typhimurium* is disrupted in mismatch-repair mutants. *Nature, London*, **342**, 396–401.

Roberts, S. J. & Macelis, D. (1993). REBASE – restriction enzymes and methylases. *Nucleic Acids Research*, **21**, 3125–37.

Rosamond, J., Endlich, B. & Lim, S. (1979). Electron microscopic studies of the mechanism of action of the restriction endonuclease of *Escherichia coli* B. *Journal of Molecular Biology*, **129**, 619–35.

Ryu, J., Rajadas, P. T. & Bullas, L. R. (1988). Complementation and hybridization evidence for additional families of type I DNA restriction and modification genes in *Salmonella* serotypes. *Journal of Bacteriology*, **170**, 5785–8.

Saitou, N. & Nei, M. (1987). The neighbor-joining method: a new method for reconstructing phylogenetic trees. *Molecular Biology and Evolution*, **4**, 406–25.

Selander, R. K., Caugant, D. A., Ochman, H., Musser, J. M., Gilmour, M. N. & Whittam, T. S. (1986). Methods of multilocus enzyme electrophoresis for bacterial population genetics and systematics. *Applied Environmental Microbiology*, **51**, 873–84.

Selander, R. K., Caugant, D. A. & Whittam, T. S. (1987). Genetic structure and variation in natural populations of *Escherichia coli*. In Escherichia coli *and* Salmonella typhimurium: *Cellular and Molecular Biology*. (Neidhart, F. C., Ingraham, J. L., Low, K. B., Magasanik, B., Schaechter, M. and Umbarger, H. E. eds.), pp. 1625–48. American Society for Microbiology, Washington, DC.

Simmon, V. F. & Lederberg, S. (1972). Degradation of bacteriophage lambda deoxyribonucleic acid after restriction by *Escherichia coli* K-12. *Journal of Bacteriology*, **112**, 161–9.

Sharp, P. M., Kelleher, J. E., Daniel, A. S., Cowan, G. M. & Murray, N. E. (1992). Roles of selection and recombination in the evolution of type I restriction-modification systems in enterobacteria. *Proceedings of the National Academy of Sciences, USA*, **89**, 9836–40.

Smith, G. M. (1991). Conjugational recombination in *E. coli*: myths and mechanisms. *Cell*, **64**, 19–27.

Somer, R. & Schaller, H. (1979). Nucleotide sequence of the recognition site of the B-specific restriction modification system in *Escherichia coli*. *Molecular and General Genetics*, **168**, 331–5.

Stahl, M. M., Kobayashi, I., Stahl, F. W. & Huntingdon, S. K. (1983). Activation of Chi, a recombinator, by the action of endonuclease at a distant site. *Proceedings of the National Academy of Sciences, USA*, **80**, 2310–13.

Stahl, F. W., Thomason, L. C., Siddiqi, I. & Stahl, M. M. (1990). Further tests of a recombination model in which χ removes the RecD subunit from the RecBCD enzyme of *Escherichia coli*. *Genetics*, **126**, 519–33.

Suri, B., Shepherd, J. C. W. & Bickle, T. A. (1984). The *Eco*A restriction and modification system of *Escherichia coli* IST⁻: enzyme structure and DNA recognition sequence. *EMBO Journal*, **3**, 575–9.

Taylor, I., Patel, J., Firman, K. & Kneale, G. (1992). Purification and biochemical characterisation of the *Eco*R124 type I modification methylase. *Nucleic Acids Research*, **20**, 179–86.

Taylor, I., Watts, D. & Kneale, G. (1993). Substrate recognition and selectivity in the type IC DNA modification methylase M.*Eco*R124I. *Nucleic Acids Research,* **21**, 4929–35.

Thaler, D. S., Stahl, M. M. & Stahl, F. W. (1987). Tests of the double-strand-break repair model for Red-mediated recombination of phage λ and plasmid λdv. *Genetics*, **116**, 501–11.

Thoms, B. & Wackernagel, W. (1982). UV-induced alleviation of λ restriction in *Escherichia coli* K-12: kinetics of induction and specificity of this SOS function. *Molecular and General Genetics*, **186**, 111–17.

Tyndall, C., Meister, J. & Bickle, T. A. (1994). The *Escherichia coli prr* region encodes a functional type IC DNA restriction system closely integrated with an anticodon nuclease gene. *Journal of Molecular Biology*, **237**, 266–74.

Van Pel, A. & Colson, C. (1974). DNA restriction and modification systems in *Salmonella*. II. Genetic complementation between the K and B systems of *Escherichia coli* and the *Salmonella typhimurium* system SB, with the same chromosomal location. *Molecular and General Genetics*, **135**, 51–60.

Waite-Rees, P. A., Keating, C. J., Moran, L. S., Slatko, B. E., Hornstra, L. J. & Benner, J. S. (1991). Characterization and expression of the *Escherichia coli* Mrr restriction system. *Journal of Bacteriology*, **173**, 5207–19.

Weiserova, M., Janscak, P., Benada, O., Hubácek, J., Zinkevich, V. E., Glover, S. W. & Firman, K. (1993). Cloning, production and characterisation of wild type and mutant forms of the R*Eco*K endonucleases. *Nucleic Acids Research*, **21**, 373–9.

Whittam, T. S., Ochman, H. & Selander, R. K. (1983). Multilocus genetic structure in natural populations of *Escherichia coli*. *Proceedings of the National Academy of Sciences, USA*, **80**, 1751–5.

Wilson, G. G. & Murray, N. E. (1991). Restriction and modification systems. *Annual Review of Genetics*, **25**, 585–627.

GENE TRANSFER BY BACTERIAL CONJUGATION: DIVERSITY OF SYSTEMS AND FUNCTIONAL SPECIALIZATIONS

BRIAN M. WILKINS

Department of Genetics, University of Leicester, Leicester LE1 7RH
UK

INTRODUCTION

Almost 50 years ago J. Lederberg and E. M. Tatum discovered genetic recombination in bacteria through the serendipitous choice of a strain of *Escherichia coli* that happened to carry the F plasmid. Since then striking advances have been made in understanding the transmission genetics of bacteria with the result that we now understand mechanistic features of diverse transformation, transduction and conjugation systems. Impressive advances have also been made in understanding mechanisms for integrating transferred genes into the recipient genome, whether by some form of recombination with resident DNA or through inheritance of transferred DNA as an autonomous plasmid.

Major impetus for these fundamental studies of bacterial genetics has come from the quest to understand ways by which bacteria acquire resistance to the onslaught of antibiotics used in clinical medicine and animal husbandry. Appearance of the same or similar resistance determinants in bacteria of different type and ecological distribution provides compelling evidence for widespread exchange of genes in nature. Bacterial conjugation is implicated as a major route for such disseminations through the frequent detection of resistance genes on conjugative elements (DeFlaun & Levy, 1989; Amábile-Cuevas & Chicurel, 1992). The importance of conjugative plasmids is emphasized further by laboratory demonstrations of conjugative DNA transfer between distantly related bacteria and from bacteria to eukaryotes (Mazodier & Davies, 1991; Courvalin, 1994), while findings that some plasmid types are maintained autonomously in a broad range of hosts have given rise to the concept of the 'promiscuous' plasmid as a specialized entity (Thomas, 1989).

A contrasting picture emerges when we consider conjugative dissemination of chromosomal genes. Despite the proven ability of the F plasmid to transport long segments of the donor chromosome to recipients in the laboratory, discovery of strong linkage disequilibrium in populations of *E. coli* implies that recombination of large chromosomal segments of this species is rare in nature (Maynard-Smith *et al.*, 1993). One possible explanation is that conjugative transmission of chromosomal genes is a rare event in the wild.

These considerations raise a number of questions to be addressed in this chapter. One concerns mechanistic similarities between the conjugation system of plasmid F and those encoded by the plethora of conjugative elements now known in Gram-positive and Gram-negative bacteria: does the F transfer system, typified by the transfer of a specific DNA strand between bacteria that are brought into contact by the extracellular conjugative pilus, continue to provide a representative model of the conjugation process? Another set of questions concerns the perceived promiscuity of conjugation: are all systems manifestly indiscriminate or is there evidence of functional and ecological specialization at such levels as donor-recipient cell interactions and the regulation of transfer activity? Conversely, have broad host-range plasmids been found to carry specialized transfer genes that aid promiscuity by widening the effective conjugation range, and have such plasmids evolved systems for evading the barrier to transfer constituted by host restriction systems? Another important question springs from the apparent distinction between the mobility of conjugative plasmids and their cargoes of specialized genes on the one hand and the relative immobility of chromosomal genes on the other. Is mobilization of chromosomal genes, a process fundamental to the discovery of genetic recombination in bacteria, a property peculiar to sex factor F or a characteristic of all conjugative plasmids?

MECHANISTIC DIVERSITY OF CONJUGATION SYSTEMS

Bacterial conjugation is defined as the transmission of genetic material from a bacterium to another cell by a process requiring cell to cell contact. These contacts are established by the 'mating apparatus' of the conjugation system and culminate in the formation of some specialized 'bridge' for gene transfer. Establishment of an effective mating contact in turn triggers a series of DNA processing reactions resulting in the transfer of the conjugative element to the recipient cell. Our understanding of these steps in conjugation comes largely from complementary studies with two types of Gram-negative bacterial plasmid, namely the F-like plasmids (Willetts & Skurray, 1980; Ippen-Ihler & Skurray, 1993) and the Birmingham IncPα plasmids consisting of the similar, if not identical, R18, R68, RK2 and RP4 elements (Guiney & Lanka, 1989; Guiney, 1993; Wilkins & Lanka, 1993). The nucleotide sequence of the transfer (Tra) regions of F and the Birmingham IncPα plasmids has recently been published (Frost, Ippen-Ihler & Skurray, 1994; Pansegrau et al., 1994).

The F and Pα Tra systems are typical of those examined in E. coli in specifying a conjugative pilus as an essential component of the mating apparatus. The precise contribution of the conjugative pilus remains unclear. As viewed through the F system, it appears that the tip of the pilus makes the initial mating contact with the recipient cell and that pilus retraction, accomplished by disassembly of pilin subunits, bring cells into surface contact as a prelude to DNA transfer (Dürrenberger, Villiger & Bächi, 1991; Frost *et al.*, 1994). Possibly some structure at the base of a pilus forms part of the bridge for DNA transport, while the pilus and/or its assembly proteins in the cell envelope contribute to the mechanism that couples conjugative processing of the plasmid to the formation of an effective mating contact. It has been reported that Hfr donors and recipient cells form recombinants when incubated for up to ten hours on opposite sides of a membrane filter with 'straight-through' pores, implying that DNA is transferrable through an extended pilus (Harrington & Rogerson, 1990). This phenomenon needs to be re-examined with donors of autonomous conjugative plasmids and small mobilizable plasmids, particularly to quantify transfer efficiencies in shorter time periods.

The general model of conjugation, developed from the study of a few *E. coli* systems, states that DNA transfer is initiated at the unique origin of transfer (*oriT*) site to cause export of a specific plasmid strand with a leading 5' terminus. Single-stranded (ss) DNA transfer is associated with synthesis of a replacement strand in the donor cell and a complementary strand in the recipient (Fig. 1; Wilkins & Lanka, 1993). An essential component of the DNA processing mechanism is the relaxosome. This structure, characterized in exceptional detail for RP4, is a natural complex of negatively supercoiled plasmid DNA and a subset of the Tra proteins involved in the DNA processing reactions. Treatment of the isolated relaxosome with a protein-denaturing agent releases an open circular form of DNA that has a strand- and site-specific cleavage at *oriT* and contains one relaxosomal protein, called relaxase, covalently linked to the 5'-phosphate of the nucleotide at the nick (Fürste *et al.*, 1989; Pansegrau, Ziegelin & Lanka, 1990). This DNA–relaxase complex is thought to model the transfer intermediate, with the covalently linked protein providing the mechanism for circularizing the transferred plasmid strand (Wilkins & Lanka, 1993).

The F and Pα Tra systems are genetically complex. F Tra is encoded by a 33.3 kb region containing *oriT*, 26 *tra* genes and >10 genes of unknown function. Only four of these genes (*traD, I, M* and *Y*) are essential for conjugative processing of DNA with *traI* specifying a multifunctional protein with relaxase and DNA helicase I activities. The remaining *tra* genes function in the cell-to-cell interactions fundamental to conjugation: 15 participate in the synthesis and assembly of the extracellular pilus, itself comprised of the processed product of a single gene (*traA*), two other *tra* genes stabilize cells in mating aggregates, while a further two determine surface/entry exclusion. The latter process blocks conjugation of cells harbouring closely related plasmids and may be important in a donor colony to prevent multiple cycles of mating and consequent membrane damage.

DONOR   **RECIPIENT**

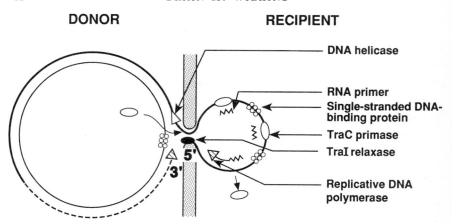

Fig. 1. Scheme of the conjugative processing of a Gram-negative bacterial plasmid exemplified by RP4. A unique DNA strand is transferred in the 5' to 3' direction following cleavage of the relaxosome at *oriT*. DNA relaxase (TraI) is linked covalently to the 5' terminus at the *oriT* nick site, and the complex is localized in the region of the DNA transport pore. The 3'OH terminus at the nick site is extended by rolling circle DNA replication in the donor. Nascent DNA is indicated by the broken line. Export of a unit length of DNA brings the reconstituted nick region into contact with the DNA-linked relaxase to allow a coupled cleavage–ligation reaction at the nick region to generate a monomeric circle of transferred DNA. DNA unwinding in the donor is thought to require a DNA helicase but it is unknown whether a plasmid- or host-encoded enzyme participates in the RP4 system. Molecules of the 118 kDa TraC DNA primase escort the transferred strand into the recipient cell to generate primers for discontinuous synthesis of the complementary strand. (Adapted from Wilkins & Lanka, 1993).

The F Tra region is located on the plasmid such that it is transferred last to the recipient cell. The complexity of conjugation is compounded by the finding that genes with ancillary functions are located in the leading region, defined as the first portion to be transferred. These ancillary genes are thought to promote establishment of the transferred plasmid DNA in the new host cell (Ippen-Ihler & Skurray, 1993; Wilkins & Lanka, 1993; Frost *et al.*, 1994).

The Pα transfer genes are arranged in two regions that are separated by a DNA sector specifying the Par partitioning–Mrs multimer resolution systems, insertion sequence IS*21* (=IS*8*) and the kanamycin-resistance (*aphA*) gene (Fig. 2). The TraI region contains *oriT* and at least 14 putative transfer genes in a segment of approximately 13 kb; this includes *traA* to *traJ*, inclusive, and *traX*, collectively located in two operons anticlockwise of *oriT* on the conventional map of the plasmid, and *traK*, *L* and *M* in the 'leader' operon on the clockwise side of *oriT*. Several of these loci function in DNA processing reactions but a core of only five genes (*F*, *G*, *I*, *J* and *K*) clustered around the essential *oriT* region is required for RP4 transfer between *E. coli* cells. TraI protein is the relaxase, TraJ acts in relaxosome formation and TraK is a specificity determinant for *oriT* function. TraG may couple the relaxosome to membrane proteins for DNA export, and TraF is thought to

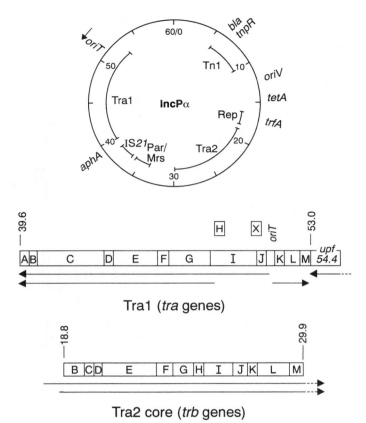

Fig. 2. Outline map of IncPα Birmingham plasmids such as RK2 and RP4. Numbers identify kilobase coordinates. The arrow at *oriT* indicates the clockwise direction of transfer. Expansions show the Tra1 (*traA* to *traM*) and Tra2 core (*trbB* to *trbL*) regions; horizontal lines indicate transcripts and arrowheads show transcriptional directions. *Abbreviations*: Rep, replication functions; *oriV*, origin of vegetative replication; *aphA*, *bla*, *tetA*, resistance to kanamycin, ampicillin and tetracycline, respectively. Other abbreviations are defined in the text. IS*21* is also designated IS*8*. (Data are from Pansegrau *et al.*, 1994).

be part of the mating apparatus that functions at the cell surface. Other genes contributing to the mating apparatus are located in the Tra2 region. The Tra2 core, transcribed as a single unit, contains 11 genes (designated *trbB* to *trbL*) required for *E. coli* conjugation. At least *trbB, C, D, E, G, H, I, J* and *L* encode essential proteins for the formation of mating aggregates, while *trbK* and the multifunctional *trbJ* gene function in entry exclusion. The putative pilin-subunit gene is *trbC* (Guiney, 1993; Lessl *et al.*, 1992, 1993; Pansegrau *et al.*, 1994; E. Lanka, personal communication).

While plasmid-encoded pili are a common feature of conjugation systems examined in Gram-negative bacteria (Frost, 1993), there is as yet no

DONOR   **RECIPIENT**

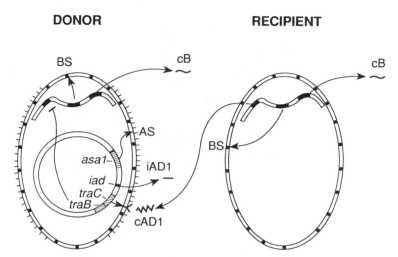

Fig. 3. Schematic representation of pheromone-induced transfer of plasmid pAD1 in conjugation of *Enterococcus faecalis*. The bacterial chromosome, represented by the linear structure,
encodes the octapeptide pheromone cAD1 (∿), a second pheromone arbitrarily called cB (~)
and binding substance (BS, ●). The latter is a normal component of the cell surface. The donor
cell harbours plasmid pAD1 (c. 60 kb, represented by the circle), which specifically responds to
cAD1 pheromone. Exogenous cAD1 is sensed by TraC (−<) functioning as a putative receptor
protein. Sensing induces expression of *tra* genes in the half of the plasmid located anti-clockwise
of *iad*. One such gene (*asa1*) determines the surface-associated aggregation substance (AS) that
binds BS on the recipient cell. Endogenous cAD1 synthesis by the donor is shut down by
plasmid TraB protein. iAD1 is the plasmid-encoded secreted octapeptide (—) that inhibits
cAD1. Further details are given by Clewell (1993*a*, *b*).

evidence that such structures participate in conjugation of Gram-positive
bacteria. One functional alternative to the pilus is the proteinaceous adhesin
determined by pheromone-responding plasmids of enterococci (Fig. 3;
Clewell, 1993*a*, *b*). The adhesin, also called 'aggregation substance' (AS),
has been identified by immunoelectron microscopy to consist of a dense
layer of protein microfibrils projecting about 30 nm from the donor cell
surface (Galli, Wirth & Wanner, 1989). Interaction between the adhesin
and its 'binding site' receptor on the recipient cell surface gives mating
contacts that are sufficiently stable to support efficient plasmid transfer
between cells in a liquid environment (10^{-2} transconjugants or more per
donor cell).

Other Gram-positive bacterial plasmids apparently specify less elaborate
mating apparatuses. A transfer region of about 15 genes has been identified
on the large staphylococcal plasmids pSK41 and pGO1, which are thought to
transfer by a pheromone-independent mechanism. There is no evidence that
these *tra* loci specify an aggregation substance or a cell-surface appendage,
which is consistent with the finding that the plasmids transfer relatively

infrequently (10^{-4} to 10^{-6} transconjugants per donor) by a surface-obligatory process (Firth *et al.*, 1993; Morton *et al.*, 1993).

One of the simplest conjugation systems is that encoded by pIJ101 (8.8 kb), discovered in *Streptomyces lividans*. Transfer of this plasmid requires very little genetic information on the element, and only one essential *tra* gene is involved. Despite this simplicity, conjugation can be very efficient. Possibly the *tra* gene specifies key DNA processing activities, while chromosomal genes determine other functions such as hyphal unions. Apart from small multicopy plasmids like pIJ101, *Streptomyces* spp. harbour integrating plasmids that insert into the host genome by site-specific recombination and 'giant' linear plasmids. Much remains to be learnt about the transfer of these plasmids but their diversity suggests that conjugation is of ecological and evolutionary importance to the host organisms (Hopwood & Kieser, 1993).

Compared to plasmids examined in *E. coli*, very little is known about origin of transfer sites on conjugative elements of Gram-positive bacteria and none has been shown to transfer in single-stranded form. However, there is indirect evidence for a ssDNA intermediate in the transfer of rolling circle staphylococcal plasmids, since these are mobilizable by a system that typifies Gram-negative counterparts in requiring a *cis*-acting *oriT* site and *mob* genes (Novick, 1989). Moreover, the staphylococcal plasmids exist as relaxosomes and the relaxase proteins possess motifs common to RP4 TraI (Lessl *et al.*, 1992). Further indirect evidence for ssDNA transfer is the similarity between the predicted TraK protein of the staphylococcal conjugative plasmids pSK41 and pGO1 and F TraD and RP4 TraG (Firth *et al.*, 1993; Morton *et al.*, 1993). Proteins of this family may facilitate transfer by coupling the relaxosome to the DNA export apparatus or by enabling ssDNA transfer through cell membranes (Lessl *et al.*, 1993; Wilkins & Lanka, 1993).

Not all conjugative elements are plasmids. Non-replicative types are illustrated by the conjugative transposons of Gram-positive bacteria, which were discovered as the disseminating agents of antibiotic resistance genes among streptococcal strains lacking detectable plasmids (Scott, 1992; Clewell & Flannagan, 1993). Particularly well studied are the conjugative transposons of a family that includes Tn*916*, originally isolated from *Enterococcus faecalis*, and Tn*1545*, first described in *Streptococcus pneumoniae*. These elements are relatively small (16–25 kb) and, like other conjugative transposons, carry a tetracycline-resistance gene of the *tetM* or a closely related type that confers resistance by ribosome protection.

Transposition of these elements is proposed to involve excision of the conjugative element to give a non-replicative, covalently closed, circular intermediate that can reinsert at a heterologous sequence in the donor genome. The joint in the intermediate is produced by a novel type of recombination that pairs non-complementary sequences derived by staggered nicking of the 'coupling' regions adjacent to the insert in the donor molecule (Caparon & Scott, 1989; Poyart-Salmeron *et al.*, 1989). Unlike other prokaryotic transposable elements, excision can be precise, thus restoring the function of the sequence originally disrupted by insertion. Conjugative behaviour would be fulfilled by transmission of the intermediate, or a derived single strand, followed by its insertion at one of many different sites in the resident DNA of the recipient cell. Transconjugants often carry multiple inserts but it is unknown whether these result from transfer of multiple copies or intracellular transposition of the original element inserted in the recipient cell. Very little is known about the cell to cell interactions supporting transfer of conjugative transposons but it has been suggested the process might cause some form of cell fusion (see a later section).

INTERRELATIONSHIPS BETWEEN CONJUGATION SYSTEMS

Nucleotide sequence comparisons have revealed that transfer systems are far more interrelated than suspected. One interesting relationship exists between the Pα Tra system and the Vir system determined by the Ti plasmid of *Agrobacterium tumefaciens*. The latter mediates transfer of the T-DNA sector to plant cells by a conjugation-like process initiated by VirD2 protein (Pansegrau *et al.*, 1993). This site-specific endonuclease nicks within the 25 bp border repeats that flank the T-DNA in the plasmid (Fig. 4). A 12-nucleotide consensus sequence has been extracted from the nick site regions in the Ti border repeats and in RP4 *oriT*. Other sequence and organizational similarities have been detected between VirD2 and the functionally related RP4 TraI (relaxase) protein, and between six of the predicted products of both the RP4 Tra2 region and the Ti VirB operon (Pansegrau & Lanka, 1991; Waters *et al.*, 1991; Lessl *et al.*, 1992). Yet a further relationship has been found between Ti VirB2 and F TraA, the pilin subunit precursor. Hence, it appears that a pilin-like protein contributes to the mating apparatus for agrobacterial DNA transfer to plant cells (Shirasu & Kado, 1993).

RP4 TraI-like sequences are also present in the naturally mobilizable pTF-FC2 plasmid isolated from a biomining bacterium, *Thiobacillus ferrooxidans*. Extensive similarities have been found between four *mob* genes surrounding pTF-FC2 *oriT* and four *tra* genes (*traI, J, K*, and *L*) surrounding RP4 *oriT*. Both sets of homologues have the same relative order. Interestingly, the minimal pTF-FC2 replicon shows a high degree of sequence similarity to the replicon of IncQ mobilizable plasmids, but similarities between these two small plasmids do not extend to their Mob regions (Rohrer & Rawlings, 1992).

Further evidence of the interchangeability of backbone portions of plasmids is provided by the hybrid nature of IncI1 conjugative plasmids. The I1 Tra system specifies two genetically and functionally discrete types of conjugative pilus, implying that the system is composed of components of

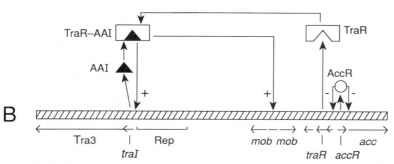

Fig. 4. Scheme of the regulation of the conjugative transfer genes on a classical nopaline-type Ti plasmid. (A) Organization of pTiC58 (c. 200 kb) presented in linear form. Regions shown are Noc, nopaline catabolism; Tra, conjugative transfer including genes for the mating apparatus; Rep, replication; Mob, conjugative transfer (mobilization) functions; Reg/Acc, regulation plus uptake and catabolism of agrocinopines A and B; Vir, virulence; T, T-region plus left (LB) and right (RB) border repeats (adapted from Farrand, 1993). The hatched segment is shown in greater detail in panel B. (B) Regulation of the pTiC58 conjugative transfer system. AccR repressor (○) regulates the leftward-transcribed *traR* gene and the rightward-transcribed *acc* genes. Agrocinopines are inducers. The *traI* locus is the first gene in the Tra3 operon and specifies production of the homoserine lactone autoinducer (▲; AAI). This interacts with TraR receptor protein to activate expression of at least *traI*, and hence Tra3, and the two divergent *mob* operons. Products of *mob* include proteins that are sequence-related to RSF1010 relaxase, F TraI DNA helicase and RP4 TraC DNA primase. *oriT* is located between the *mob* operons and resembles that of RSF1010 (Cook & Farrand, 1992). The Figure is based on unpublished concepts kindly provided by Dr Stephen Farrand.

two different ancestral lines (Rees, Bradley & Wilkins, 1987). One sub-region of I1 Tra resembles the Pα TraI region in that there are sequence and organizational similarities between *oriT* sites, relaxosome loci and DNA primase genes (Furuya & Komano, 1991; Furuya, Nisioka & Komano, 1991; Pansegrau & Lanka, 1991; Strack *et al.*, 1992). In contrast, the leading region and the replication region of I1 plasmids, which together flank the Tra region, are related to equivalent portions of F-like plasmids (Couturier *et al.*, 1988; Jones, Barth & Wilkins, 1992).

In summary, eubacteria harbour a wide variety of conjugative and mobilizable elements that are transferred by a mechanistically diverse range of systems. There is increasing evidence of unsuspected networks of relationships between plasmids, showing that during their evolution, these organisms have exchanged genetic modules for such fundamental activities as conjugative transfer and vegetative replication.

PROTEIN TRANSFER, CELL FUSION AND RETROTRANSFER

Bacterial conjugation is viewed classically as a process causing unidirectional DNA transfer from a donor to a recipient cell. While there is no extensive mixing of cell contents in Gram-negative bacterial conjugation, some systems do support unidirectional transfer of specific Tra proteins that promote DNA transport and initiation of DNA synthesis on the transferred plasmid strand. One example of these transmissible proteins is the DNA primases encoded by the *sog* and *traC* genes of IncI1 and IncPα plasmids, respectively (Rees & Wilkins, 1989; 1990; also see a later section). A further example of a transmissible protein is VirE2 of the Ti plasmid. This ssDNA-binding protein is thought to coat the intracellular T-strand and facilitate its transport into the plant cell (Citovsky *et al.*, 1992). Another potentially transmissible protein is the relaxase molecule linked to the 5′ end of the transferring DNA strand. This protein is thought to mediate circularization of the transferred strand by an event that reverses the *oriT*-nicking reaction, but it unknown whether the DNA-associated relaxase remains in the donor in the vicinity of the DNA transport pore or enters the recipient cell as a pilot to the transferring strand (Fig. 1).

Cell fusion has been proposed to occur in some conjugation systems. The finding that transfer of conjugative transposons correlates with extensive recombination of chromosomal genes of both mating partners has been interpreted to indicate formation of transient diploids (Torres *et al.*, 1991). It is unknown how cell fusion might be achieved but one speculation invokes formation of a variably sized DNA transport bridge that at one extreme allows bidirectional exchange of cell contents (Clewell & Flannagan, 1993).

An aspect of Gram-negative bacterial conjugation that is formally similar to cell fusion in its genetic outcome is retrotransfer. This causes genetic markers located on the chromosome or a non-conjugative plasmid of the original recipient to be recovered in the original donor of the conjugative plasmid. It has been proposed that retrotransfer is a specialized feature of the conjugation systems of some broad host-range plasmids, conferring on established plasmid-containing cells the advantage of capturing genes from other organisms, and reflects some form of hermaphroditism mediated by functions supplied by the original donor to allow this bacterium to donate and receive genetic information in a mating. Results of mass-action mathematical modeling support the notion that retrotransfer is a one-step process, mediated for example by the mating apparatus determined by the original donor (Mergeay *et al.*, 1987; Top *et al.*, 1992).

Other studies suggest that retrotransfer is a two-step process. This is implied by the delay found between forward and reverse transfers (Blanco *et al.*, 1991). Further arguments are the requirement for protein synthesis in the original recipient and the inverse correlation between retrotransfer efficiency and strength of the entry exclusion barrier determined by the conjugative plasmid. Coupled with detection of retrotransfer in F-mediated conjugation, these findings suggest the more orthodox explanation that retrotransfer is newly initiated conjugation of transconjugants and original donor cells (Heinemann & Ankenbauer, 1993). Clearly, further studies are required to define the mechanistic principles of the phenomenon.

BROAD TRANSFER RANGES

Many plasmids featuring in early studies of conjugation originated from *E. coli* and its close relatives. It now appears that most of these elements, which include F and the colicinogenic plasmid ColIb-P9 (IncI1), are maintained autonomously only in genera of the Enterobacteriaceae or even a sub-division of this family (Jacob *et al.*, 1977; Datta & Hughes, 1983; Tardif & Grant, 1983). Such plasmids are described as having a 'narrow' host range. In contrast, some of the antibiotic-resistance plasmids isolated in the last 25 years have a much broader host range. Classic examples are conjugative plasmids of the IncP group, which have been found in natural isolates of enterobacterial and other Gram-negative genera including *Pseudomonas aeruginosa*, *Alcaligenes eutrophus* and *Bordetella bronchiseptica*. More-over, IncP plasmids are maintained in almost all Gram-negative bacterial species tested in the laboratory (Smith & Thomas, 1989). Another set of broad host-range plasmids are the mobilizable plasmids of the IncQ group. This group includes RSF1010 and R300B, which were originally isolated in enterobacteria and *P. aeruginosa* but are now known to replicate in Gram-positive bacteria (Barth & Grinter, 1974; Guerry, van Embden & Falkow, 1974; see next section). Plasmids of Gram-positive bacteria have also been shown to have a remarkably broad host range; for example, staphylococcal rolling circle plasmids can replicate not only in phylogenetically distant Gram-positive organisms but also in *E. coli* and yeast (Goursot *et al.*, 1982; Novick, 1989).

While many plasmids have the potential to transfer widely, attention must be given to transfer efficiencies and to the transmission range in nature. Little is known about the latter and caution must be exercised in assuming that host range identified through laboratory tests reflects the breadth of the plasmid's distribution in nature. Furthermore, we possibly overestimate the prevalence of broad host-range plasmids by focusing on drug-resistance plasmids, chosen for their easily traceable phenotypic trait. This point is illustrated by the characterization of plasmid types present in a collection of enterobacteria established by E.D.G. Murray in the 'preantibiotic' years from 1917 to 1954 (Datta & Hughes, 1983). Most of the strains were *Salmonella*, *Shigella*, *Escherichia* and *Klebsiella* spp. Of 84 conjugative plasmids isolated from these strains, none specified antibiotic resistance and most belonged to the F complex (IncFII and FIV), the I complex (IncB, I1, I2 and K) and the IncN and IncX groups. Members of these Inc groups are often isolated nowadays as resistance plasmids, presumably reflecting the transposable mobility of resistance genes and the intense selection pressure exerted for their carriage by the use of antibiotics. While contemporary enterobacterial resistance plasmids include members of the IncA/C, H, M, P and W groups, no such plasmids were present in the Murray collection (Jacob *et al.*, 1977; Datta & Hughes, 1983). One possible explanation is that the distribution of the latter groups, which include some broad host-range plasmids (IncA/C, P and W), has expanded opportunistically in response to the widespread use of antibiotics.

It has frequently been found that the transfer potential of a plasmid is broader than its host range. One method demonstrating this point tests for the recovery of a plasmid-borne transposon or the plasmid itself as an insertion in a replicon native to the recipient cell following a cycle of authentic conjugation. Another approach uses naturally mobilizable or *oriT*-bearing shuttle plasmids as reporters of the transfer range of the mobilizing conjugative plasmid. Details are given elsewhere (Simon, 1989; Mazodier & Davies, 1991).

These approaches have shown that the Pα Tra system causes productive conjugation of *E. coli* and microorganisms outside the Gram-negative group. For example, RP4/RK2 mobilized RSF1010 at relatively high frequencies to the actinomycetes *Streptomyces lividans* and *Mycobacterium smegmatis* and to a cyanobacterium. RSF1010 was maintained in such hosts for many generations without selection, illustrating the remarkably broad maintenance range of small IncQ plasmids (Kreps *et al.*, 1990; Gormley & Davies, 1991). It has further been demonstrated with a mobilizable shuttle vector that an IncPβ plasmid (R751) causes *E. coli* to conjugate with the yeast *Saccharomyces cerevisiae*. However, the frequency of effective trans-kingdom transfer per recipient cell was low, ranging from 10^{-4} to 10^{-7} of the value for *E. coli* conjugation (Heinemann & Sprague, 1989). There is no evidence that IncP plasmids replicate in yeast but it has been reported that one of the rolling circle staphylococcal plasmids has this capacity (Goursot *et al.*, 1982).

The F transfer system also causes conjugation of *E. coli* and yeast but again the efficiency was found to be low, in the region of 10^{-7} transconjugants per donor (Heinemann & Sprague, 1989). In bacterial conjugation, F Tra appears to be less well adapted than Pα Tra for inter-familial transfers. This conclusion stems from the finding that F was 10^4-fold less efficient than RK2 in directing conjugation of *E. coli* and *P. aeruginosa*, as judged by mobilization of a broad host-range vector carrying a cognate *oriT*, whereas both plasmids were equally proficient in causing conjugation of *E. coli* strains (Guiney, 1982). Thus, some conjugation systems do appear better suited than others for promoting promiscuous gene transfer. Possibly the low rates of gene exchange detected between distantly related organisms reflect a background level of conjugation that is non-specific in terms of cell-to-cell interactions. Furthermore, such transfers may be of no consequence without some mechanism to ensure stable inheritance of the transferred DNA plus strong selection and even mutational change to optimise expression of the transferred genes in the new cell line.

Conjugative transposons transfer rather inefficiently with frequencies in the range of 10^{-9} to 10^{-5} per donor cell (Clewell & Flannagan, 1993). However, these elements transfer widely as independent units and as components of other conjugative elements. For example, Tn916 located on a self-transmissible plasmid in *E. faecalis* donors was found to transfer in laboratory conditions to the Gram-negative organisms *A. eutrophus*, *Citrobacter freundii* and *E. coli* and to move from an artificial plasmid construct in *E. coli* donors to low G+C content Gram-positive bacteria (Bertram, Strätz & Dürre, 1991). The last few years have seen a dramatic increase in the number of clinical isolates of Gram-positive and Gram-negative genera that have apparently acquired antibiotic resistance via conjugative transposition. This is scarcely surprising since the elements are widely transmissible, their maintenance requirements are simplified by transposition into a host replicon and their *tetM* genes are expressed in many Gram-positive and Gram-negative bacteria. Furthermore, transfer frequencies can be enhanced significantly by sub-inhibitory concentrations of tetracycline in the medium (see Poyart-Salmeron *et al.*, 1992). Although unproven, it appears that the natural distribution of conjugative transposons has enlarged in response to their acquisition of resistance genes and the widespread use of antibiotics (Roberts & Lansciardi, 1990; Poyart-Salmeron *et al.*, 1992; Swartley *et al.*, 1993).

SPECIFICITY OF CELL SURFACE INTERACTIONS

The broad transfer range of many conjugation systems implies that the contribution of chromosomally encoded components is relatively non-specific. Attempts to identify an essential 'mating receptor' in the envelope of the recipient cell have focused on the isolation of ConF$^-$ mutants defective in conjugation with F donors. Such an approach has given equivocal data, since mutants isolated for conjugation-deficiency in a liquid medium were generally found to conjugate normally on a surface. Several such ConF$^-$ mutants were defective in OmpA, a finding generally interpreted to indicate a role for this major outer membrane protein in the stabilization of cells in mating aggregates. Involvement of OmpA is remarkably system-specific, since other F-like plasmids can transfer normally to ConF$^-$ *ompA* mutants, as can IncI1 plasmids. Lipopolysaccharide (LPS) mutants were also recovered as ConF$^-$ mutants, but this phenotype may reflect inappropriate insertion of OmpA in the outer membrane of the mutants (Manning & Achtman, 1979; Willetts & Skurray, 1980). Possibly LPS has a more direct role in F-mediated conjugation, suggested by the recent finding that formation of mating pairs involves the pyrophosphory-lethanolamine residue on the first heptose of the inner core of the lipopoly-saccharide (L. S. Frost, personal communication). LPS appears to fulfil a central role in mating interactions in the I1 system, since ConI⁻ mutants were found to be LPS mutants and normal LPS was observed to inhibit conjugation (Havekes *et al.*, 1977).

While the contribution of host-encoded components to enterobacterial conjugation remains elusive, there is an essential role for the plasmid-encoded pilus. Conjugative pili can be distinguished into three morphologi-cal forms, described as rigid, thin and flexible, and thick and flexible (Bradley, Taylor & Cohen, 1980, Frost, 1993). Some forms show apparent environmental adaptation. For example, rigid pili are associated with a surface-obligatory system that operates >2000-fold more efficiently when *E. coli* cells are on a surface rather than in a liquid. This type of pilus is determined by plasmids of the IncN, P and W groups, which are some of the groups described as having a broad host range. Plasmids of the I complex (IncB, I1, I2, I5, K and Z) also specify a rigid pilus; this organelle is fundamental to the mating apparatus but functions as part of a two-pilus system that is 'universal' in operating equally effectively in liquid and on a surface. The second type of I pilus is thin and flexible and is thought to stabilize rigid pilus-mediated contacts in liquid (Bradley, 1984; Rees *et al.*, 1987). The thick and flexible type of pilus, as encoded by F-like plasmids, also contributes to a universal mating system but as a single organelle.

Clearly, a number of systems appear to be adapted through the mating apparatus for conjugation of cells on a surface. While it is attractive to speculate that the ecological role of the universal system of F-like plasmids is to cause conjugation of bacteria in liquid, which is the traditional medium used in laboratory studies, it has been argued from other evidence that the F Tra system is adapted to surface habitats (see next section).

REGULATORY SYSTEMS INDICATING SPECIALIZATION

There is increasing evidence that conjugation systems are regulated to promote transfer in particular ecological situations. For example, it has been argued that the FinOP fertility inhibition system of F-like plasmids is an adaptation to surface-dwelling hosts. FinOP regulates expression of the *tra* genes through an antisense RNA-repression system targeted against the production of TraJ protein, the transcriptional activator of the transfer operon. Operationally, *finP* transcript interacts with *traJ* mRNA to block translation of the latter and hence accumulation of the activator. Repression breaks down spontaneously in a small fraction of cells carrying an estab-lished Fin⁺ plasmid, thereby allowing occasional transfers. However, the Fin⁺ phenotype is only slowly established in the transconjugants, which therefore enjoy transient high fertility. Consequently once transfer starts, a

Fin$^+$ plasmid can spread epidemically through a population of recipient bacteria. Despite this spread, the original donor population remains much less fertile in broth culture than one harbouring a plasmid genetically derepressed (*drd*) for transfer functions (Willetts & Skurray, 1980; Dempsey, 1993).

A conventional view is that fertility inhibition is a price paid for host fitness, achieved by limiting the transcriptional and translational load of *tra* gene expression and the susceptibility to infection by pilus-specific phages. However unlike matings in liquid, a *drd* mutant enjoys little advantage over a Fin$^+$ plasmid when donor and recipients are incubated on a surface from a low plating density. Presumably conjugation becomes possible when a donor and recipient colony meet by convergent growth and, once transfer has been initiated from a physiologically derepressed cell at the margin of the donor colony, the plasmid can spread epidemically through the recipient colony. The transfer dynamics lead to the intriguing speculation that the Fin system of F-like plasmids has evolved as an adaptation to their natural hosts living in surface habitats (Simonsen, 1990).

Other transfer systems show more obvious evidence of specialization in being turned on by specific diffusible signals that sense environmental conditions suitable for conjugation. A classic example is the pheromone-responding system of enterococcal plasmids, whereby expression of conjugation genes involved in cell aggregation, DNA transfer and surface exclusion is induced by small octa- or heptapeptides secreted into the medium by potential enterococcal recipients (Fig. 3). Enterococci can specify several pheromones but any one responding plasmid shows specificity for a particular type and secretion of that pheromone is shut down following acquisition of the plasmid. In addition, such plasmids specify a small secreted peptide that competitively inhibits the cognate pheromone at low concentration. This type of inhibition may function to prevent induction of the conjugation system by suboptimal levels of pheromone, as would occur when the recipient population is insufficiently dense to support efficient aggregation (Clewell, 1993a,b). Such a signalling system indicates a high degree of discrimination in the conjugation process. It is unknown how these peptides evolved as important regulatory components of conjugation. Interestingly, the cAD1 pheromone and the precursor of iAD1 inhibitor (described in Fig. 3) are related by amino acid sequence to the signal peptide of the TraH protein of the staphylococcal pSK41 plasmid, supporting the notion that pheromones are derived from the signal sequences of larger proteins (Firth *et al.*, 1993).

The Ti plasmid of agrobacteria determines two transfer systems (Vir and Tra) that respond to specific low molecular weight signals. The Vir system mediates the conjugation-like transfer of T-DNA from agrobacteria to plant cells, and its expression is induced transcriptionally by phenolic compounds released by wounded plant cells. These signals are recognized by a two-component signal transduction system involving VirA and VirG. VirA is a member of the histidine protein kinase family of sensors, while VirG belongs to the class of response regulators whose promoter-binding properties are affected by N-terminal phosphorylation (Winans, 1992).

Following its integration into the genome of the plant cell, T-DNA determines a class of low molecular weight carbon compounds called opines. These are taken up and consumed as nutrients by pTi-containing agrobacteria colonizing the plant tumour. Opines can also induce expression of a Tra system that mediates conjugative transfer of the plasmid between bacteria (Farrand, 1993; Fuqua, Winans & Greenberg, 1994). Figure 4 illustrates the scheme thought to hold for the nopaline-type plasmid pTiC58. Catabolism of agrocinopines is mediated by the *acc* gene cluster, which is regulated at the transcriptional level by the product of an adjacent repressor gene called *accR*. AccR repressor also governs expression of a nearby master regulatory gene (*traR*) encoding the transcriptional activator protein of the Tra system. The AccR-regulated genes are induced by agrocinopines, explaining how opines control conjugative transmission of the plasmid (von Bodman, Hayman & Farrand, 1992; Farrand, 1993).

Transcriptional activation of the Tra system requires a second plasmid-determined signal that is specified by the *traI* gene and called conjugation factor (CF) or *Agrobacterium* autoinducer (AAI). This compound is a homoserine lactone derivative that is diffusible between cells. AAI interacts with the TraR receptor protein to form the activator for transcription of at least the Tra3 operon, which includes *traI* itself, and the *mob* operons. The intriguing finding is that TraR-AAI is homologous to the well-characterized LuxR-AI autoinduction system that effects cell density-dependent expression of bioluminescence genes in the marine bacterium *Vibrio fischeri* (Zhang *et al.*, 1993; Piper, von Bodman & Farrand, 1993). By analogy with AI activity, AAI is predicted to stimulate the frequency of conjugation in a manner that senses the density of agrobacterial donors under conditions where diffusion of the autoinducer is impeded. These would include the plant tumour. Clearly, the Vir and Tra systems of the Ti plasmid provide important and mechanistically intriguing precedents for induction of Gram-negative bacterial conjugation by specific environmental signals.

Induction of the conjugative transfer system of chromosomal resistance elements provides yet another example of environment sensing. The large (>70 kb) tetracycline-resistance (Tc[r]) elements of *Bacteroides* species mediate their own transfer to the recipient cell where they are recoverable in the chromosome as conjugative transposons. Tc[r] elements are also able to cause excision and circularization of mobilizable insertion elements called non-replicating *Bacteroides* units (NBUs). NBU excision-circularization and self-transfer of the Tc[r] elements are remarkable in being stimulated at least 100-fold by brief preexposure of cells to tetracycline. This antibiotic exerts its controlling effect through transcriptional induction of an operon containing a *tetQ* resistance gene and two downstream regulatory loci that activate a cascade of secondary regulators (Stevens *et al.*, 1993).

'PROMISCUITY' DETERMINANTS

The concept of the promiscuous plasmid raises the question of whether appropriate transfer systems include genes that potentiate conjugation of an extended range of species. This notion provides a possible explanation of the finding that only five of the 14 *tra* genes assigned to the Pα Tra1 region are essential for conjugation of *E. coli* (Lessl *et al.*, 1993). Promoters governing expression of Pα Tra1 varied widely in strength when tested in five Gram-negative genera, which might reflect a strategy to vary relative levels of gene expression according to needs in different genetic backgrounds (Greener, Lehman & Helinski, 1992).

Two RP4 *tra* genes have been implicated through transposon mutagenesis to function as 'promiscuity' genes. One such gene is *traC* (=*pri*; Fig. 2), encoding the plasmid DNA primase (Wilkins & Lanka, 1993). Primases are a class of enzymes that initiate DNA synthesis on ssDNA by synthesizing the oligonucleotide primers necessary for *de novo* starts by a DNA polymerase. Genetic evidence indicates that RP4 TraC primase is required in some recipient species for productive transmission of the plasmid and that the enzyme fulfilling this function is supplied by the donor cell (Lanka & Barth, 1981; Merryweather, Barth & Wilkins, 1986; Krishnapillai, 1988). The prediction that enzyme molecules are transferred between cells, apparently as DNA-binding protein, has been confirmed (Fig. 1; Rees & Wilkins, 1990).

Plasmid primases are thought to extend the conjugation range of a plasmid by rendering conjugative DNA synthesis on the transferred strand independent of host primer-generating systems. In the absence of the plasmid protein, the host priming mechanisms of some recipient species can substitute to allow productive conjugation. One complicating observation is that a *traC*::Tn7 mutation reduced the transfer efficiency of R18 from *P. aeruginosa* to *Pseudomonas stutzeri* but had little effect on transfers between *P. stutzeri* strains. Hence, TraC may have an additional role to priming, one possibility being that the transferred protein helps the immigrant plasmid to evade restriction.

While RP4 *traC* can be classified as a promiscuity determinant, there is no obvious correlation on plasmids between the carriage of a primase gene and the breadth of maintenance host range. The evidence is that primase genes have been detected on plasmids representing approximately one-third of the Inc groups studied in *E. coli* but the list includes narrow host-range plasmids of the I complex and lacks broad host-range plasmids of the IncN and W groups (Wilkins & Lanka, 1993).

The second Pα promiscuity gene is *traN*, which maps on the *oriT*-distal side of the leader operon. The gene is now called *upf54.4* (Fig. 2; *upf* = unknown plasmid function), recognizing the possibility that it may not be a transfer determinant (Pansegrau *et al.*, 1994). The phenotype of appropriate transposon mutants was found to differ according to the position of the insertion in the gene, with one class of mutants transferring normally from *P. aeruginosa* to *P. stutzeri* but deficiently to *E. coli* C and other recipient species and a second class showing deficient transfer only to *P. stutzeri*. The finding that one class of mutants was rescued by the provision of the appropriate normal DNA in the recipient whereas the other class was rescued through the donor cell suggests that Upf54.4 may have more than one function (Krishnapillai, 1988).

RESTRICTION-EVASION SYSTEMS RAISE EVOLUTIONARY CONSIDERATIONS

The concept of the promiscuous plasmid raises the further question of whether broad host-range plasmids have evolved mechanisms to bypass host restriction systems. Such systems have been detected in approximately one quarter of all bacterial isolates and more than 200 different types of specificity site have been identified (Wilson & Murray, 1991; Barcus & Murray, this volume). Hence restriction systems are likely to be encountered frequently not only in interspecific conjugations but also in transfers between strains of the same species.

One passive process allowing plasmids to bypass restriction is elimination of restriction sites by selection. This should operate each time a plasmid transfers in the unmodified state to a restricting host to favour recovery of mutants altered in the target sites. The IncPα plasmid 'backbone' structure, defined here as the transfer and replication-maintenance regions, is remarkably deficient of cleavage sites for many type II restriction enzymes (Smith & Thomas, 1989), fostering the view that broad host-range plasmids have lost restriction sites during an evolutionary history of promiscuous transfers. This explanation appears to be insufficient in the light of our analysis of oligonucleotide frequencies in the published nucleotide sequence of the Birmingham IncPα plasmids (M. Pocklington & B.M. Wilkins, in preparation). Specifically the backbone structure of these plasmids contains an abundance of the tetranucleotide pair CGGC/GCCG, present on average about one per 30 nucleotides (Table 1). The high incidence of the tetranucleotide suggests that the plasmid structure has experienced some form of motif-specific mutational pressure during its evolution. Assuming that the mechanism responsible was bacterial rather than plasmid-determined, it follows that IncPα plasmid backbone segments evolved as part of the genome of a preferred host type and that the chromosomal DNA of this bacterial type should be marked with a similar nucleotide signature. *P.*

Table 1. *DNA characteristics of known natural hosts of IncPα plasmids and of regions of RP4*[a]

Hosts[b] and plasmid components[c]	G + C (mol%)	Abundance of CGGC/GCCG
Pseudomonas aeruginosa	64.3	1/31
Serratia marcescens	57.5–60	1/36
Klebsiella pneumoniae	56–58	1/42
Proteus mirabilis	39.3	1/300
Providencia sp.	39–42	n.d.
Birmingham IncPα plasmids	61.8	1/31
Tra region of RP4	63.7	1/27
Tn*1* of RP4	49.8	1/110
IS*8* of RP4	52.3	1/125

[a]Values computed by Dr Michael Pocklington.
[b]Hosts of early isolates of IncPα plasmid-containing strains, of which *P. aeruginosa* was the most frequent (Smith & Thomas, 1989).
[c]Values derived from the nucleotide sequence data of Pansegrau *et al.* (1994).

aeruginosa is a possible candidate from the hosts of the first IncPα plasmid-containing strains to be isolated: as judged from a panel of nucleotide sequences taken from databases, the chromosomal DNA of this organism is similar by G+C content and CGGC/GCCG frequency to the Birmingham group of plasmids (Table 1).

The finding that IncPα plasmid backbone regions have a similar nucleotide signature to the chromosomal DNA of *P. aeruginosa* provides a basis for examining whether type II restriction targets are depleted from the plasmid. To this end, the frequency of each of the 64 possible palindromic hexanucleotides was computed for the plasmid and bacterial DNA. Frequency trends in the two molecules were similar with a bias towards high G+C sequences incorporating the tetranucleotide pair. However, approximately 24 of the hexanucleotides were clearly deficient on the plasmid, presumably reflecting selection. Interestingly, selection could have occurred during historic transfers between pseudomonads, since *P. aeruginosa* strains are the source of about 30 differently named restriction enzymes that collectively recognise ten different target sequences (listed in REBASE; Roberts & Macelis, 1993) and nine of these sequences are severely depleted from the backbone structure of the plasmid. Taken together, the findings suggest that the backbone portion of IncPα plasmids evolved in a limited set of hosts and, while deficiencies of restriction sites may favour the promiscuity of contemporary plasmids, it should not be assumed that the depletions were selected during an evolutionary history of conjugative promiscuity.

In contrast to the backbone regions of IncPα plasmids, there is a relative abundance of restriction sites in regions containing antibiotic resistance genes and transposable elements (Smith & Thomas, 1989). Table 1 shows that the latter (Tn*1* and IS*8*) have a different nucleotide signature from a representative portion of the RP4 backbone, namely the Tra region. Presumably the resistance determinants and transposable elements were acquired relatively recently in the evolution of this group of plasmids.

Plasmids can also evade restriction by a process dependent on conjugative transfer of more than one copy of the plasmid to the recipient cell. The process is non-specific in acting against at least type I and type II restriction systems and presumably involves breakdown of the restriction mechanism in response to an early event in conjugation. Possibly transfer of enzyme-saturating amounts of DNA is the responsible factor (Read, Thomas & Wilkins, 1992). Other physiological explanations should be considered in view of the recent finding that restriction (R) but not modification (M) activity is delayed extensively in transconjugant cells following acquisition of type I R–M genes (Prakash-Cheng & Ryu, 1993).

Carriage of an anti-restriction gene constitutes yet another type of restriction-evasion mechanism. One such gene is located on the IncI1 plasmid ColIb-P9 (Delver *et al.*, 1991). The gene, called *ardA*, specifically alleviates restriction of DNA by all three families of type I restriction enzyme. Genetic evidence shows that carriage of *ard in cis* allows un-modified ColIb to evade restriction following its transfer by conjugation to recipients specifying the type I enzyme *Eco*K. However, no such protection is conferred on ColIb DNA transferred by transformation. The anti-restriction determinant maps in the ColIb leading region, which is the expected location of a gene that is expressed early in the transconjugant cell to promote installation of the immigrant plasmid (Read *et al.*, 1992).

Tests with representatives of 23 different incompatibility groups of conjugative plasmid have detected *ardA*-hybridizing sequences on members of the IncI1-B-K subset of the I complex of plasmids, on an IncFV plasmid of the F complex, and on IncN group plasmids. The IncB and FV hybridizing sequences map to the leading region and are functional anti-restriction genes with at least 63% nucleotide sequence identity to ColIb *ardA* (P.M. Chilley & B.M. Wilkins, submitted). The same properties have been reported for *ardA* on an IncN plasmid (Belogurov, Delver & Rodzevich, 1992). Why should *ard* homologues be present on plasmids of the I complex, F complex and the IncN group, which are inferred from structural consider-ations to be a motley collection of plasmids representing different evolution-ary lineages? These plasmids, together with members of the IncX group, are authentic plasmids of *Escherichia coli* and close relatives, as judged by the plasmids present in the Murray collection of bacteria described earlier. A similar cluster of enterobacteria is a known source of type I restriction enzymes (Wilson & Murray, 1991). Hence, carriage of *ardA* appears to be an adaptation allowing conjugative plasmids to escape a particular type of restriction system determined by their natural hosts, rather than a general mechanism for restriction evasion.

Conjugative transposons are remarkably resistant to restriction when transferred by conjugation. Resistance is demonstrated by the finding that the large (>60kb) Tn5253 element is refractory to DpnII-restriction following transfer in unmodified form to S. pneumoniae recipients and yet the DNA is sensitive to in vitro cleavage by the enzyme. Since conjugative plasmids transferred from the same donors were DpnII-sensitive, general breakdown of restriction in conjugating cells is not the explanation (Guild, Smith & Shoemaker, 1982). The molecular basis of the resistance is unknown and explanation is likely to require a better understanding of the structure of the transfer intermediate. Some ideas have been discussed by Scott (1992).

CONJUGATIVE TRANSFER OF CHROMOSOMAL GENES

In addition to supporting transmission of the conjugative element, conjugation systems can cause transfer of segments of the bacterial chromosome by the process of mobilization. Chromosome mobilization is thought to require the coupling of the donor cell chromosome to a functional $oriT$, achieved most simply by covalent linkage of the chromosome to a conjugative plasmid. The resulting cointegrate is a fusion of two replicons and, if stable, gives rise to an Hfr donor strain. Chromosome transfer occurs unidirectionally from the integrated $oriT$ of such cells, giving a high frequency of recombinants for bacterial genes that are transferred early.

Understanding of the molecular basis of chromosome mobilization is most advanced for plasmid F. Cultures of F^+ cells give recombinants with frequencies of about 10^{-5} per donor cell. This low level of fertility is partly due to stable cointegrates (Hfr chromosomes) in the population but mostly due to transient interactions between the plasmid and the chromosome (Curtiss & Stallions, 1969). These interactions are undefined but it is assumed here that they involve replicon fusion rather than the fortuitous recognition of chromosomal sequences that happen to resemble $oriT$ of the plasmid. One mechanism known to support cointegration is RecA-mediated recombination of one of the three insertion sequences (IS2, IS3a or IS3b) on the native F plasmid and one of the several homologues in the bacterial chromosome. The resulting cointegrates can be remarkably stable, even in rec^+ cells. Transmissable cointegrates can be generated likewise by recombination of Tn1000 ($\gamma\delta$) on F and a copy of the transposon on another replicon. Tn1000 is a member of the Tn3 family of transposons, which transpose by a replicative mechanism that cointegrates donor and target replicons as a transient intermediate. If such an intermediate is formed at the time of conjugation, the target replicon will be transferred to the recipient cell behind the leading portion of the F plasmid (Low, 1987).

The considerable value of F-based Hfr strains to genetic studies of *E. coli* has stimulated many attempts to isolate equivalent strains in other bacterial species. A range of plasmids has been shown to promote chromosomal mobilization but transfer generally occurs with a lower frequency and from fewer sites than for F-containing cells. Many studies have focused on IncPα plasmids, chosen for their broad host-range maintenance properties, but the chromosome-mobilizing ability (Cma) of these plasmids is generally very inefficient, as documented by Holloway (1979) and Reimmann & Haas (1993).

One explanation of the low Cma of IncPα plasmids is the infrequent formation of cointegrates and the instability of these replicon fusions. IncPα plasmids carry a copy of Tn*1* (Fig. 2), which transposes by a cointegrate-forming mechanism similar to that of Tn*1000*. Even when breakdown of cointegrates is impeded by mutation of both the *tnpR* resolvase gene of the transposon and the RecA-dependent system of the host, stable cointegrates are found only very rarely (Reimmann & Haas, 1986). The probability of cointegration with the bacterial chromosome is limited further by the fact that transposons of the Tn*1* type transpose preferentially to plasmid rather than chromosomal sites.

Spontaneous IncPα-derived plasmids showing enhanced Cma have been isolated. One, R68.45, carries a tandem duplication of IS*21* (or IS*8*) normally present in a single copy on these plasmids (Fig. 2). IS*21* tandems are quite stable and cause cointegrates in *E. coli* at a frequency of 10^{-2} to 10^{-5}, apparently by a non-replicative pathway. The tandem duplication of the element enhances replicon fusion by creating a new promoter for expression of the transposase/cointegrase gene and by providing closely linked IS*21* ends which are thought to interact strongly with the cointegrating enzyme (Reimmann & Haas, 1990; 1993).

An important factor influencing the stability of an Hfr strain may be whether or not the plasmid component retains the potential to initiate DNA replication. Several lines of evidence point to this conclusion. One is the frequent detection of stable Hfrs among IncP-10 plasmid transconjugants formed in a *Pseudomonas putida* strain incapable of supporting autonomous replication of the plasmid (Strom *et al.*, 1990). A complementary finding is that IncPα plasmids recovered after forced chromosomal integration were frequently replication-defective; in some cases the mutation responsible was shown to be in *trfA*, an essential plasmid replication gene (Grinter, 1984; Reimmann & Haas, 1986). Similarly, Hfr strains selected for carriage of an integrated temperature-sensitive pSC101 plasmid replicon were found to grow poorly on return to a temperature permissive for plasmid replication (François, Conter & Louarn, 1990).

How can these observations be reconciled with the apparent stability and fitness of some F-mediated Hfr strains? One explanation centres on whether or not the replication origin of the plasmid is active in the cointegrate. F has a copy number of about one per bacterial chromosome and initiation of replication from the F origin is inhibited in at least some Hfrs where the integration site is sufficiently near the chromosomal origin to maintain the concentration of F above the critical level that triggers plasmid replication (Chandler *et al.*, 1976; Lycett & Pritchard, 1986). No such replication constraint will operate for plasmids with significantly higher copy numbers, such as IncPα plasmids and pSC101 which collectively have a copy number in the range of four to eight per chromosome (Kornberg & Baker, 1992). In these cases, replication initiated from the integrated plasmid origin might derange cellular physiology by altering relative gene concentrations, increasing gene dosage and creating topological problems at cell division. Such adverse effects on fitness have been observed when replication of the *E. coli* chromosomes is initiated from an integrated, temperature-dependent, runaway-replication derivative of the IncFII plasmid R1, whose copy number increases at high temperature (Bernander, Merryweather & Nordström, 1989). Yet another potential complication is that a number of plasmids replicate unidirectionally, whereas bacterial chromosome replication from the native origin is bidirectional (Kornberg & Baker, 1992). Unidirectional replication of the bacterial chromosome may cause complications formally similar to those resulting from large inversions of the bacterial chromosome, a topic reviewed elsewhere (Krawiec & Riley, 1990).

CONCLUDING REMARKS

Conjugative plasmids and chromosomal elements have been discovered in a rich array of Gram-positive and Gram-negative bacteria. The transfer systems of these elements are mechanistically diverse, especially with respect to the mating apparatus for establishing contacts between cells. There is increasing evidence that conjugation systems are adapted through their mating apparatus and regulatory controls to promoting gene transfer between particular groups of organism in preferred ecological conditions. Plasmids described as having a broad host range do not appear to be exceptional: there is no compelling evidence that transfer genes allowing promiscuity are restricted to these elements and backbone portions of some broad host-range plasmids have apparently evolved in a closely related set of hosts. Laboratory studies have shown that conjugative elements are transmissible between distantly related organisms but these transfers may reflect background conjugation that is non-specific at the level of cellular interactions and inconsequential without selection. Some transmissible plasmids cause efficient mobilization of large segments of the bacterial chromosome to the recipient cell but this is not a general property of conjugative elements.

ACKNOWLEDGEMENTS

This chapter is dedicated to the memory of Bill Hayes (1913–1994), pioneer of plasmid and bacterial genetics and an inspirational teacher. I am indebted to Paul Chilley, Patrice Courvalin, Stephen Farrand, Laura Frost, Karin Ippen-Ihler, Erich Lanka and Michael Pocklington for ideas and unpublished data.

REFERENCES

Amábile-Cuevas, C. F. & Chicurel, M. E. (1992). Bacterial plasmids and gene flux. *Cell*, **70**, 189–99.

Barth, P. T. & Grinter, N. J. (1974). Comparison of the deoxyribonucleic acid molecular weights and homologies of plasmids conferring linked resistance to streptomycin and sulfonamides. *Journal of Bacteriology*, **120**, 618–30.

Belogurov, A. A., Delver, E. P. & Rodzevich, O. V. (1992). IncN plasmid pKM101 and IncI1 plasmid ColIb-P9 encode homologous antirestriction proteins in their leading regions. *Journal of Bacteriology*, **174**, 5079–85.

Bernander, R., Merryweather, A. & Nordström, K. (1989). Overinitiation of replication of the *Escherichia coli* chromosome from an integrated runaway-replication defective derivative of plasmid R1. *Journal of Bacteriology*, **171**, 674–83.

Bertram, J., Strätz, M. & Dürre, P. (1991). Natural transfer of conjugative transposon Tn*916* between gram-positive and gram-negative bacteria. *Journal of Bacteriology*, **173**, 443–8.

Blanco, G., Ramos, F., Medina, J. R., Gutierrez, J. C. & Tortolero, M. (1991). Conjugal retrotransfer of chromosomal markers in *Azotobacter vinelandii*. *Current Microbiology*, **22**, 241–6.

Bradley, D. E. (1984). Characteristics and function of thick and thin conjugative pili determined by transfer-derepressed plasmids of incompatability groups I_1, I_2, I_5, B, K and Z. *Journal of General Microbiology*, **130**, 1489–502.

Bradley, D.E., Taylor, D.E. & Cohen, D.R. (1980). Specification of surface mating systems among conjugative drug resistance plasmids in *Escherichia coli* K-12. *Journal of Bacteriology*, **143**, 1466–70.

Caparon, M. G. & Scott, J. R. (1989). Excision and insertion of the conjugative transposon Tn*916* involves a novel recombination mechanism. *Cell*, **59**, 1027–34.

Chandler, M., Silver, L., Roth, Y. & Caro, L. (1976). Chromosome replication in an Hfr strain of *Escherichia coli*. *Journal of Molecular Biology*, **104**, 517–23.

Clewell, D. B. (1993*a*). Bacterial sex pheromone-induced plasmid transfer. *Cell*, **73**, 9–12.

Clewell, D. B. (1993*b*). Sex pheromones and the plasmid-encoded mating response in *Enterococcus faecalis*. In *Bacterial Conjugation*, Clewell, D.B., ed., pp. 349–67. Plenum Press, New York.

Clewell, D. B. & Flannagan, S. E. (1993). The conjugative transposons of gram-positive bacteria. In *Bacterial Conjugation*, Clewell, D. B., ed., pp. 369–93.

Citovsky, V., Zupan, J., Warnick, D. & Zambryski, P. (1992). Nuclear localization of *Agrobacterium* VirE2 protein in plant cells. *Science*, **256**, 1802–5.

Cook, D. M. & Farrand, S. K. (1992). The *oriT* region of *Agrobacterium tumefaciens* Ti plasmid pTiC58 shares DNA sequence identity with the transfer origins of RSF1010 and RK2/RP4 and with T-region borders. *Journal of Bacteriology*, **174**, 6238–46.

Courvalin, P. (1994). Transfer of antibiotic resistance genes between Gram-positive and Gram-negative bacteria. *Antimicrobial Agents and Chemotherapy*, **38**, 1447–51.

Couturier, M., Bex, F., Berguist, P. L. & Maas, W. K. (1988). Identification and classification of bacterial plasmids. *Microbiological Research*, **52**, 375–95.

Curtiss III, R. & Stallions, D. R. (1969). Probability of F integration and frequency of stable Hfr donors in F$^+$ populations of *Escherichia coli* K-12. *Genetics*, **63**, 27–38.

Datta, N. & Hughes, V. M. (1983). Plasmids of the same Inc groups in Enterobacteria before and after the medical use of antibiotics. *Nature, London*, **306**, 616–17.

DeFlaun, M. F. & Levy, S. B. (1989). Genes and their varied hosts. In *Gene Transfer in the Environment*, Levy, S. B. & Miller, R. V., eds., pp. 1–32. McGraw-Hill, New York.

Delver, E. P., Kotova, V. U., Zavilgelsky, G. B. & Belogurov, A. A. (1991). Nucleotide sequence of the gene (*ard*) encoding the antirestriction protein of plasmid CoIIb-P9. *Journal of Bacteriology*, **173**, 5887–92.

Dempsey, W. B. (1993). Key regulatory aspects of transfer of F-related plasmids. In *Bacterial Conjugation*, Clewell, D. B., ed., pp. 53–73. Plenum Press, New York.

Dürrenberger, M. B., Villiger, W. & Bächi, T. (1991). Conjugational junctions: morphology of specific contacts in conjugating *Escherichia coli* bacteria. *Journal of Structural Biology*, **107**, 146–56.

Farrand, S. K. (1993). Conjugal transfer of *Agrobacterium* plasmids. In *Bacterial Conjugation*, Clewell, D.B., ed., pp. 255–91. Plenum Press, New York.

Firth, N., Ridgway, K. P., Bryne, M. E., Fink, P. D., Johnson, L., Paulsen, I. T. & Skurray, R. A. (1993). Analysis of a transfer region from the staphylococcal conjugative plasmid pSK41. *Gene*, **136**, 13–25.

François, V., Conter, A. & Louarn, J.-M, (1990). Properties of new *Escherichia coli* Hfr strains constructed by integration of pSC101-derived conjugative plasmids. *Journal of Bacteriology*, **172**, 1436–40.

Frost, L. S. (1993). Conjugative pili and pilus-specific phages. In *Bacterial Conjugation*, Clewell, D. B., ed. pp. 189–221. Plenum Press, New York.

Frost, L. S., Ippen-Ihler, K. & Skurray, R. A. (1994). Analysis of the sequence and gene products of the transfer region of the F sex factor. *Microbiological Reviews*, **58**, 162–210.

Fuqua, W. C., Winans, S. C. & Greenberg, E. P. (1994). Quorum sensing in bacteria: the LuxR-LuxI family of cell density-responsive transcriptional regulators. *Journal of Bacteriology*, **176**, 269–75.

Fürste, J. P., Pansegrau, W., Zeigelin, G., Kröger, M. & Lanka, E. (1989). Conjugative transfer of promiscuous IncP plasmids: Interactions of plasmid-encoded products with the transfer origin. *Proceedings of the National Academy of Sciences, USA*, **86**, 1771–5.

Furuya, N. & Komano, T. (1991). Determination of the nick site at *oriT* of IncI1 plasmid R64: global similarity of *oriT* structures of IncI1 and IncP plasmids. *Journal of Bacteriology*, **173**, 6612–17.

Furuya, N., Nisioka, T. & Komano, T. (1991). Nucleotide sequence and functions of the *oriT* operon in IncI1 plasmid R64. *Journal of Bacteriology*, **173**, 2231–7.

Galli, D., Wirth, R. & Wanner, G. (1989). Identification of aggregation substances of *Enterococcus faecalis* cells after induction by sex pheromones. An immunological and ultrastructural investigation. *Archives of Microbiology*, **151**, 486–90.

Gormley, E. P. & Davies, J. (1991). Transfer of plasmid RSF1010 by conjugation from *Escherichia coli* to *Streptomyces lividans* and *Mycobacterium smegmatis*. *Journal of Bacteriology*, **173**, 6705–8.

Goursot, R., Goze, A., Niaudet, B. & Ehrlich, S. D. (1982). Plasmids from *Staphylococcus aureus* replicate in yeast *Saccharomyces cerevisiae*. *Nature, London*, **298**, 488–90.

Greener, A., Lehman, S. M. & Helinski, D. R. (1992). Promoters of the broad host range plasmid RK2: analysis of transcription (initiation) in five species of Gram-negative bacteria. *Genetics*, **130**, 27–36.

Grinter, N. J. (1984). Replication defective RP4 plasmids recovered after chromosomal integration. *Plasmid*, **11**, 65–73.

Guerry, P., van Embden, J. & Falkow, S. (1974). Molecular nature of two nonconjugative plasmids carrying drug resistance genes. *Journal of Bacteriology*, **117**, 619–30.

Guild, W. R., Smith, M. D. & Shoemaker, N. B. (1982). Conjugative transfer of chromosomal R determinants in *Streptococcus pneumoniae*. In *Microbiology-1982*. Schlessinger, D., ed., pp. 82–92. American Society for Microbiology, Washington, DC.

Guiney, D. G. (1982). Host range of conjugation and replication functions of *Escherichia coli* sex factor F *lac*: comparison with the broad host range plasmid RK2. *Journal of Molecular Biology*, **162**, 699–703.

Guiney, D. G. (1993). Broad host range conjugative and mobilizable plasmids in Gram-negative bacteria. In *Bacterial Conjugation*, Clewell, D. B., ed., pp. 75–103. Plenum Press, New York.

Guiney, D. G. & Lanka, E. (1989). Conjugative transfer of IncP plasmids. In *Promiscuous Plasmids of Gram-negative Bacteria*. Thomas, C. M., ed., pp. 27–56. Academic Press, London.

Harrington, L. C. & Rogerson, A. C. (1990). The F pilus of *Escherichia coli* appears to support stable DNA transfer in the absence of wall-to-wall contact between cells. *Journal of Bacteriology*, **172**, 7263–4.

Havekes, L., Tommassen, J., Hoekstra, W. & Lugtenberg, B. (1977). Isolation and characterization of *Escherichia coli* K-12 F⁻ mutants defective in conjugation with an I-type donor. *Journal of Bacteriology*, **129**, 1–8.

Heinemann, J. A. & Ankenbauer, R. G. (1993). Retrotransfer in *Escherichia coli* conjugation: bidirectional exchange or de novo mating? *Journal of Bacteriology*, **175**, 583–8.

Heinemann, J. A. & Spraque Jr G. F. (1989). Bacterial conjugative plasmids mobilize DNA transfer between bacteria and yeast. *Nature, London*, **340**, 205–9.

Holloway, B. W. (1979). Plasmids that mobilize bacterial chromosome. *Plasmid*, **2**, 1–19.

Hopwood, D. A. & Keiser, T. (1993). Conjugative plasmids of *Streptomyces*. In *Bacterial Conjugation*, Clewell, D. B., ed., pp. 293–311. Plenum Press, New York.

Ippen-Ihler, K. & Skurray, R. A. (1993). Genetic organization of transfer-related determinants on the sex factor F and related plasmids. In *Bacterial Conjugation*, Clewell, D. B., ed., pp. 23–52. Plenum Press, New York.

Jacob, A. E., Shapiro, J. A., Yamamoto, L., Smith, D. I., Cohen, S. N. & Berg, D. (1977). Table 1b. Plasmids studied in *Escherichia coli* and other enteric bacteria. In *DNA Insertion Elements, Plasmids and Episomes*, Bukhari, A. I., Shapiro, J. A. & Adhya, S. L., eds., pp. 607–38. Cold Spring Harbor Laboratory, Cold Spring Harbor.

Jones, A. L., Barth, P. T. & Wilkins, B. M. (1992). Zygotic induction of plasmid *ssb* and *psiB* genes following conjugative transfer of IncI1 plasmid CoIIb-P9. *Molecular Microbiology*, **6**, 605–13.

Kornberg, A. & Baker, T. A. (1992). *DNA Replication*, 2nd end. W.H. Freeman, New York.

Krawiec, S. & Riley, M. (1990). Organization of the bacterial chromosome. *Microbiological Reviews*, **54**, 502–39.

Kreps, S., Ferino, F., Mosring, C., Gerits, J., Mergeay, M. & Thuriaux, P. (1990). Conjugative transfer and autonomous replication of a promiscuous IncQ plasmid in the cyanobacterium *Synechocystis* PCC 6803. *Molecular and General Genetics*, **221**, 129–33.

Krishnapillai, V. (1988). Molecular genetic analysis of bacterial plasmid promiscuity. *FEMS Microbiology Reviews*, **54**, 223–38.

Lanka, E. & Barth, P. T. (1981). Plasmid RP4 specifies a deoxyribonucleic acid primase involved in its conjugal transfer and maintenance. *Journal of Bacteriology*, **148**, 769–81.

Lederberg, J. & Tatum, E. L. (1946). Gene recombination in *Escherichia coli*. *Nature, London*, **158**, 558.

Lessl, M., Balzer, D., Pansegrau, W. & Lanka, E. (1992). Sequence similarities between RP4 Tra2 and the Ti VirB region strongly support the conjugation model for T-DNA transfer. *Journal of Biological Chemistry*, **267**, 20471–80.

Lessl, M., Balzer, D., Weyrauch, K. & Lanka, E. (1993). The mating pair formation system of plasmid RP4 defined by RSF1010 mobilization and donor-specific phage propagation. *Journal of Bacteriology*, **175**, 6415–25.

Low, K. B. (1987). Hfr strains of *Escherichia coli* K-12. In Escherichia coli *and* Salmonella typhimurium: *Cellular and Molecular Biology*, Neidhardt, F. C., Ingraham, J. L., Low, K. B., Magasanik, B., Schaechter, M. & Umbarger, H. E., eds., pp. 1134–7. American Society for Microbiology, Washington DC.

Lycett, G. W. & Pritchard, R. H. (1986). Functioning of the F plasmid origin of replication in an *Escherichia coli* K12 Hfr strain during exponential growth *Plasmid*, **16**, 168–74.

Manning, P. A. & Achtman, M. (1979). Cell-to-cell interactions in conjugating *Escherichia coli*: the involvement of the cell envelope. In *Bacterial Outer Membranes: Biogenesis and Functions*, Inouye, M., ed., pp. 409–47. John Wiley, New York.

Maynard Smith, J., Smith, N. H., O'Rourke, M. & Spratt, B. G. (1993). How clonal are bacteria? *Proceedings of the National Academy of Sciences, USA*, **90**, 4284–8.

Mazodier, P. & Davies, J. (1991). Gene transfer between distantly related bacteria. *Annual Review of Genetics*, **25**, 147–71.

Mergeay, M., Lejeune, P., Sadouk, A., Gerits, J. & Fabry, L. (1987). Shuttle transfer (or retrotransfer) of chromosomal markers mediated by plasmid pULB113. *Molecular and General Genetics*, **209**, 61–70.

Merryweather, A., Barth, P. T. & Wilkins, B. M. (1986). Role and specificity of plasmid RP4-encoded DNA primase in bacterial conjugation. *Journal of Bacteriology*, **167**, 12–7.

Morton, T. M., Eaton, D. M., Johnston, J. L. & Archer, G.L. (1993). DNA sequence and units of transcription of the conjugative transfer gene complex (*trs*) of *Staphylococcus aureus* plasmid pGO1. *Journal of Bacteriology*, **175**, 4436–47.

Novick, R. P. (1989). Staphylococcal plasmids and their replication. *Annual Review of Microbiology*, **43**, 537–65.

Pansegrau, W. & Lanka, E. (1991). Common sequence motifs in DNA relaxases and nick regions from a variety of DNA transfer systems. *Nucleic Acids Research*, **19**, 3455.

Pansegrau, W., Lanka, E., Barth, P. T., Figurski, D. H., Guiney, D. G., Haas, D., Helinski, D. R., Schwab, H., Stanisich, V. A. & Thomas, C. M. (1994). Complete nucleotide sequence of Birmingham IncPα plasmids: compilation and comparative analysis. *Journal of Molecular Biology*, **239**, 623–63.

Pansegrau, W., Schoumacher, F., Hohn, B. & Lanka, E. (1993). Site-specific cleavage and joining of single-stranded DNA by VirD2 protein of *Agrobacterium tumefaciens* Ti plasmid: analogy to bacterial conjugation. *Proceedings of the National Academy of Sciences, USA*, **90**, 11538–42.

Pansegrau, W., Ziegelin, G. & Lanka, E. (1990). Covalent association of the *traI* gene product of plasmid RP4 with the 5′-terminal nucleotide at the relaxation nick site. *Journal of Biological Chemistry*, **265**, 10637–44.

Piper, K. R., von Bodman, S. B. & Farrand, S. K. (1993). Conjugation factor of *Agrobacterium tumefaciens* regulates Ti plasmid transfer by autoinduction. *Nature, London*, **362**, 448–50.

Prakash-Cheng, A. & Ryu, J. (1993). Delayed expression of in vivo restriction activity following conjugal transfer of *Escherichia coli* hsd_K (restriction-modification) genes. *Journal of Bacteriology*, **175**, 4905–6.

Poyart-Salmeron, C., Trieu-Cuot, P., Carlier, C. & Courvalin, P. (1989). Molecular characterization of two proteins involved in the excision of the conjugative transposon Tn*1545*: homologies with other site-specific recombinases. *EMBO Journal*, **8**, 2425–33.

Poyart-Salmerson, C., Trieu-Cuot, P., Carlier, C., MacGowan, A., McLauchlin, J. & Courvalin, P. (1992). Genetic basis of tetracycline resistance in clinical isolates of *Listeria monocytogenes*. *Antimicrobial Agents and Chemotherapy*, **36**, 463–6.

Read, T. D., Thomas, A. T. & Wilkins, B. M. (1992). Evasion of type I and type II restriction systems by IncI1 plasmid ColIb-P9 during transfer by bacterial conjugation. *Molecular Microbiology*, **6**, 1933–41.

Rees, C. E. D. & Wilkins, B. M. (1989). Transfer of *tra* proteins into the recipient cell during bacterial conjugation mediated by plasmid ColIb-P9. *Journal of Bacteriology*, **171**, 3152–7.

Rees, C. E. D. & Wilkins, B. M. (1990). Protein transfer into the recipient cell during bacterial conjugation: studies with F and RP4. *Molecular Microbiology*, **4**, 1199–205.

Rees, C. E. D., Bradley, D. E. & Wilkins, B. M. (1987). Organization and regulation of the conjugation genes of IncI₁ plasmid ColIb-P9. *Plasmid*, **18**, 223–36.

Reimmann, C. & Haas, D. (1986). IS*21* insertion in the *trfA* replication control gene of chromosomally integrated plasmid RP1: a property of stable *Pseudomonas auruginosa* Hfr strains. *Molecular and General Genetics*, **203**, 511–19.

Reimmann, C. & Haas, D. (1990). The *istA* gene of insertion sequence IS*21* is essential for cleavage at the inner 3′ ends of tandemly repeated IS*21* elements *in vitro*. *EMBO Journal*, **9**, 4055–63.

Reimmann, C. & Haas, D. (1993). Mobilization of chromosomes and nonconjugative plasmids by cointegrative mechanisms. In *Bacterial Conjugation*, Clewell, D. B., ed., pp. 137–88. Plenum Press, New York.

Roberts, M. C. & Lansciardi, J. (1990). Transferable TetM in *Fusobacterium nucleatum*. *Antimicrobial Agents and Chemotherapy*, **34**, 1836–8.

Roberts, R. J. & Macelis, D. (1993). REBASE-restriction enzymes and methylases. *Nucleic Acids Research*, **21**, 3125–37.

Rohrer, J. & Rawlings, D. E. (1992). Sequence analysis and characterization of the mobilization region of a broad-host-range plasmid, pTF-FC2, isolated from *Thiobacillus ferrooxidans*. *Journal of Bacteriology*, **174**, 6230–7.

Scott, J. R. (1992). Sex and the single circle: conjugative transposition. *Journal of Bacteriology*, **174**, 6005–10.

Shirasu, K. & Kado, C. I. (1993). Membrane location of the Ti plasmid VirB proteins involved in the biosynthesis of a pilin-like conjugative structure on *Agrobacterium tumefaciens*. *FEMS Microbiology Letters*, **111**, 287–94.

Simon, R. (1989). Transposon mutagenesis in non-enteric Gram-negative bacteria. In *Promiscuous Plasmids of Gram-negative Bacteria*. Thomas, C. M., ed., pp. 207–28. Academic Press, London.

Simonsen, L. (1990). Dynamics of plasmid transfer on surfaces. *Journal of General Microbiology*, **136**, 1001–7.

Smith, C. A. & Thomas, C. M. (1989). Relationships and evolution of IncP plasmids. In *Promiscuous Plasmids of Gram-negative Bacteria*. Thomas, C. M., ed., pp. 57–77. Academic Press, London.

Stevens, A. M., Shoemaker, N. B., Li, L-Y. & Salyers, A. A. (1993). Tetracycline regulation of genes on *Bacteroides* conjugative transposons. *Journal of Bacteriology*, **175**, 6134–41.

Strack, B., Lessl, M., Calendar, R. & Lanka, E. (1992). A common sequence motif, -E-G-Y-A-T-A, identified within the primase domains of plasmid-encoded I- and P-type DNA primases and the α protein of the *Escherichia coli* satellite phage P4. *Journal of Biochemical Chemistry*, **267**, 13062–72.

Strom, A. D., Hirst, R., Petering, J. & Morgan, A. (1990). Isolation of high frequency of recombination donors from Tn5 chromosomal mutants of *Pseudomonas putida* PPN and recalibration of the genetic map. *Genetics*, **126**, 495–503.

Swartley, J. S., McAllister, C. F., Hajjen, R. A., Heinrich, D.W. & Stephens, D. S. (1993). Deletions of Tn*916*-like transposons are implicated in *tetM*-mediated resistance in pathogenic *Neisseria*. *Molecular Microbiology*, **10**, 299–310.

Tardif, G. & Grant, R. B. (1983). Transfer of plasmids from *Escherichia coli* to *Pseudomonas aeruginosa*: characterization of a *Pseudomonas aeruginosa* mutant with enhanced recipient ability for enterobacterial plasmids. *Antimicrobial Agents and Chemotherapy*, **24**, 201–8.

Thomas, C. M. (ed.) (1989). *Promiscuous Plasmids of Gram-negative Bacteria*. Academic Press, London.

Top, E., Vanrolleghem, P., Mergeay, M. & Verstraete, W. (1992). Determination of the mechanisms of retrotransfer by mechanistic mathematical modeling. *Journal of Bacteriology*, **174**, 5953–60.

Torres, O. R., Korman, R. Z., Zahler, S. A. & Dunny, G. M. (1991). The conjugative transposon Tn925: enhancement of conjugal transfer by tetracycline in *Enterococcus faecalis* and mobilization of chromosomal genes in *Bacillus subtilis* and *E. faecalis*. *Molecular and General Genetics*, **225**, 395–400.

von Bodman, S. B., Hayman, G. T. & Farrand, S. K. (1992). Opine catabolism and conjugal transfer of the nopaline Ti plasmid pTiC58 are coordinately regulated by a single repressor. *Proceedings of the National Academy of Sciences, USA*, **89**, 643–7.

Waters, V. L., Hirata, K. H., Pansegrau, W., Lanka, E. & Guiney, D. G. (1991). Sequence identity in the nick regions of IncP plasmid transfer origins and T-DNA borders of *Agrobacterium* Ti plasmids. *Proceedings of the National Academy of Sciences, USA*, **88**, 1456–60.

Wilkins, B. & Lanka, E. (1993). DNA processing and replication during plasmid transfer between Gram-negative bacteria. In *Bacterial Conjugation*, Clewell, D. B., ed., pp. 105–36, Plenum Press, New York.

Willetts, N. & Skurray, R. (1980). The conjugation system of F-like plasmids. *Annual Review of Genetics*, **14**, 41–76.

Wilson, G. G. & Murray, N. E. (1991). Restriction and modification systems. *Annual Review of Genetics*, **25**, 585–627.

Winans, S. C. (1992). Two-way chemical signaling in *Agrobacterium*-plant interactions. *Microbiological Reviews*, **56**, 12–31.

Zhang, L., Murphy, P. J., Kerr, A. & Tate, M. E. (1993). *Agrobacterium* conjugation and gene regulation by *N*-acyl-L-homoserine lactones. *Nature, London*, **362**, 446–8.

INSERTION SEQUENCE (IS) ELEMENTS AS NATURAL CONSTITUENTS OF THE GENOMES OF THE GRAM-NEGATIVE *RHIZOBIACEAE* AND THEIR USE AS A TOOL IN ECOLOGICAL STUDIES

SELBITSCHKA WERNER, DORIS JORDING, REINHARD SIMON & ALFRED PÜHLER

Universität Bielefeld, Fakultät für Biologie, Lehrstuhl für Genetik, Postfach 100131, D-33501, Bielefeld, Germany

INTRODUCTION

Bacterial insertion sequence (IS) elements are genetic entities which are able to translocate to new genetic locations either within a replicon or between different replicons of their host cell, a process called transposition (for transposition mechanism see Galas & Chandler, 1989). Typically, IS elements are 0.7–2.5 kilobases (kb) in size and are bordered by perfect or nearly perfect inverted terminal repeat sequences. Insertion sequence elements encode only genetic functions related to their transposition activity, such as the enzyme transposase which mediates transposition. This feature distinguishes IS elements from transposons, another class of mobile genetic elements which usually carry selectable traits. With few exceptions, IS elements generate the duplication of a characteristic number of nucleotides upon transposition into a target site. Thereby, after insertion, the target duplication borders the IS termini as a direct repeat.

Insertion sequence elements were first discovered as the causative agent of spontaneous mutations in *Escherichia coli* (Jordan, Saedler & Starlinger, 1968, Shapiro, 1969). It was shown that transposition of an IS element into a structural gene resulted in the physical disruption of the gene and consequently, could lead to the biosynthesis of a mutated, non-functional protein. Insertion sequences often exert a polar effect on transcription of downstream located genes if they transpose into a transcriptional unit. Some IS elements are able to activate cryptic genes upon insertion, or mediate the overexpression of nearby genes (see Galas & Chandler, 1989 and references therein). Moreover, IS elements were shown to contribute significantly to the genetic plasticity of their host cell's genome. Major changes like deletion formation, inversion or replicon fusion may occur as a direct consequence of IS transposition. Alternatively, since IS elements constitute regions of portable homology, similar changes are induced if the host recombinational apparatus uses IS elements as a substrate for homologous recombination.

The ability of IS elements to transpose into transmissible elements like plasmids or phages ensures their distribution within bacterial populations (for a review on the population dynamics of transposable elements see Ajioka & Hartl, 1989). Moreover, since pairs of IS elements are able to transpose any DNA sequence that lies between them, they participate actively in the horizontal transmission of genetic information among bacterial cells. An excellent example of the involvement of IS elements in genetic exchange provides the broad-host-range IncP plasmid R68.45 which is able to mobilize chromosomal markers between conjugating cells at high frequency (for a review see Reimmann & Haas, 1993).

A yet open question concerns the origin of insertion sequences as well as the evolutionary forces which act upon their long-term maintenance in genomes. It remains to be determined whether IS elements are maintained primarily in genomes due to their ability to propagate by transposition in host cells, i.e. to behave as selfish DNA (Doolittle & Sapienza, 1980; Orgel & Crick, 1980) or because they provide some adaptive role to their host (Campbell, 1981). This topic is discussed in detail in recent reviews (Syvanen, 1984; Hartl et al., 1986; Arber, 1991).

Insertion sequences were identified in numerous bacterial species and until now the detection of numerous IS elements has been reported. Although IS elements were isolated from diverse species belonging to the eubacterial as well as the archaebacterial urkingdoms, this class of mobile genetic elements has been most thoroughly studied in E. coli. The review will focus on indigenous IS elements of the Gram-negative, plant-interacting Rhizobiaceae. In addition, the application of IS elements for the unequivocal identification of bacterial strains as well as their potential use to assess the genetic diversity of natural populations will be discussed.

IDENTIFICATION OF IS ELEMENTS WITHIN THE RHIZOBIACEAE

The bacterial family Rhizobiaceae is taxonomically subdivided into the genera Rhizobium, Bradyrhizobium and Azorhizobium as well as the genera Phyllobacterium and Agrobacterium. Bacterial species belonging to the former genera are able to fix atmospheric nitrogen in symbiosis with leguminous plants. Owing to this unique property, these plant-beneficial microorganisms are of enormous agricultural value. Certain species of Agrobacterium are also of agricultural interest since they are the causative agent of plant diseases.

Like in E. coli, the occurrence of IS elements within the Rhizobiaceae was first discovered by their mutational activity. IS elements ISR1, ISRm1, ISRm2 and IS66 of the Rhizobiaceae were detected owing to the high instability of the resistance plasmid RP4 in *B. lupini* (Priefer *et al.*, 1981), by their mutational insertion into symbiotic genes of *R. meliloti* (Ruvkun *et al.*, 1982; Dusha *et al.*, 1987) or into the Ti plasmid of *A. tumefaciens* (Binns, Sciaky & Wood, 1982). Whereas the previous detection and isolation of IS elements was basically by chance, the development of new techniques for the detection of mutational events, e.g. insertions by a positive selection procedure greatly facilitated the identification of bacterial IS elements. The *Bacillus subtilis sacRB* gene, encoding the enzyme levansucrase has proven to be a valuable tool for such purposes (Gay *et al.*, 1985). Gram-negative bacteria expressing the *sacRB* gene are effectively killed when grown in the presence of sucrose. Therefore, the inactivation of the *sacRB* gene is positively selectable. Another approach used the *E. coli rpsL* gene which encodes the ribosomal S12 protein. If the wild-type allele is provided on a plasmid to a spontaneous streptomycin-resistant mutant, the bacterial cells become streptomycin-sensitive, i.e. streptomycin-sensitivity is trans-dominant over streptomycin-resistance (Dean, 1981). Inactivation of the *rpsL* gene, e.g. by an insertion, results in the recovery of streptomycin resistance, and again this event is positively selectable. Finally, a third approach used the *E. coli pheS* gene which encodes the α-subunit of the phenylalanine-t-RNA synthetase. Spontaneous mutants of *E. coli* resistant to the phenylalanine analogon, para-fluor phenylalanine (pfp), regain sensitivity if they carry the wild type allele on a plasmid and are therefore killed in the presence of the compound (Hennecke, Günther & Binder, 1982). The identification of plasmids carrying an inactivated *pheS* gene is based on the back-crossing of such a plasmid from the bacterial species to be analysed for the presence of IS elements, into the pfp-resistant *E. coli* strain and subsequent selection for pfp-resistance.

Simon *et al.* (1991) extended the applicability of the above described marker genes for the isolation of IS elements by inserting them into the broad-host-range vector pSUP104. Due to the replication functions of RSF1010 carried by pSUP104, this vector is stably maintained in a wide range of Gram-negative bacterial species. The newly constructed entrapment vectors were shown to be advantageous to perform a systematic search for IS elements in various Gram-negative bacteria, and were especially successfully employed for this purpose in *R. meliloti*. Another vector system which was developed in our group is suitable to search for a special class of insertion sequences, ISp elements, which carry outreading promoters and therefore activate the transcription of adjacent genes. There are outwardly directed promoters (pOUT) located within the element or, alternatively, hybrid promoters that are created upon insertion into the target sequence to direct transcription (see also Galas & Chandler, 1989). Positive selection vectors based on the broad-host-range vector pSUP104 were constructed to carry a promotorless tetracycline or neomycin resistance gene as selectable markers (Labes & Simon, 1990). The system was successfully employed for the isolation of *R. meliloti* ISp elements. Fig. 1 shows the restriction maps of the various positive selection vectors described above.

CHARACTERISTICS OF IS ELEMENTS OF THE *RHIZOBIACEAE*

Rhizobium leguminosarum biovar *viciae*

Using the above-mentioned entrapment vectors, four transposable elements comprising one transposon and three IS elements have been isolated

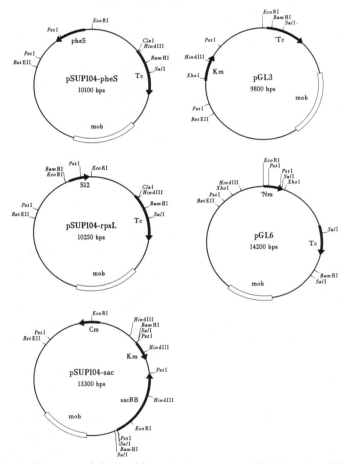

Fig. 1. Restriction maps of the positive selection vectors pSUP104-*rpsL*, pSUP104-*sac*, pSUP104-*pheS* and the probe vectors pGL3 and pGL6. Bold arrows indicate the positive selection markers *rpsL* (encoding the ribosomal S12 protein), *sac* and *pheS* or the promotorless tetracyclin ('Tc) and neomycin ('Nm) resistance genes, respectively. Shaded arrows indicate the antibiotic resistance markers of the plasmids. Mob indicates the mobilization functions of RSF1010.

from *R. leguminosarum* bv. *viciae* strains. Transposon Tn*163* is a class II transposon and carries an open reading frame with the deduced amino acid sequence showing homology to the trifolitoxin resistance protein TfxG (Ulrich & Pühler, 1994). Insertion sequence IS*RlF7-2* (Simon *et al.*, 1991) derives from strain F7, a wild-type isolate from a *Vicia faba* root nodule. Strain F7 harbours a genetically unstable cryptic plasmid which frequently undergoes deletion formation, replicon fusion or is completely lost by its host cell (Selbitschka & Lotz, 1991). DNA sequence analysis revealed that IS*RLF7-2* is, with the exception of two base substitutions identical to IS*RlD1*, isolated from a French *R. leguminosarum* bv. *viciae* strain (S. Mazurier pers. comm.). Comparison of the IS*RlF7-2* nucleotide sequence with data base entries indicate that IS*RlF7-2*/IS*RlD1* belong to the IS*5* group of the IS*4* family (Rezsöhazy *et al.*, 1993) and are closely related to the IS elements IS*Rm4* of *R. meliloti* as well as IS*1031A*, *C*, *D*, (Coucheron, 1993) and IS*1032* of *Acetobacter xylinum* (our unpublished results). This close relationships among IS elements could be indicative of a horizontal gene transfer event. In addition, the putative transposases encoded by IS*RlF7-2*/IS*RlD1*, IS*Rm4* and IS*1031* show significant homology to a hypo-

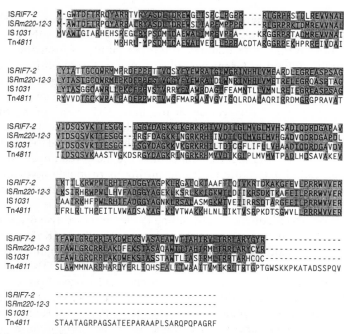

Fig. 2. Alignment of the putative transposases of IS elements IS*RIF7-2* of *R. leguminosarum* bv. *viciae*, IS*Rm220-12-3* of *R. meliloti*, IS*1031* of *A. xylinum* and transposon Tn*4811* of *S. lividans*. Identical residues are underlaid with black boxes. If at a given position two couples of identical amino acid residues occur, identity is also indicated with grey boxes.

thetical protein encoded by ORF3 of the *Streptomyces lividans* transposon Tn*4811* (Chen *et al.*, 1992) (Fig. 2).

The third IS element of *R. leguminosarum* bv. *viciae*, ISR*l1* whose nucleotide sequence has been deposited in the gene bank data base (O'Brien *et al.*, 1993 unpublished observations) shows neither at the DNA sequence nor at the deduced amino acid sequence level significant homology to the above mentioned elements. However, ISR*l1* shares an interesting feature with agrobacterial IS elements as well as some other pro- and eukaryotic DNA sequences. This refers to a plasmid-borne open reading frame called ORF104, identified in a pathogenic *E. coli* isolate (Knoop & Brennicke, 1994). Interestingly, the nucleotide sequence immediately upstream of ORF104 shows the structural characteristics of eukaryotic group II intron domaines. Further upstream of this sequence there is an ORF situated with the deduced amino acid sequence showing homology to an intron encoded maturase. Open reading frames displaying significant homology to ORF104 are also present in agrobacterial IS elements IS*66*, IS*866* and IS*1131*. The functional importance of ORF104 remains to be established.

Bradyrhizobium *spp.*

The first insertion sequence of the *Rhizobiaceae* to be reported, ISR*1* of *B. lupini*, was analysed in detail by Priefer *et al.* (1981, 1989). DNA sequence analysis showed that ISR*1* belongs to the IS*3* family of insertion elements which are distributed in Gram-negative as well as Gram-positive bacterial species (Schwartz, Kröger & Rak, 1988; Rubens, Heggen & Kuypers, 1989). Insertion sequence element ISR*1* carries two open reading frames with the translational stop codon of the first ORF overlapping the start codon of the second ORF in a -1 frame. Immediately in front of this structure there is a poly A stretch situated, a signal which is considered to facilitate the slippage of the translating ribosome into the new reading frame. This mechanism called 'translational frameshifting' is probably involved in the regulation of transpositional activity of some IS elements (Chandler & Fayet, 1993).

Repetitive DNA sequences resembling IS elements are also present in the genome of *B. japonicum* strains, the symbionts of soybean (Kaluza, Hahn & Henneke, 1985; Rodriguez-Quinones *et al.*, 1992). Although transpositional activity has not been demonstrated yet, these repeated DNA sequences (RS sequences) display typical structural characteristics of prokaryotic IS elements (Kaluza *et al.*, 1985; Judd & Sadowski, 1993). Five different repeated sequences designated RSα, RSβ, RSγ, RSδ, and RSε were found in the *B. japonicum* strain 110 genome with the copy number ranging between four and twelve copies (Kaluza *et al.*, 1985; Hahn & Hennecke, 1987). The hyperreiterated sequence HRS1 which is found in *B. japonicum* serocluster 123 strains (Rodriguez-Quinones *et al.*, 1992) is unrelated to the repeated sequences mentioned above. The HRS1 sequence lacks terminal inverted repeats, however comparison of the deduced amino acid sequence of an open reading frame of HRS1 revealed a high degree of sequence homology to an open reading frame of the *Acetobacter pasteurianus* IS element IS*1380* (Judd & Sadowsky, 1993).

Agrobacterium spp.

So far, the nucleotide sequences of seven agrobacterial IS elements are available including those of IS*66*, IS*426* (identical to IS*136*), IS*427*, IS*866*, IS*869*, IS*870* and IS*1131* (see Table 1). All IS elements originate from *A. tumefaciens* strains except IS*870* which was found in *Agrobacterium vitis*. Moreover, an IS-like DNA sequence was identified in the Ri plasmid of

Table 1. *Molecular characteristics of IS elements of the* Rhizobiaceae *with known nucleotide sequence. Sizes in parenthesis indicate ambiguous length*

IS element	Size (bp)	Target duplication (bp)	Inverted repeats (bp)	Source	Reference
ISR1F7-2 ISR1D1	932	3	17	*R. leguminosarum* bv. *viciae*	unpublished
ISR1I	2495	8	13	*R. leguminosarum* bv. *viciae*	O'Brien *et al.*, 1993 (unpublished)
ISR1	1260	4	13	*B. lupini*	Priefer *et al.*, 1989
RSRjα	(1126)	(5)	(4)	*B. japonicum*	Kaluza, Hahn & Hennecke, 1985; Ramseier & Göttfert 1991
HRS1	2070	5	–	*B. japonicum*	Judd & Sadowsky, 1993
IS66	2548	8	18/20	*A. tumefaciens*	Machida *et al.*, 1984
IS426	1313	9	32/30	*A. tumefaciens*	Vanderleyden *et al.*, 1986
IS427	1271	2	16	*A. tumefaciens*	DeMeirsman *et al.*, 1990
IS866	2716	8	27	*A. tumefaciens*	Bonnard, Vincent & Otten, 1989
IS869	847	4	15	*A. tumefaciens*	Paulus *et al.*, 1991
IS870	(1129)	(4)	(9)	*A. vitis*	Fournier, Paulus & Otten, 1993
IS1131	2773	8	12	*A. tumefaciens*	Wabiko, 1992
ISRm1	1319	5	31/32	*R. meliloti*	Watson & Wheatcroft, 1991
ISRm3	1298	8–9	30	*R. meliloti*	Wheatcroft & Laberge, 1991; Soto *et al.*, 1992*a*
ISRm4	933	3	17	*R. meliloti*	Soto *et al.*, 1992*b*
ISRm220-12-3	934				
ISRm2011-2	1050	2	19	*R. meliloti*	unpublished

Agrobacterium rhizogenes (Slightom *et al.*, 1986; Ramseier & Göttfert, 1991). Although agrobacterial IS elements are both, plasmid-borne and resident in the chromosome of their host cells, they often show characteristic associations with Ti plasmids (Paulus, Ride & Otten, 1989; Otten *et al.*, 1992). Consequently, information about the presence of IS elements in Ti plasmids has been used to infer the evolution of octopine Ti plasmids (Paulus *et al.*, 1991). Comparison of the nucleotide sequences as well as the deduced amino acid sequences of potential ORFs of these IS elements with data base entries reveal homologies among agrobacterial IS elements as well as homologies of agrobacterial IS elements to mobile genetic elements from other bacterial species (see Table 2). If the relationships among IS elements of *Agrobacterium* are considered, the agrobacterial IS elements roughly can be allocated to three homology groups: (i) IS elements IS*427* and IS*869* are closely related, (ii) IS elements IS*66*, IS*426*, IS*866* and IS*1131* are distantly related and (iii) IS*870* is related to the IS-like element situated in the Ri plasmid of *A. rhizogenes* as well as to repeated sequences RFRS of *R. fredii* (Krishnan & Pueppke, 1993) and RSRj of *B. japonicum* (Ramseier & Göttfert, 1991). Interestingly, as a unique feature of agrobacterial IS elements, IS*870* requires a 5'-CTAG-3' target sequence to generate a stop codon for its ORF1 which encodes the putative transposase.

R. meliloti

All IS elements of *R. meliloti* which have been sequenced to date are unrelated among each other but show striking homology to IS elements of the *Rhizobiaceae* or other bacterial species (see Table 2). Insertion sequence element IS*Rm1*, initially discovered as the causative agent of symbiotic mutations of *R. meliloti* was the first *R. meliloti* IS element to be sequenced. IS*Rm1* seems to be identical to ISp*Rm2011-2* described by Labes and Simon (1990) as well as to IS*Rm2011-1* described by Simon *et al.* (1991). This conclusion is based on the comparable length of the IS elements, their restriction sites as well as their characteristic distribution in the genome of *R. meliloti* 2011. The structural features of IS*Rm1* allocate this IS element to the IS*3* family of IS elements. Homologues of IS element IS*Rm3*, another *R. meliloti* IS element were isolated independently from strains 102F70 (Wheatcroft & Laberge, 1991) and GR4 (Soto *et al.*, 1992*a*), respectively. Both IS elements differ in their nucleotide sequence at seven positions and hence, the overall homology is 99.5%. Similarly, homologues of IS*Rm4* identified in *R. meliloti* strains GR4 (Soto *et al.*, 1992*b*) and 220-13 (Simon *et al.*, 1991), differ in 11 positions of their nucleotide sequence and show an overall homology of 99% (our unpublished results). These data demonstrate the high degree of nucleotide sequence conservation between IS elements isolated from different strains of the same rhizobial species. Analogous results for homologues of IS elements isolated from different *E.*

Table 2. *Homologies between IS elements of the* Rhizobiaceae *and IS elements isolated from various bacterial species*

IS element	Source	Homology to ORFs of	Source	Accession no.
IS*RIF7-2*	*R. leguminosarum*	IS*Rm4*	*R. meliloti*	JQ1940
IS*RID1*	bv. *viciae*	IS*1031*A, D, C	*A. xylinum*	S35003-5
		IS*1032*	*A. xylinum*	U02294
		Tn*4811*	*S. lividans*	S19841
		IS*427*	*A. tumefaciens*	JQ1762
		IS*869*	*A. tumefaciens*	S14988
		IS*402*	*P. cepacia*	P24537
		IS*6501*	*B. ovis*	X71024
		IS?	*M. tubercolosis*	X65618
IS*Rl1*	*R. Leguminosarum*	IS*Rm1*	*R. meliloti*	S37715
	bv. *viciae*	IS*1131*	*A. tumefaciens*	JC1151
		IS*426*	*A. tumefaciens*	S37713
		IS*2*	*E. coli*	P19776
		IS*1163*	*L. sake*	X75164
		IS*904*	*L. lactis*	D00696
		Tn*4521*	*E. coli*	M35123
		IS*L1*	*L. casei*	X02734
IS*R1*	*B. lupini*	IS*476*	*X. campestris*	P25438
		IS*407*	*P. cepacia*	P24580
		IS*2*	*E. coli*	P19777
		IS*1163*	*L. sake*	X75164
RSRjα	*B. japonicum*	IS*870*	*A. tumefaciens*	Z18270
		RFRS9	*R. fredii*	A48935
HR S1	*B. japonicum*	IS1380	*A. pasteurianus*	B41313
IS66	*A. tumefaciens*	IS*1131*	*A. tumefaciens*	JC1151
		IS*Rm1*	*R. meliloti*	S37715
		IS*426*	*A. tumefaciens*	S37713
		IS*2*	*E. coli*	P19776
IS*426*	*A. tumefaciens*	IS*2*	*E. coli*	P19777
		IS*Rm1*	*R. meliloti*	S37716
		Tn*4521*	*E. coli*	M35123
		IS*407*	*P. cepacia*	P24577
		Tn*552*	*S. aureus*	P18416
		IS*150*	*E. coli*	P19769
		IS*476*	*X. campestris*	P25438
		IS*L1*	*L. casei*	X02734
		IS*904*	*L. lactis*	P35878
IS*427*	*A. tumefaciens*	IS*869*	*A. tumefaciens*	S14986
		IS?	*M. tubercolosis*	S21394
		IS*6501*	*B. ovis*	X71024
		Tn*4811*	*S. lividans*	S19841
		IS*402*	*P. cepacia*	P24537
		IS*1031*	*A. xylinum*	S35003-5
		IS*Rm4*	*R. meliloti*	JQ1940
		IS*1106*	*N. meningitidis*	Q00840
IS*866*	*A. tumefaciens*	IS*1131*	*A. tumefaciens*	JC1151
IS*869*	*A. tumefaciens*	IS*427*	*A. tumefaciens*	JQ1762
		IS?	*M. tuberculosis*	S21394
		IS*6501*	*B. ovis*	X71024
		Tn*4811*	*S. lividans*	S19841

Continued

Table 2. *Continued*

IS element	Source	Homology to ORFs of	Source	Accession no.
		IS*1031*	*A. xylinum*	S35003-5
		IS*Rm4*	*R. meliloti*	JQ1940
IS*870*	*A. vitis*	RSRj9	*B. japonicum*	A48358
		RFRS9	*R. fredii*	A48935
IS*1131*	*A. tumefaciens*	IS*866*	*A. tumefaciens*	JC1151
IS*Rm1*	*R. meliloti*	IS*426*	*A. tumefaciens*	S37714
		IS2	*E. coli*	P19777
		Tn*4521*	*E. coli*	M35123
		IS*476*	*X. campestris*	P25438
IS*Rm3*	*R. meliloti*	IS*1191*	*S. thermophilus*	S37549
		IS*905*	*L. lactis*	P35881
		IS*1201*	*L. helveticus*	P35880
		IS*256*	*S. aureus*	P19775
		IS*406*	*P. cepacia*	P24575
		IS*6120*	*M. smegmatis*	P35883
		IS*1081*	*M. bovis*	X61270
		IST2	*T. ferrooxidans*	P35884
IS*Rm4*	*R. meliloti*	IS*RIF7-2/ISRlD1*	*R. leguminosarum*	
IS*Rm220-12-3*			bv. *viciae*	
		IS*1031*	*A. xylinum*	S35003-5
		Tn*4811*	*S. lividans*	S19841
		IS*427*	*A. tumefaciens*	JQ1762
		IS*1032*	*A. xylinum*	U02294
		IS*869*	*A. tumefaciens*	S14988
		IS*402*	*P. cepacia*	P24537
		IS*6501*	*B. ovis*	X71024
		IS?	*M. tuberculosis*	X65618
IS*Rm2011-2*	*R. meliloti*	–	–	–

coli strains were reported by Lawrence, Ochman & Hartl (1992). The authors concluded a high turnover and rapid movement of IS elements among *E. coli* hosts. Insertion sequence element IS*Rm2011-2* which was isolated by Simon *et al.* (1991) shows neither at the nucleotide sequence level nor at the deduced amino acid sequence level significant homology to any IS element sequenced so far (our unpublished results). The homologies found between *R. meliloti* IS elements and IS elements of other bacterial species are indicated in Table 2.

OCCURRENCE AND DISTRIBUTION OF IS ELEMENTS IN POPULATIONS OF *RHIZOBIUM MELILOTI*

As shown in the last paragraph, a substantial number of IS elements and IS-like DNA sequences of the *Rhizobiaceae* have been analysed. In most cases these mobile genetic elements were shown to be polymorphic with respect to copy number and genomic location (e.g. Wheatcroft & Watson, 1988*b*;

Simon *et al.*, 1991). However, there are striking differences between strains belonging to different species or genera, e.g. IS elements seem to be abundant in strains of *R. meliloti* whereas few transposable elements exist in *R. leguminosarum* bv. *viciae* strains (Ulrich & Pühler, 1994). This refers to the number of unrelated IS elements as well as the copy number of a given IS element present in a strain's genome.

In general, according to Southern hybridization analyses, IS elements of the *Rhizobiaceae* are confined to the species from which they were isolated. However, sequences exhibiting partial homology to IS elements isolated from *R. meliloti* were detected in strains of *R. leguminosarum* biovars *phaseoli* (Wheatcroft & Watson, 1988*a*; Wheatcroft & Laberge, 1991), *trifolii* (Dusha *et al.*, 1987) and *viciae* (Dusha *et al.*, 1987; Wheatcroft & Laberge, 1991).

Insertion sequence elements were found to reside on plasmids as well as on the chromosome of strains of the *Rhizobiaceae* (Hartmann & Amarger, 1991, Wheatcroft & Watson, 1988*a*; Labes & Simon, 1990). Hence, the spread of IS elements within bacterial populations via plasmid transfer is possible. We have addressed the question of the content of various unrelated IS elements within the genomes of a strain collection (reference strains) as well as strains of natural populations of *R. meliloti*. The group of the reference strains consisted of ten strains deriving from various geographical origins. As natural *R. meliloti* populations strains sampled from sites in Spain (22 strains) and Germany (78 strains) were taken.

Although some differences were observed in the percentage of strains of the various populations harbouring a given IS element, the distribution of IS elements within the three populations was essentially the same. The percentage of IS-positive strains in natural populations or reference strains of *R. meliloti* varied between 40% and 100% (see Fig. 3). IS element ISRm1 was reported to be present in approx. 65% of *R. meliloti* strains investigated (Watson & Wheatcroft, 1991). Since ISRm2011-1 which was used by Kosier, Pühler & Simon (1993) is identical to ISRm1, the percentage of ISRm2011-1 in natural populations of Spain (our unpublished results) and Germany (Kosier *et al.*, 1993) confirms the observation of these authors. Especially striking is the abundance of IS element ISRm2011-2 which it is present in virtually every *R. meliloti* strain tested so far. Based on its widespread occurrence we conclude that ISRm2011-2 is an ancestral IS element. Another IS element, ISRm3, which was not included in our study was reported to be present in 91% of the *R. meliloti* strains analysed (Wheatcroft & Laberge, 1991).

Depending on the strain/IS element combination the copy number of IS elements ranged between zero and 16 copies per genome in strains of the Spanish local population (see Fig. 4) as well as the reference collection (Simon *et al.*, 1991). *R. meliloti* strain NRG185 which displayed a complex banding pattern with eight out of nine IS elements tested (Simon *et al.*, 1991) was chosen to calculate the proportion of IS element-specific DNA within its genome. The amount of IS element-specific genetic information present in strain NRG185 was determined to be approximately 110 kb. Assuming that the size of the genome is in the range of 8000 kb, the proportion of IS-specific DNA is in the order of at least 1.4% per genome. This value is significantly larger than that calculated for strains of *E. coli* (approx. 1%, Arber 1991).

INSERTION SEQUENCE TYPING AS A TOOL FOR STRAIN CHARACTERIZATION

The unequivocal identification of bacteria at the strain level is of crucial importance in many areas of microbiology and consequently various methods were developed to differentiate strains of a given species. There is agreement that methods which type a strain according to its genotype (at the DNA sequence level) are superior to methods that group strains on basis of phenotypic traits. Restriction fragment length polymorphism (RLFP) analysis of chromosomal genes is considered to represent the most convenient of the reliable strain characterization methods. Thereby, repetitive DNA sequences that are polymorphic with respect to copy number and chromosomal location in a strain's genome are of particular value.

The potential of IS elements as probes to unambiguously characterize *E.*

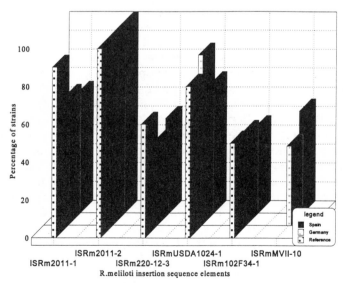

Fig. 3. Occurrence of *R. meliloti* IS elements IS*Rm2011-1* (IS*Rm1*), IS*Rm2011-2*, IS*Rm220-12-3* (IS*Rm4*), IS*RmUSDA1024-1*, IS*Rm102F34-1* and IS*RmMVII-10* in reference strains of various geographical origins and in two local *R. meliloti* populations isolated in Germany and Spain. The percentage of strains of each population harbouring a given IS element is shown. The presence of IS*RmMVII-10* in strains of the reference collection was not determined.

coli strains was first recognized by Green *et al*. (1984). Meanwhile, the IS typing method has become of widespread use in the epidemiological study of various pathogenic bacteria like *Salmonella* species (Ezquerra *et al*., 1993; Stanley *et al*., 1992), *Staphylococcus aureus* (Tenover *et al*., 1994), *Mycobacterium tubercolosis* (van Embden *et al*., 1993) as well as *Leptospira borgpetersenii* (Zuerner *et al*., 1993). Moreover, IS typing has been proposed to study the genetic variability of the plant-pathogenic *Xanthomonas* bacteria (Berthier *et al*., 1994).

Since strains of *Bradyrhizobium/Rhizobium* are used extensively throughout the world as seed inoculants to increase crop yield, there is an obvious need for reliable strain differentiation methods in order to be able to select superior strains for such purposes. Moreover, the potential application of genetically engineered strains of *Bradyrhizobium/Rhizobium* in the open environment requires reliable monitoring methods as well. The use of repetitive DNA sequences to characterize strains of *Bradyrhizobium* was first proposed by Kaluza *et al*. (1985). Similarly, Wheatcroft & Watson

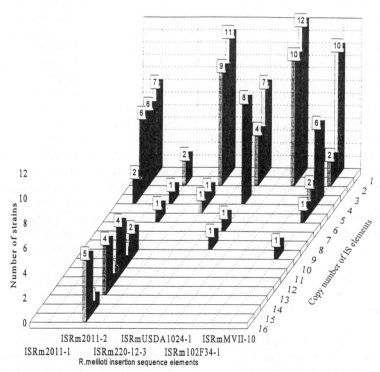

Fig. 4. Correlation of the proportion of strains of a natural spanish *R. meliloti* population with the numbers of IS element copies per strain of IS*Rm2011-1* (IS*Rm1*), IS*Rm2011-2*, IS*Rm220-13-2* (IS*Rm4*), IS*RmUSDA1024-1*, IS*Rm102F34-1* and IS*RmMVII-10*.

(1988*a*) suggested the IS typing technique as a positive strain identification method for *R. meliloti*. In a previous paper the authors had shown that strains of *R. meliloti* were polymorphic for IS element IS*Rm1* (Wheatcroft & Watson, 1988*b*). A modification of the proposal of Wheatcroft and Watson was introduced by Simon *et al.* (1991). These authors proposed the use of different, unrelated IS elements as probes to genotype *R. meliloti* strains and thereby to characterize them unequivocally.

As a prerequisite for the meaningful application of the IS typing technique for strain characterization, the fingerprints of strains generated with IS probes should be sufficiently stable. This seems to be the case with *R. meliloti* and *B. japonicum* IS/RS fingerprints. Wheatcroft and Laberge (1991) estimated the frequency of nonlethal transposition of IS elements IS*Rm1* and IS*Rm3* in *R. meliloti* strain SU47 grown in liquid culture to be 2 $\times 10^{-4}$ and 4×10^{-5} per generation per cell, respectively. Simon *et al.* (1991) noted the apparent stability of the IS*Rm1* fingerprints of strains *R. meliloti* 2011 and 1021. Both strains are derivatives of *R. meliloti* SU47 and faced different conditions for a number of years in different laboratories. Kosier *et al.* (1993) investigated the stability of IS*Rm2011-2* fingerprints of nine different field isolates over a period of two years. Despite the fact that the strains encountered various stress conditions during storage and repeated root nodule passages, only one change in the location of an IS element could be detected, demonstrating the remarkable stability of the IS pattern. Even more conclusive than the above-described laboratory-based results, are those results previously reported by Barran *et al.* (1991). These authors investigated the genetic stability of *R. meliloti* SU47 isolates which had been recovered from the field. IS*Rm1* fingerprints of root nodule isolates from alfalfa plants grown at a site which had been inoculated with *R. meliloti* SU47 (Bromfield, Sinha & Wolynetz, 1986) were compared to the IS fingerprint patterns of single colony isolates from the peat-based inoculum. The authors reported no significant difference in the frequency of changes in fingerprint pattern among nodule isolates or single colonies grown from the inoculum. Although transposition rate of IS*Rm1* was drastically enhanced in *R. meliloti* SU47 reisolates from the field site compared to cultivation of the same strain in liquid culture, changes in fingerprint pattern of single isolates occurred in the range of 2–4%. Although not thoroughly tested, repeated sequences RSα and RSβ seem to be stable in strains of *B. japonicum* in root nodules as well as during growth in liquid culture (Kaluza *et al.*, 1985; Minamisawa *et al.*, 1992).

Laboratory-based microcosm analyses to assess the impact of genetically engineered microorganisms (GEMs) on *R. meliloti* populations, require fast and reliable methods for strain differentiation. Since the IS fingerprint method is too laborious for the characterization of large quantities of bacterial strains, we developed the much simpler IS handprint method (Kosier *et al.*, 1993). The IS handprint method is based on the characterization of a given *R. meliloti* strain due to the presence or absence of various indigenous IS elements. The main advantage of the IS handprint method is that the *R. meliloti* strains used in such studies need no further modification (e.g. introduction of marker genes, chromosomal mutations) for their unambiguous identification. Therefore, populations consisting of various wild-type strains of *R. meliloti* can be reconstituted in model-ecosystems in order to evaluate a GEM's impact.

INSERTION SEQUENCE TYPING AS A TOOL IN POPULATION GENETICS

Since strains of *Rhizobium/Bradyrhizobium* belong to those bacteria, the genetically modified variants of which are used for environmental releases, an assessment of the genetic structure of populations (at the local as well as at the worldwide scale) is of importance for several reasons. First, knowledge of the natural diversity of *Rhizobium/Bradyrhizobium* is important in choosing suitable strain genotypes as host for genetic improvement of symbiotic performance. Secondly, in order to assess the likelyhood of the dissemination of the genetically engineered sequences, it is important to get an impression of the extent to which gene exchange occurs in natural populations. Thirdly, the impact of any genetically engineered microorganisms released into the environment on the genetic structure of local populations may be of importance, e.g. if poorly nitrogen fixing strains which may be present in the soil together with good nitrogen fixers would be selected via interstrain competition processes.

Various processes have been employed in the study of the genetic diversity of bacterial populations including serology, phage typing, intrinsic antibiotic resistance (IAR) spectra, plasmid profiling as well as RFLP analysis (see Brockman & Bezdicek, 1989, and references therein). Multilocus enzyme electrophoresis (MLEE), first applied to bacterial population genetic studies in *E. coli*, is now frequently used to assess the genotypic diversity within populations of various bacterial species (Lenski, 1993), including populations of *Rhizobium/Bradyrhizobium* (e.g. Leung, Wanjage & Bottomley, 1994; Bottomley, Cheng & Strain, 1994, see also references therein). Results obtained from population genetic studies of *E. coli* indicated that IS elements could be useful genetic markers in the study of strains whose close genetic relatedness has been established, e.g. by MLEE since the distribution of IS elements in genomes changed much faster than the electrophoretic mobility of enzymes (Green *et al.*, 1984). Studies with *E. coli* revealed that strains which were electrophoretically identical in 35 enzyme loci shared on average only 43% of their insertion sequence bands. Strains that differed in the electrophoretic mobility of two or more enzymes in most cases had no insertion sequence band in common (Sawyer *et al.*, 1987). Thus, it was concluded that IS typing was only suitable to further discriminate among closely related strains (Sawyer *et al.*, 1987; Lawrence *et al.*, 1989).

Recently, Hartmann and Amarger (1991) compared the resolutive power of IS typing vs. total DNA restriction fingerprinting and plasmid profiling for the characterization of *R. meliloti* isolates. One hundred and twenty-five strains reisolated from alfalfa root nodules sampled from one site were analysed in order to assess the genetic diversity of a natural *R. meliloti* population. Thirty-two isolates taken from twelve distinct plasmid profile groups yielded 20 and 16 groups with restriction fingerprinting and IS typing, respectively. Classification of strains obtained with each of the three methods correlated well. In another study, a local *R. meliloti* population isolated from a mining recultivation field was characterized by Kosier *et al.* (1993) also applying the IS fingerprint method. Using IS*Rm2011-2* as a probe a total of 78 isolates of the population yielded 42 different fingerprint groups again stressing the high resolutive power of IS fingerprinting. A detailed analysis of six strains belonging to the same IS*Rm2011-2* fingerprint class showed that although identical in their distribution of IS*Rm2011-2* the strains differed in their contents of the unrelated IS element IS*Rm220-12-3*. By using a third IS element (IS*RmUSDA1024-1*) the genetic relatedness of strains could be elucidated unambiguously.

Using different DNA probes, among them repeated sequences RSα and RSβ, Minamisawa *et al.* (1992) assessed the genetic structure of a field population of *B. japonicum* by RFLP analysis. The authors obtained 33 different fingerprint groups of 44 isolates. Grouping of strains according to RS fingerprints correlated well with grouping generated by *nif* and *hup* specific hybridization pattern as well as phenotypic characteristics like exopolysaccharide type, production of rhizobitoxine (RT) symptoms with soybean or indol-3-acetic acid (IAA) production. Thus, fingerprinting of strains with repeated sequences proved to be suitable for population genetic studies of *B. japonicum*.

CONCLUSIONS

Within bacterial species belonging to the *Rhizobiaceae* IS elements seem to be most abundant in *R. meliloti* with respect to the number of unrelated IS elements as well as the IS copy number occurring in a strain's genome. Analyses of the distribution of IS elements in natural populations of *R. meliloti* indicates global occurrence of all mobile elements identified so far. Moreover, there seems to be little variation in the occurrence of IS elements within strains of various natural populations. Insertion sequence fingerprinting is suitable to establish a strain's identity unequivocally. This is an important issue in the monitoring of genetically engineered *R. meliloti* strains deliberately released into the environment. In addition, IS typing is a valuable tool in molecular ecology. Using IS elements as probes, a detailed analysis of the genetic relatedness among strains of natural populations is possible. In conjunction with other genotyping methods like RFLP analysis of chromosomal markers, PCR amplification of repetitive sequences with REP (repetitive extragenic palindromic), or ERIC (enterobacterial repetitive intergeneric consensus) sequence primers, or the MLLE method, IS fingerprinting should provide a detailed insight into the genetic structure of natural populations of *R. meliloti*.

ACKNOWLEDGEMENTS

Part of this work was supported by the Bundesministerium für Forschung und Technologie (BMFT).

REFERENCES

Ajioka, J.W. & Hartl, D.L. (1989). Population dynamics of transposable elements. In: Berg, D.E. & Howe, M.M., eds. *Mobile DNA*, pp. 939–958. American Society for Microbiology, Washington DC.

Arber, W. (1991). Elements in microbial evolution. *Journal of Molecular Evolution*, **33**, 4–12.

Barran, L.R., Bromfield, E.S.P., Rastogi, V., Withwill, S.T. & Wheatcroft, R. (1991). Transposition and copy number of insertion sequence IS*Rm1* are not correlated with symbiotic performance of *Rhizobium meliloti* from two field sites. *Canadian Journal of Microbiology*, **37**, 576–9.

Berthier, Y., Thierry, D., Lemattre, M. & Guesdon, J.-L. (1994). Isolation of an insertion sequence (IS*1051*) from *Xanthomonas campestris* pv. *dieffenbachiae* with potential use for strain identification and characterization. *Applied and Environmental Microbiology*, **60**, 377–84.

Binns, A., Sciaky, N. & Wood, H.N. (1982). Variation in hormone autonomy and regenerative potential of cells transformed by strain A66 of *Agrobacterium tumefaciens*. *Cell*, **31**, 605–12.

Bonnard, G., Vincent, F. & Otten, L. (1989). Sequence and distribution of IS*866*, a novel T region-associated insertion sequence from *Agrobacterium tumefaciens*. *Plasmid*, **22**, 70–81.

Bottomley, P.J., Cheng, H.-H. & Strain, S.R. (1994). Genetic structure and symbiotic characteristics of a *Bradyrhizobium* population recovered from a pasture soil. *Applied and Environmental Microbiology*, **60**, 1754–61.

Brockman, E.J. & Bezdicek, D.F. (1989). Diversity within serogroups of *Rhizobium leguminosarum* biovar *viciae* in the palouse region of eastern Washington as indicated by plasmid profiles, intrinsic antibiotic resistance and topography. *Applied and Environmental Microbiology*, **55**, 109–15.

Bromfield, E.S.P., Sinha, I.B. & Wolynetz, M.S. (1986). Influence of location, host cultivar, and inoculation on the composition of naturalized populations of *Rhizobium meliloti* in *Medicago sativa* nodules. *Applied and Environmental Microbiology*, **51**, 1077–84.

Campbell, A. (1981). Evolutionary significance of accessory DNA elements in bacteria. *Annual Review of Microbiology*, **35**, 55–83.

Chandler, M. & Fayet, O. (1993). Translational frameshifting in the control of transposition in bacteria. *Molecular Microbiology*, **7**, 497–503.

Chen, C.W., Yu, T.-W., Chung, H.-M. & Chou, C.-F. (1992). Discovery and characterization of a new transposable element, Tn*4811*, in *Streptomyces lividans* 66. *Journal of Bacteriology*, **174**, 7762–9.

Choucheron, D.H. (1993). A family of IS*1031* elements in the genome of *Acetobacter xylinum*: nucleotide sequences and strain distribution. *Molecular Microbiology*, **9**, 211–18.

Dean, D. (1981). A plasmid cloning vector for the direct selection of strains carrying recombinant plasmids. *Gene*, **15**, 99–102.

De Meirsman, C., van Soom, C., Verreth, C., van Gool, A. & Vanderleyden, J. (1990). Nucleotide sequence analysis of IS*427* and its target sites in *Agrobacterium tumefaciens* T37. *Plasmid*, **24**, 227–34.

Doolittle, W.F. & Sapienza, C. (1980). Selfish genes, the penotype paradigm and genome evolution. *Nature*, **284**, 601–3.

Dusha, I, Kovalenko, S., Banfalvi, Z. & Kondorosi, A. (1987). *Rhizobium meliloti* insertion element IS*Rm2* and its use for identification of the *fixX* gene. *Journal of Bacteriology*, **169**, 1403–9.

Ezquerra, E., Burnens, A., Jones, C. & Stanley, J. (1993). Genotypic typing and phylogenetic analysis of *Salmonella paratyphi* B and *S. java* with IS*200*. *Journal of General Microbiology*, **139**, 2409–14.

Fournier, P., Paulus, F. & Otten, L. (1993). IS*870* requires a 5'-CTAG-3' target sequence to generate the stop codon for its large ORF1. *Journal of Bacteriology*, **175**, 3151–60.

Galas, D.J. & Chandler, M. (1989). Bacterial insertion sequences. In: Berg, D.E. & Howe, M.M., eds. Mobile DNA, pp. 109–162. American Society for Microbiology, Washington DC.

Gay, P., Le Coq, D., Steinmetz, M., Berkelman, T. & Kado, C.I. (1985). Positive selection procedure for entrapment of insertion sequence elements in gram-negative bacteria. *Journal of Bacteriology*, **164**, 918–21.

Green, L., Miller, R.D., Dykhuizen, D.E. & Hartl, D.L. (1984). Distribution of DNA insertion element IS*5* in natural isolates of *Escherichia coli*. *Proceedings of the National Academy of Sciences, USA*, **81**, 4500–4.

Hahn, M. & Hennecke, H. (1987). Mapping of a *Bradyrhizobium japonicum* DNA region carrying genes for symbiosis and an asymetric accumulation of reiterated sequences. *Applied and Environmental Microbiology*, **53**, 2247–52.

Hartl, D.L., Medhora, M., Green, L. & Dykhuizen, D.E. (1986). The evolution of DNA sequences in *Escherichia coli*. *Philosophical Transactions of The Royal Society of London*, **312**, 191–204.

Hartmann, A. & Amarger, N. (1991). Genotypic diversity of an indigenous *Rhizobium meliloti* field population assessed by plasmid profiles, DNA fingerprinting, and insertion sequence typing. *Canadian Journal of Microbiology*, **37**, 600–8.

Hennecke, H., Günther, I. & Binder, F. (1982). A novel cloning vector for the direct selection of recombinant DNA in *E. coli*. *Gene*, **19**, 231–4.

Jordan, E., Saedler, H. & Starlinger, P. (1968). O-zero and strong polar mutations in the *gal* operon are insertions. *Molecular and General Genetics*, **102**, 353–63.

Judd, A.K. & Sadowsky, M.J. (1993). The *Bradyrhizobium japonicum* serocluster 123 hyperreiterated DNA region, HRS1, has DNA and amino acid sequence homology to IS*1380*, an insertion sequence from *Acetobacter pasteurianus*. *Applied and Environmental Microbiology*, **59**, 1656–61.

Kaluza, K., Hahn, M. & Hennecke, H. (1985). Repeated sequences similar to insertion elements clustered around the *nif* region of the *Rhizobium japonicum* genome. *Journal of Bacteriology*, **162**, 535–42.

Knoop, V. & Brennicke, A. (1994). Evidence for a group II intron in *Escherichia coli* inserted into a highly conserved reading frame associated with mobile DNA sequences. *Nucleic Acids Research,* **22**, 1167–71.

Kosier, B., Pühler, A. & Simon, R. (1993). Monitoring the diversity of *Rhizobium meliloti* field and microcosm isolates with a novel rapid genotyping method using insertion elements. *Molecular Ecology,* **2**, 35–46.

Krishnan, H.B. & Pueppke, S.G. (1993). Characterization of RFRS9, a second member of the *Rhizobium fredii* repetitive sequence family from the nitrogen-fixing symbiont *R. fredii* USDA257.

Labes, G. & Simon, R. (1990). Isolation of DNA insertion elements from *Rhizobium meliloti* which are able to promote transcription of adjacent genes. *Plasmid,* **24**, 235–9.

Lawrence, J.G., Dykhuizen, D.E., DuBose, R.F. & Hartl, D.L. (1989). Phylogenetic analysis using insertion sequence fingerprinting in *Escherichia coli*. *Molecular Biology and Evolution,* **6**, 1–14.

Lawrence, J.G., Ochman, H. & Hartl, D.L. (1992). The evolution of insertion sequences within enteric bacteria. *Genetics,* **131**, 9–20.

Lenski, R.E. (1993). Assessing the genetic structure of microbial populations. *Proceedings of the National Academy of Sciences, USA,* **90**, 4334–6.

Leung, K., Wanjage, F.N. & Bottomley, P.J. (1994). Symbiotic characteristics of *Rhizobium leguminosarum* bv. *trifolii* isolates which represent major and minor nodule-occupying chromosomal types of field-grown subclover (*Trifolium subterraneum* L.). *Applied and Environmental Microbiology,* **60**, 427–33.

Machida, Y., Sakurai, M., Kiyokawa, S., Ubasawa, A., Suzuki, Y. & Ikeda, J.E. (1984). Nucleotide sequence of the insertion sequence found in the T-DNA region of mutant Ti plasmid pTiA66 and distribution of its homologues in octopine Ti plasmids. *Proceedings of the National Academy of Sciences, USA,* **81**, 7495–9.

Minamisawa, K., Seki, T., Onodera, S. Kubota, M. & Asami, T. (1992). Genetic relatedness of *Bradyrhizobium japonicum* field isolates as revealed by repeated sequences and various other characteristics. *Applied and Environmental Microbiology,* **58**, 2832–9.

Orgel, L.E. & Crick, F.H.C. (1980). Selfish DNA: the ultimate parasite. *Nature,* **284**, 604–7.

Otten, L., Canaday, J., Gerard, J.-C., Fournier, P., Crouzet, P. & Paulus, F. (1992) Evolution of agrobacteria and their Ti plasmids – a review. *Molecular Plant–Microbe Interactions,* **5**, 279–87.

Paulus, F., Ride, M. & Otten, L. (1989). Distribution of two *Agrobacterium tumefaciens* insertion elements in natural isolates: evidence for stable association between Ti plasmids and their bacterial host. *Molecular and General Genetics,* **219**, 145–52.

Paulus, F., Canaday, J., Vincent, F., Bonnard, G., Kares, C. & Otten, L. (1991). Sequence of the *iia* and *ipt* region of different *Agrobacterium tumefaciens* biotype III octopine strains: reconstruction of octopine Ti plasmid evolution. *Plant Molecular Biology,* **16**, 601–14.

Priefer, U.B., Burkhardt, H.J., Klipp, W. & Pühler, A. (1981). IS*R1*: an insertion element isolated from the soil bacterium *Rhizobium lupini*. *Cold Spring Harbor Symposia on Quantitative Biology,* **45**, 87–91.

Priefer, U.B., Kalinowski, J., Rüger, B., Heumann, W. & Pühler, A. (1989). ISR1, a transposable DNA sequence resident in *Rhizobium* class IV strains, shows structural characteristics of classical insertion elements. *Plasmid,* **21**, 120–8.

Ramseier, T.M. & Göttfert, M. (1991). Codon usage and G + C content in *Bradyrhizobium japonicum* genes are not uniform. *Archives of Microbiology,* **156**, 270–6.

Reimmann, C. & Haas, D. (1993). Mobilization of chromosomes and nonconjugative plasmids by cointegrative mechanisms. In: Clewell, D.B., ed. *Bacterial Conjugation*, pp. 137–188. Plenum Press, New York.

Rezsöhazy, R., Hallet, B., Delcour, J. & Mahillon, J. (1993). The IS*4* family of insertion sequences: evidence for a conserved transposase motif. *Molecular Microbiology*, 9, 1283–95.

Rodriguez-Quinones, F., Judd, A.K., Sadowsky, M.J., Lui, R.-L. & Cregan, P.B. (1992). Hyperreiterated DNA regions are conserved among *Bradyrhizobium japonicum* serocluster 123 strains. *Applied and Environmental Microbiology*, 58, 1878–85.

Rubens, C.E., Heggen, L.M. & Kuypers, J.M. (1989). IS*861*, a group B streptococcal insertion sequence related to IS*150* and IS*3* of *Escherichia coli*. *Journal of Bacteriology*, 171, 5531–5.

Ruvkun, G.B., Long, S.R., Meade, H.M., van den Bos, R.C. & Ausubel, F.M. (1982). IS*Rm1*: a *Rhizobium meliloti* insertion sequence that transposes preferentially into nitrogen fixation genes. *Journal of Molecular and Applied Genetics*, 1, 405–18.

Sawyer, S.A., Dykhuizen, D.F., DuBose, R.F., Green, L., Mutangadura-Mhlanga, T. Wolczyk, D.E. & Hartl, D.L. (1987). Distribution and abundance of insertion sequences among natural isolates of *Escherichia coli*. *Genetics*, 115, 51–63.

Schwartz, E., Kröger, M. & Rak, B. (1988). IS*150*: distribution, nucleotide sequence and phylogenetic relationships of a new *E. coli* insertion element. *Nucleic Acids Research*, 16, 6789–802.

Selbitschka, W. & Lotz, W. (1991). Instability of cryptic plasmids affects the symbiotic effectivity of *Rhizobium leguminosarum* bv. *viciae* strains. *Molecular Plant–Microbe Interactions*, 4, 608–18.

Shapiro, J.A. (1969). Mutations caused by the insertion of genetic material into the galactose operon of *Escherichia coli*. *Journal of Molecular Biology*, 40, 93–105.

Simon, R., Hötte, B., Klauke, B. & Kosier, B. (1991). Isolation and characterization of insertion sequence elements from Gram-negative bacteria by using new broad-host-range, positive selection vectors. *Journal of Bacteriology*, 173, 1502–8.

Slightom, J.L., Durand-Tardif, M., Jouanin, L. & Tepfer, D. (1986). Nucleotide sequence analysis of TL-DNA of *Agrobacterium rhizogenes* agropine type plasmid: identification of open reading frames. *Journal of Biological Chemistry*, 261, 108–21.

Soto, J.M., Zorzano, A., Olivares, J. & Toro, N. (1992*a*). Nucleotide sequence of *Rhizobium meliloti* GR4 insertion sequence IS*Rm3* linked to the nodulation competitiveness locus *nfe*. *Plant Molecular Biology*, 20, 307–9.

Soto, J.M., Zorzano, A., Olivares, J. & Toro, N. (1992*b*). Sequence of IS*Rm4* from *Rhizobium meliloti* strain GR4. *Gene*, 120, 125–6.

Stanley, J., Burnens, A., Powell, N., Chowdry, N. & Jones, C. (1992). The insertion sequence IS*200* fingerprints chromosomal genotypes and epidemiological relationships in *Salmonella heidelberg*. *Journal of General Microbiology*, 138, 2329–36.

Syvanen, M. (1984). The evolutionary implications of mobile genetic elements. *Annual Review of Genetics*, 18, 271–93.

Tenover, F.C., Arbeit, R., Archer, G., Biddle, J., Byrne, S., Goering, R. Hancock, G., Hebert, G.A., Hill, B., Hollis, R., Jarvis, W.R., Kreiswirth, B., Eisner, W., Maslow, J., McDougal, L.K., Miller, J.M., Mulligan, M. & Pfaller, M.A. (1994). Comparison of traditional and molecular methods of typing isolates of *Staphylococcus aureus*. *Journal of Clinical Microbiology*, 32, 407–15.

Ulrich, A. & Pühler, A. (1994). The new class II transposon Tn*163* is plasmid-borne in two unrelated *Rhizobium leguminosarum* biovar *viciae* strains. *Molecular and General Genetics*, **242**, 505–16.

Vanderleyden, J., Desair, J., De Meirsman, C., Michiels, K., Van Gool, A., Chilton, M.-D. & Jen, G.C. (1986). Nucleotide sequence of an insertion sequence (IS) element identified in the T-DNA region of a spontaneous variant of the Ti-plasmid pTiT37. *Nucleic Acids Research*, **16**, 6699–709.

van Embden, J.D.A., Cave, M.D., Crawford, J.T., Dale, J.W., Eisenach, K.D., Gicquel, B., Hermans, P., Martin, C., McAdam, R., Shinnick, T.M. & Small, P.M. (1993). Strain identification of *Mycobacterium tuberculosis* by DNA finger-printing: recommendations for a standardized methodology. *Journal of Clinical Microbiology*, **31**, 406–9.

Wabiko, H. (1992). Sequence analysis of an insertion element, IS*1131*, isolated from the nopaline-type Ti plasmid of *Agrobacterium tumefaciens*. *Gene*, **114**, 229–33.

Watson, R.J. & Wheatcroft, R. (1991). Nucleotide sequence of *Rhizobium meliloti* insertion sequence IS*Rm1*: homology to IS*2* from *Escherichia coli* and IS*426* from *Agrobacterium tumefaciens*. *Journal of DNA Sequencing and Mapping*, **2**, 163–72.

Wheatcroft, R. & Watson, R.J. (1988a). A positive strain identification method for *Rhizobium meliloti*. *Applied and Environmental Microbiology*, **54**, 574–6.

Wheatcroft, R. & Watson, R.J. (1988b). Distribution of insertion sequence IS*Rm1* in *Rhizobium meliloti* and other Gram-negative bacteria. *Journal of General Microbiology*, **134**, 113–21.

Wheatcroft, R. & Laberge, S. (1991). Identification and nucleotide sequence of *Rhizobium meliloti* insertion sequence IS*Rm3*: similarity between the putative transposase encoded by IS*Rm3* and those encoded by *Staphylococcus aureus* IS*256* and *Thiobacillus ferrooxidans* IS*T2*. *Journal of Bacteriology*, **173**, 2530–8.

Zuerner, R.L., Ellis, W.A., Bolin, C.A. & Montgomery, J.M. (1993). Restriction fragment length polymorphism distinguish *Leptospira borgpetersenii* serovar hardjo type hardjo-bovis isolates from different geographical locations. *Journal of Clinical Microbiology*, **31**, 578–83.

EVOLUTION OF GENE SEQUENCES BETWEEN AND WITHIN SPECIES OF *BACILLUS*

PAUL M. SHARP,[1] NIAMH C. NOLAN[2] and KEVIN M. DEVINE[2]

[1]*Department of Genetics, University of Nottingham, Queens Medical Centre, Nottingham NG7 2UH, UK*
[2]*Department of Genetics, Trinity College, Dublin 2, Ireland*

INTRODUCTION

The study of population genetics and evolution in prokaryotes has generally lagged far behind similar work on multicellular eukaryotes, but in recent years this situation has begun to change. First, the evolutionary relationsips of a very large number of bacterial species have been elucidated through the comparative analysis of 16S rRNA sequences (for review see Olsen, Woese & Overbeek, 1994). Secondly, multilocus enzyme electrophoresis (MLEE) studies, and latterly DNA sequence comparisons, of strains or closely related species have started to shed light on the population genetic structure of bacteria. However, extensive population genetic studies, particularly those examining DNA sequence data, have so far been largely limited to the closely related gram negative bacteria *Escherichia coli* and *Salmonella typhimurium*. Given the enormous evolutionary diversity within the eubacteria, it is not clear to what extent these results can be generalised to other species.

The major questions in bacterial population genetics concern not only the extent and nature of genetic diversity within and among species, but also whether 'species' even exist, at least in the terms that, for example, a population geneticist working on animals would recognize. Thus, much recent work has focussed on whether bacteria have clonal population structures (see, e.g. Maynard Smith *et al.*, 1993): what is the extent of recombination, and over what range can such genetic exchange occur? The population genetics and evolution of *E. coli* have been most intensively studied utilizing the ECOR collection, a set of 72 natural isolates of diverse origins (Ochman & Selander, 1984), and by comparison with *S. typhimurium* (Sharp, 1991). MLEE studies of *E. coli* have revealed very strong linkage disequilibrium among different genes, indicating that this species is largely clonal (Selander & Levin, 1980; Selander, Caugant & Whittam, 1987). In contrast, DNA sequence studies have provided evidence that genetic exchanges among strains of *E. coli* have occurred: phylogenies based on different genes, or even different regions within a gene are not always concordant (DuBose, Dykhuizen & Hartl, 1988; Bisercic, Feutrier & Reeves, 1991; Dykhuizen & Green, 1991). These seemingly discrepant observations have been reconciled by suggesting that occasional exchanges of quite short genetic fragments have not been sufficient to destroy the overall clonal frame of the chromosome (Milkman & Bridges, 1990; 1993). Indeed, some genes (Hall & Sharp, 1992; Nelson & Selander, 1992; Boyd *et al.*, 1994) do produce phylogenies not dissimilar to that derived from MLEE (Herzer *et al.*, 1990). It has been suggested that the genes which have been exchanged frequently are mainly those that may confer some immediate adaptive advantage, such as those encoding cell-surface proteins or restriction-modification systems, in contrast to 'housekeeping' genes, such as those encoding the enzymes studied by MLEE (Nelson & Selander, 1992).

From the earliest electrophoretic studies it became clear that *E. coli* populations exhibit much more biochemical diversity than higher eukaryotes (Milkman, 1973; Selander & Levin, 1980). Sequence comparisons provide more detailed information on the levels of genetic diversity within and among bacterial species. Genes compared among the ECOR collection typically differ at 1–2% of nucleotides (see, e.g. Hall & Sharp, 1992). Some genes exhibit (sometimes much) higher diversity, although again this appears to be explicable in terms of particular loci being subject to diversifying selection and recombination (Dykhuizen & Green, 1991; Sharp *et al.*, 1992*a*). The value of 1–2% is indeed rather higher than those seen within, for example, humans (<0.1%; Li & Sadler, 1991), but not much greater than seen in *Drosophila* (up to 1%; Hey & Kliman, 1993). Genes compared between *E. coli* and *S. typhimurium* differ on average at 16% of nucleotides (Sharp, 1991). This value is higher than those seen in comparisons between animal species perhaps considered to be similarly closely related, such as human and chimpanzee (1–2%; Li, Tanimura & Sharp, 1987), or mouse and rat (7%; Wolfe & Sharp, 1993). The extent of divergence between *E. coli* and *S. typhimurium* differs among genes, and among sites within genes, although it appears that this can largely be explained in terms of different levels of selective constraint, i.e. it is not necessary to invoke exchange between these species to explain the lower divergence of some genes. In fact, *E. coli* and *S. typhimurium* appear to be effectively isolated (Rayssiguier, Thaler & Radman, 1989).

After *E. coli*, the most intensively studied bacterial species in terms of general molecular genetics is *Bacillus subtilis*. It has a well-defined genetic and physical map (Piggot *et al.*, 1990; Amjad *et al.*, 1990; Itaya & Tanaka, 1991) and is now the subject of a genome sequencing project (Kunst & Devine, 1991; Glaser *et al.*, 1993; Sorokin *et al.*, 1993; Ogasawara, Nakai & Yoshikawa, 1994). It might be anticipated that the population genetics and evolution of *Bacilli* are very different from *E. coli*. These species are very distantly related (Olsen, Woese & Overbeek, 1994): their common ancestry might date back to 1500 Myr ago (Ochman & Wilson, 1987). Furthermore, *B. subtilis* and *E. coli* have very different lifestyles, and habitats (soil, and the mammalian intestine, respectively). Importantly, many wild strains of *B. subtilis* are competent for transformation (Cohan, Roberts & King, 1991), meaning that this species is potentially far more likely to undergo recombination among different strains, or even exchange genetic material among different species. Istock and colleagues have found little evidence of linkage disequilibrium among various allozyme markers in a wild population of *B. subtilis*, indicating that this species is far less clonal than *E. coli* (Istock *et al.*, 1992). The same authors have even documented the formation of hybrids, and recombination, between *B. subtilis* and a related though clearly distinct species *B. licheniformis* in soil cultures (Duncan *et al.*, 1989). However, population genetic and evolutionary studies of *B. subtilis* and its relatives have not yet focussed on DNA sequence data. We have begun such a study (Sharp *et al.*, 1992*b*). Here we attempt to place this work in context, by reviewing what is known so far about the extent and nature of sequence divergence among *B. subtilis* strains, and among *B. subtilis* and its close relatives, and contrasting these results with those for *E. coli*.

INTERPRETATION OF COMPARATIVE SEQUENCE DATA

The rate and manner in which the DNA sequences of genes evolve are expected to be influenced by a number of factors. The first is the underlying mutation rate, which is often implicitly assumed to be similar across genes, although it is possible that the rate may vary around the bacterial chromosome (Chilton & McCarthy, 1969; Sharp *et al.*, 1989). Natural selection is the second factor, which will affect genes, and sites within genes, in different ways. In the main, selection is viewed as being a constraint on sequence change (Kimura, 1983), although clearly it can increase diversity and accelerate evolution under certain circumstances. In comparative sequence analyses, sites within genes must be separated into those where synonymous (silent) nucleotide changes can occur, and those where mutations are non-synonymous, i.e. cause amino acid replacements. Synonymous changes generally occur at much faster rates than non-synonymous changes, because of the relative lack of constraint on the former.

Nevertheless, even synonymously variable sites do not evolve in the same way in different genes. Thus, the usage of alternative synonymous codons in both *E. coli* (Gouy & Gautier, 1982) and *B. subtilis* (Shields & Sharp, 1987) is quite non-random, and varies among genes. Highly expressed genes have a very biased pattern of codon usage, with a strong preference towards one or two translationally optimal codons for each amino acid; lowly expressed genes have more uniform codon usage, apparently primarily influenced by mutation patterns (see Sharp *et al.*, 1993). This constraint on codon usage in highly expressed genes has reduced the rate at which silent sites have diverged between *E. coli* and *S. typhimurium* (Sharp, 1991).

Superimposed upon the classical forces of mutation and selection (and random genetic drift) is the effect of recombination. Recombination can be detected when the relationships among strains inferred from different genes (or regions within genes) vary, implying that different parts of the genome have different evolutionary histories. In the absence of phylogenetic information, recombination may be evident from different degrees of relative divergence for different genes (or parts of genes) between a particular pair of strains (see, e.g. Maynard Smith, 1992).

At all but the lowest levels of divergence it is necessary to make some correction of the observed sequence difference, in an attempt to take account of superimposed substitutions (multiple hits). We have used the method of Li, Wu & Luo (1985), as modified by Li (1993), which distinguishes between synonymous and nonsynonymous substitutions, and also allows for different rates of transitions and transversions. This method produces estimates of K_S and K_A, the numbers of synonymous and nonsynonymous substitutions per site, respectively.

To investigate the divergence among *Bacillus* sequences, in the tables that follow, we present the following information for each gene that can be compared between different strains or species:

Map: the map position, in degrees, of the gene in *B. subtilis* 168, where known (Piggot *et al.*, 1990).

Codons: the number of codons compared, with + indicating that the comparison is across partial sequences.

F_{op}: the 'frequency of optimal codons' in the *B. subtilis* gene, measuring the strength of codon usage bias, calculated according to Sharp *et al.* (1990).

%DNA: the percentage nucleotide difference between the two sequences.

%AA: the percentage amino acid difference between the two sequences.

K_A: the estimated number of nonsynonymous nucleotide substitutions per site.

K_S: the estimated number of synonymous nucleotide substitutions per site, both calculated by the method of Li (1993).

Strain: the strain designation for the *B. amyloliquefaciens* and *B. licheniformis* genes (where known).

DIVERGENCE AMONG '*BACILLUS SUBTILIS*' STRAINS

Most molecular genetic studies of *B. subtilis* have examined strain 168, or derivatives of it. However, some work has been done using strain W23, and a number of genes can be compared between these two strains (Table 1). These include sequences obtained for comparative studies, around *divIB* (Harry *et al.*, 1994) and *terC* (*rtp*; Ahn & Wake, 1991), as well as three genes investigated in our own diversity project (*bgl*, *rplX* and *spo0A*). We have also included the *thyB* gene from ATCC 6633, since at *rplX*, strains W23 and ATCC 6633 are very similar, and approximately equidistant from 168 (Sharp *et al.*, 1992*b*).

Across these genes, the weighted average DNA sequence difference between 168 and W23 is 5.6%. This is remarkably similar to the estimate of 7%, made 25 years ago, from the thermal stability of DNA from a number of loci (Chilton & McCarthy, 1969). The levels of DNA sequence divergence vary from 1.6% to 7.5% among the genes in Table 1. As expected, the extent of synonymous substitution is always much higher than non-synonymous substitution, even in the three short partial open reading frames (*orfs*) of unknown function. Much of the variation among genes in DNA sequence divergence is due to different extents of protein sequence conservation. Among the complete genes, *rtp* has only silent changes between 168 and W23, whereas the xylose repressor protein (encoded by *xylR*) differs at 7% of residues. Only one of these genes (*rplX*, encoding ribosomal protein L24) has strong codon usage bias. The number of synonymous substitutions per site (K_S) at *rplX* between 168 and W23 is much lower than the average (0.22), indicating that codon selection constrains the rate of evolution at silent sites in this gene (Sharp *et al.*, 1992*b*). The divergence values for *thyB*

Table 1. *Divergence between* B. subtilis *strains 168 and W23*

Gene	Map	Codons	F_{op}	%DNA	%AA	K_A	K_s	References
rplX	12	104	0.67	1.6	1.0	0.005	0.07	1,2
xylR	48	385	0.29	6.6	7.0	0.035	0.22	3,4
orf2	135	43+	0.40	2.3	2.4	0.012	0.09	5,6
divIB	135	264	0.37	4.8	1.1	0.004	0.22	5,6
orf4	135	31+	0.33	6.5	3.2	0.017	0.22	5,6
orf238	180	29+	0.37	5.7	0.0	0.000	0.39	7,8
rtp	180	123	0.36	6.0	0.0	0.000	0.26	7,9
orf257	180	258	0.25	5.2	3.9	0.017	0.18	8,9
thyB	200	265	0.28	7.4	3.4	0.016	0.36	10,11
spo0A	217	137+	0.18	4.4	0.7	0.004	0.15	12,13
bgl	334	89+	0.35	7.5	10.1	0.049	0.21	13

References: 1. Henkin *et al.* (1989); 2. Sharp *et al.* (1992*b*); 3. Hastrup (1987); 4. Kreuzer *et al.* (1989); 5. Harry & Wake (1989); 6. Harry *et al.* (1994); 7. Carrigan *et al.* (1987); 8. Ahn & Wake (1991); 9. Lewis & Wake (1989); 10. Iwakura *et al.* (1988); 11, Montorsi & Lorenzetti (1993); 12. Ferrari *et al.* (1985); 13. Nolan, N.C. (unpublished).

(from ATCC 6633) are high, but within the range of other genes, seeming to justify its inclusion in the comparison.

Roberts and Cohan (1993) used the presence and absence of restriction sites at the *rpoB* locus to estimate DNA sequence divergence among a number of *B. subtilis* strains. Their approach sampled a total of 364 bp dispersed across a 3.3 kb region. They reported 3.1% difference between strains placed in 168 and W23 'subgroups'. The *rpoB* product, the beta subunit of RNA polymerase, is a highly conserved protein. In addition, it is highly expressed, and has strong codon usage bias ($F_{op} = 0.51$). Thus, both synonymous and non-synonymous substitutions would be expected to be constrained in *rpoB*, and its lower than average divergence is consistent with this.

Both *rplX* and *rpoB* map to the region around 12 degrees on the *B. subtilis* chromosome (Piggot *et al.*, 1990). From the relative frequencies of homologous and heterologous transformation, Chilton & McCarthy (1969) deduced that sequences in this region were much less divergent (between *B. subtilis* 168, W23 and *B. globigii*) than for markers from other regions of the chromosome. Sequence comparisons suggest that this can be explained in terms of constraints on protein sequence and codon usage. Nevertheless, as more data become available, it will be interesting to question whether chromosomal location also influences evolutionary rate.

Although the overall level of DNA sequence divergence between 168 and W23 is relatively low, some regions of the genome are quite different in these two strains. For example, in the region around *terC*, the replication terminus at 180 degrees, W23 contains an insertion of approximately 1300 bp including an open reading frame of 405 codons, between *orf238* and *rtp* (Ahn & Wake, 1991). In addition, a major biochemical difference between 168 and W23 concerns the teichoic acids found in the cell wall: the major anionic polymer in 168 is polyglycerol phosphate, whereas the cell wall of W23 contains polyribitol phosphate. It has been found that the *tag* and *tar* genes involved in biosynthesis of these teichoic acids, although located at the same map position in the two strains, exhibit no homology in Southern hybridization experiments (Young *et al.*, 1989).

PBSX is a defective prophage located around map position 112 on the *B. subtilis* chromosome. The sequence of a 1200 bp fragment of PBSX containing a repressor gene (*xre*) has been determined from two strains, SO113 and IA4201 (Wood, Devine & McConnell, 1990). Both strains were thought to be derived from 168, but it was found that the two strains differed at 5.5% of nucleotides. Although IA4201 had been subjected to EMS mutagenesis, this treatment would be most unlikely to have caused so many changes, and so it was concluded that the SO113 sequence was probably not derived from strain 168 (Wood *et al.*, 1990). Indeed, this level of divergence is very similar to that found above between 168 and W23, and so it is interesting to examine the sequence differences between the two strains in

Table 2. *Comparison of PBSX sequences*

Gene	Codons	F_{op}	%DNA	%AA	K_A	K_s
orf60	60	0.21	5.6	5.1	0.022	0.23
xre	114	0.29	4.1	2.7	0.013	0.19
orf53	53+	0.35	7.5	1.9	0.010	0.41

xre, and in two flanking open reading frames. These differences are concentrated at silent sites, consistent with the theory that most had accumulated during 'natural' evolution rather than in the laboratory, and the numbers of synonymous substitutions (Table 2) are within the range seen between 168 and W23 (Table 1).

In their study of *rpoB* cited above, Roberts and Cohan (1993) also derived a phylogeny for various strains. These included members of three sub-groups, centred on strains 168, W23 and ROH1. The latter subgroup was considered to be a separate species. The divergence between ROH1 and 168 (4.6%) was higher than between W23 and 168 (3.1%), but W23 and ROH1 formed a monophyletic cluster distinct from 168. This may indicate that W23 also warrants separate species status. To place the divergence of 168 and W23 in perspective, note that their average DNA sequence difference (5.6%) is rather higher than the values of 1–2% seen between *E. coli* strains, but much lower than the average difference (16%) between *E. coli* and *S. typhimurium*.

In contrast to the apparent genetic distinctiveness of the two laboratory strains of *B. subtilis* (168 and W23), some wild strains are very similar at the DNA level. For example, among strains isolated from the Mojave Desert, one was found to be indistinguishable from 168, and two identical to W23, in a four-cutter restriction analysis of the *polC* gene (Cohan *et al.*, 1991). We have reported a strain originally isolated in the Gobi Desert (BSG67), which was found to be identical to 168 in its *rplX* sequence (Sharp *et al.*, 1992*b*).

DIVERGENCE AMONG *BACILLUS* SPECIES

Among species closely related to *B. subtilis* only *B. amyloliquefaciens* and *B. licheniformis* have been the subject of much molecular genetic analysis. *B. subtilis* and *B. amyloliquefaciens* are thought to be sister species, with *B. licheniformis* somewhat more distantly related (Fig. 1). This has emerged from a numerical taxonomy based on numerous morphological and bio-chemical traits (Priest, Goodfellow & Todd, 1988), and from sequence comparisons of 16S rRNA sequences (Ash *et al.*, 1991). In fact, *B. subtilis* and *B. amyloliquefaciens* are so similar phenotypically that the distinct species status of the latter has only recently been reintroduced (Priest *et al.*, 1987).

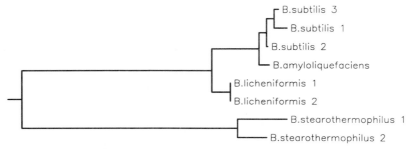

Fig. 1. Phylogenetic relationships among *B. subtilis*, *B. amyloliquefaciens*, *B. licheniformis* and *B. stearothermophilus*, derived from 16S rRNA sequences. The tree was derived by the neighbor-joining method (Saitou & Nei, 1987) applied to a matrix of pairwise distances corrected for multiple hits by the 2-parameter method (Kimura, 1980). Horizontal branch lengths are drawn to scale, vertical separation is for clarity only. The *B. subtilis* sequences are all from strain 168: 1, from the *rrnB* operon at 280 degrees; 2, from the *rrnO* operon at 0 degrees; 3, was sequenced as RNA. The other strains are *B. amyloliquefaciens* ATCC 23350, *B. licheniformis* DSM 13 and NCDO 1772 (sequences identical), and *B. stearothermophilus* NCDO 1768 (1) and T-10 (2). All sequences were taken from the GenBank/EMBL/DDBJ database.

We have collated data for 11 genes sequenced in both *B. subtilis* and *B. amyloliquefaciens* (Table 3). Across these genes, the weighted average DNA sequence difference is 22%. This value is somewhat higher than that (16%) between *E. coli* and *S. typhimurium* (Sharp, 1991), confirming the distinct nature of *B. subtilis* and *B. amyloliquefaciens* at the genetic level. None of the genes examined has very strong codon usage bias, nor do any encode particularly highly conserved proteins. The majority of genes have K_s values in excess of one synonymous substitution per site, and one (for the 3'*orf* at *sipS*) is so high as to be unestimable. Two genes have much lower K_s values. That for *sacQ* may simply be a sampling effect, given its rather short sequence. However, the value for *bgl*, encoding endo-beta-1,3-1,4-glucanase, is quite surprising. The high level of divergence for *comQ* is due

Table 3. *Divergence between* B. subtilis *and* B. amyloliquefaciens

Gene	Map	Codons	F_{op}	%DNA	%AA	K_A	K_s	Strain
aprE	84	382	0.39	19.6	13.9	0.090	0.97	ATCC 23844
bgl	334	240	0.35	9.9	10.0	0.045	0.31	ATCC 15841
comQ	279	97+	0.21	42.3	51.6	0.419	2.38	F
nprE	127	521	0.40	24.2	19.4	0.121	1.24	ATCC 23350
pcp	–	216	0.23	30.6	27.9	0.209	1.88	
3'*orf*	–	56+	0.07	26.2	21.8	0.123	1.53	
sacB	296	473	0.49	17.7	10.4	0.057	1.05	ATCC 23844
sacQ	279	47	0.39	14.2	10.9	0.068	0.67	F
sipS	210	184	0.39	22.3	15.8	0.090	1.49	
3'*orf*	210	37+	0.09	29.7	16.2	0.108	**	
spoIID	316	343	0.21	25.7	21.6	0.138	1.23	Torrey 10785

All sequences were taken from the GenBank/EMBL/DDBJ database.

Table 4. *Divergence between* B. subtilis *and* B. licheniformis
(*and* B. amyloliquefaciens *and* B. licheniformis)

Gene	Map	Codons	F_{op}	%DNA	%AA	K_A	K_s	Strain
adk	11	24+	0.23	13.9	12.5	0.052	0.67	DSM13
amy	?	512	–	25.0	22.7	0.142	1.23	(*Ba* v *Bl*)
aprE	84	377	0.39	32.4	33.2	0.246	1.56	NCIB 6816
		378	0.39	33.9	34.0	0.261	2.04	(*Ba* v *Bl*)
bgl	334	243	0.35	20.9	13.6	0.077	1.49	
		240	0.30	22.1	15.9	0.089	1.77	(*Ba* v *Bl*)
comQ	279	97+	0.21	43.3	49.5	0.383	**	
				42.3	53.6	0.423	**	(*Ba* v *Bl*)
cwlA	228	88+	0.39	29.5	23.0	0.155	**	FD0120
				34.1	28.7	0.215	**	MC14
cwlC	?	253	0.17	31.4	27.4	0.192	2.47	FD0120
orf2	135	185+	0.31	25.8	19.6	0.126	1.83	5A2
divIB	135	263	0.37	32.4	32.8	0.247	2.03	5A2
orf4	135	187+	0.32	35.3	34.2	0.231	**	5A2
rplO	11	73+	0.60	13.7	8.3	0.040	0.56	DSM13
rpmG	11	50	0.32	18.0	18.4	0.095	0.80	FD01
sacQ	279	46	0.39	25.4	31.1	0.178	**	71I001
				23.2	26.7	0.143	1.74	(*Ba* v *Bl*)
secE	11	60	0.38	17.8	16.9	0.072	1.08	FD01
secY	11	432	0.33	22.7	16.2	0.103	1.14	DSM13
5'orf	221	58	0.35	39.1	49.1	0.384	**	FD0120
sin	221	112	0.34	11.0	7.2	0.039	0.46	FD0120
3'orf	221	39	0.53	20.5	23.7	0.165	0.57	FD0120
sipS	210	184	0.39	31.9	30.6	0.203	**	ATCC 9789
sipB	?	187	–	26.7	26.3	0.168	1.17	(*Ba* v *Bl*)
spo0H	11	219	0.26	17.5	6.0	0.033	1.08	FD01
orf389	211	49+	0.40	25.9	31.3	0.260	0.58	6346
spoIIAA	211	118	0.27	24.9	16.2	0.096	1.52	6346
spoIIAB	211	147	0.30	20.4	11.0	0.062	1.49	6346
spoIIAC	211	256	0.32	23.7	10.2	0.068	1.16	6346
xpaB	228	88	0.29	29.9	32.2	0.248	1.21	FD0120
				35.2	37.9	0.278	**	MC14

All sequences were taken from the GenBank/EMBL/DDBJ database.

to both synonymous and non-synonymous changes. This gene is located approximately 200 bp 3' to *sacQ*, which shows only a typical degree of divergence, and so this cannot be explained by either a location-specific effect, or by the particular strain sequenced. Excluding *comQ*, the weighted average DNA sequence difference is reduced to 21%.

Rather more genes can be compared between *B. subtilis* and *B. licheniformis* (Table 4). If, as expected, *B. subtilis* and *B. amyloliquefaciens* are more closely related to each other than either is to *B. licheniformis*, then comparisons between *B. amyloliquefaciens* and *B. licheniformis* (indicated by *Ba* v *Bl* in Table 4) can also be included since they represent divergence over the same evolutionary time scale. Indeed, in the case of three genes (*aprE*, *bgl*, *sacQ*) where comparisons can be made across all three species, *B. licheniformis* is always considerably more divergent: using K_A values, the ratio of the *B. subtilis* (or *B. amyloliquefaciens*) vs. *B. licheniformis* divergence over the *B. subtilis* vs. *B. amyloliquefaciens* divergence ranges between 1.7 and 2.9. (This is not the case at *comQ*, where the three sequences appear to be approximately equidistant: the unusually high divergence of *comQ* between *B. subtilis* and *B. amyloliquefaciens* has been remarked upon above, and even the values involving *B. licheniformis* are rather high by comparison with other genes.)

The signal peptidase (*sip*) genes warrant some remarks. Two have been sequenced from each of *B. subtilis* (*sipS* and *sipP*) and *B. amyloliquefaciens* (*sipA* and *sipB*), and one from *B. licheniformis*. Among these, *B. subtilis* *sipS* and *B. amyloliquefaciens* *sipA* encode the most similar proteins, and have homologus downstream open reading frames; they are inferred to be orthologous, and were included in Table 3 where their divergence values are typical. However, in a phylogenetic analysis of the five *sip* products, using *B. caldolyticus sipC* as an outgroup, the second *sip* genes from *B. subtilis* and *B. amyloliquefaciens* are far more divergent and do not cluster (not shown); therefore we cannot be sure that these second copies are orthologues, and they have not been included. In comparison with the *B. licheniformis sip* gene, *sipS* and *sipB* are the more similar of the genes from *B. subtilis* and *B. amyloliquefaciens*, respectively, and these have been included in Table 4; their values are not unusual.

Thus 26 genes were compared between *B. subtilis*, *B. amyloliquefaciens* and *B. licheniformis* (Table 4). The weighted average DNA sequence difference between *B. subtilis* and *B. licheniformis* is 26%. This value does not seem much higher than that for *B. subtilis* and *B. amyloliquefaciens*, but two factors can explain this. First, total DNA sequence divergence for any gene increases in a logarithmic fashion over evolutionary time, owing to the increased frequency of superimposed substitutions, and because silent sites become saturated with changes. Secondly, different genes, with different evolutionary rates, appear in Tables 3 and 4. As noted above, when (corrected) substitution rates at nonsynonymous sites in homologous sequences are compared, *B. licheniformis* seems to be about two times as divergent from *B. subtilis* and *B. amyloliquefaciens* as they are from each other. Individual divergence values for *B. subtilis* and *B. licheniformis* genes vary quite remarkably, from just over 10%, to more than 35%. This reflects considerable variation in the extent of protein sequence conservation, and in silent site divergence. Many of the K_s values are over 1.5, indicating that silent sites in these genes are nearly saturated with changes, and some genes are so divergent that K_s is simply not estimable. However, the extent of silent substitution in some genes is relatively low. For *rplO* and the 3'*orf* at *sin*, this may be explained in terms of their strong codon usage bias (high F_{op} values). The gene (*rpmG*) encoding ribosomal protein L33 is interesting. This short open reading frame had gone unnoticed in the original sequence study of *B. licheniformis* (Sharp, 1994). Ribosomal proteins are highly expressed, and so are expected to have strong codon usage bias and exhibit low rates of silent site divergence (as for *rplO*). However, while *rpmG* has a low K_s value, its F_{op} value is not high. For *rpmG*, as well as *adk* and *orf389* (upstream of *spoIIA*), these low K_s values may again simply reflect the small number of silent sites examined. However, the somewhat longer *sin* gene is also interesting in having a low K_s value despite rather low codon usage bias.

Three of these genes with low K_s values are located at 11 degrees, and the estimates of K_s for three other genes from the same region (*secE*, *secY* and *spoOH*) are also lower than average. Again, we are reminded of the suggestion by Chilton & McCarthy (1969) that sequences in this region may have evolved more slowly for some reason connected with their location.

The various *B. licheniformis* sequences come from a number of different strains (Table 4). Another explanation for different levels of sequence divergence would be that some sequences had evolved over a different time scale, i.e. that perhaps not all of these *B. licheniformis* strains come from a single monophyletic group. Alternatively, it is also possible that some strains have not been accurately identified. For example, the sequence of the *aprJ* gene from *B. stearothermophilus* has been reported (Jang *et al.*, 1992). *B. stearothermophilus* is relatively phylogenetically distant from *B. subtilis*, *B. amyloliquefaciens* and *B. licheniformis* (Priest *et al.*, 1988; Ash *et al.*, 1991; see Fig. 1). The *aprE* gene is not unusually highly conserved among the latter three species. However, the *aprJ* sequence differs from the *aprE* sequence of *B. subtilis* 168 (Stahl & Ferrari, 1984) at only 1.4% of nucleotides, i.e. less than any of the genes compared between *B. subtilis* strains 168 and W23. While this might be indicative of a recent horizontal transfer from 168 to *B. stearothermophilus*, it seems more likely that the source of the *aprJ* sequence has been misidentified.

We note that the average values given above for DNA sequence divergence among *B. subtilis*, *B. amyloliquefaciens* and *B. licheniformis* are much lower than those generally quoted from DNA–DNA hybridization studies. For example, *B. subtilis* and *B. amyloliquefaciens* have been reported to show less than 40% 'DNA relatedness' (Nakamura, 1987), and *B. subtilis* and *B. licheniformis* less than 20% 'DNA homology' (Seki, Oshima & Oshima, 1975). These low values can be explained (only) in part by the fact they include total genomic DNA, with perhaps greater divergence in non-coding sequences, and the presence of non-homologous insertions and deletions. However, it is worth emphasizing that values obtained from such DNA–DNA hybridization studies, at least among more distant strains, do not reflect the actual divergence seen in gene sequences.

CONCLUSIONS

We have reviewed what is known about the levels of DNA sequence divergence seen among strains of *B. subtilis*, and its close relatives *B.*

amyloliquefaciens and *B. licheniformis*. These *Bacillus* values have been contrasted with those for *E. coli* and *S. typhimurium*, since to date these represent the most extensively studied bacterial species, with respect to both intraspecific and interspecific divergence. The divergence between *B. subtilis* and *B. amyloliquefaciens* is comparable to, though somewhat greater than, that between *E. coli* and *S. typhimurium*, while *B. licheniformis* is substantially more distant. The three species of *Bacillus* examined all belong to just one subgroup within this genus (Priest *et al.*, 1988; Ash *et al.*, 1991), confirming the view that this genus is genetically very diverse (Rössler *et al.*, 1991; Ash *et al.*, 1991). The *B. subtilis* strains 168 and W23 seem to be rather more divergent than natural isolates of *E. coli*, but it may be that 168 and W23 deserve separate species status (see above).

When *Bacillus* genes are compared between a single pair of strains or species, differences in the level of divergence (for most genes) appear to be largely explicable in terms of varying selective constraints on protein sequence or codon usage. Thus, it does not seem necessary to invoke frequent recombination. This may seem surprising: *Bacilli* are renowned for their ability to undergo transformation, and even *B. subtilis* and *B. licheniformis* can partake in genetic exchange in simulated natural conditions (Duncan *et al.*, 1989). Roberts and Cohan (1993) examined in detail sexual isolation (as reflected in reduced transformation frequency) among these strains and species. Sexual isolation was found to increase exponentially with sequence divergence. For *rpoB*, a rather highly conserved gene, the frequency of transformation between *B. subtilis* and *B. licheniformis* was reduced by a factor of about 10^3 compared to the rate for homeologous sequences. However, between 168 and W23, the rate was reduced by less than a factor of 10. Thus, it seems clear that the potential for recombination is not matched by its realized frequency, at least as measured by those events yielding genomes fit enough to compete in the wild. Cohan (1994) has reviewed theoretical arguments suggesting that bacterial strains can diverge even with relatively high rates of recombination. Since there is clearly (at least some) exchange among bacterial strains (in both *E. coli* and *B. subtilis*), and yet strains placed (largely by traditional methods) in different species seem to undergo very little recombination, it has been suggested that bacterial species are often not very different from the traditional biological species definition used in eukaryotes (Dykhuizen & Green, 1991; but see also Cohan, 1994).

ACKNOWLEDGEMENTS

This work was supported in part by a grant from EOLAS, the Irish Science and Technology Agency, and by the University of Nottingham.

REFERENCES

Ahn, K.S. & Wake, R.G. (1991). Variations and coding features of the sequence spanning the replication terminus of *Bacillus subtilis* 168 and W23 chromosomes. *Gene,* **98**, 107–12.

Amjad, M., Castro, J.M., Sandoval, H., Wu, J.-J., Yang, M., Henner, D.J. & Piggot, P.J. (1990). An *Sfi*I restriction map of the *Bacillus subtilis* 168 genome. *Gene,* **101**, 15–21.

Ash, C., Farrow, J.A.E., Wallbanks, S. & Collins, M.D. (1991). Phylogenetic heterogeneity of the genus *Bacillus* revealed by comparative sequence analysis of small-subunit-ribosomal RNA sequences. *Letters in Applied Microbiology,* **13**, 202–6.

Bisercic, M., Feutrier, J.Y. & Reeves, P.R. (1991). Nucleotide sequences of the *gnd* genes from nine natural isolates of *Escherichia coli*: evidence for intragenic recombination as a contributing factor in the evolution of the polymorphic *gnd* locus. *Journal of Bacteriology,* **173**, 3894–900.

Boyd, E.F., Nelson, K., Want, F.-S., Whittam, T.S. & Selander, R.K. (1994). Molecular genetic basis of allelic polymorphism in malate dehydrogenase (*mdh*) in natural populations of *Escherichia coli* and *Salmonella enterica*. *Proceedings of the National Academy of Sciences, USA,* **91**, 1280–4.

Carrigan, C.M., Haarsma, J.A., Smith, M.T. & Wake, R.G. (1987). Sequence features of the replication terminus of the *Bacillus subtilis* chromosome. *Nucleic Acids Research,* **15**, 8501–9.

Chilton, M.-D. & McCarthy, B.J. (1969). Genetic and base sequence homologies in Bacilli. *Genetics,* **62**, 697–710.

Cohan, F.M. (1994). Genetic exchange and evolutionary divergence in prokaryotes. *Trends in Ecology and Evolution,* **9**, 175–80.

Cohan, F.M., Roberts, M.S. & King, E.C. (1991). Potential for genetic exchange by transformation within a natural population of *Bacillus subtilis*. *Evolution,* **45**, 1383–421.

DuBose, R.F., Dykhuizen, D.E. & Hartl, D.L. (1988). Genetic exchange among natural isolates of bacteria: recombination within the *phoA* gene of *Escherichia coli*. *Proceedings of the National Academy of Sciences, USA,* **85**, 7036–40.

Duncan, K.E., Istock, C.A., Graham, J.B. & Ferguson, N. (1989). Genetic exchange between *Bacillus subtilis* and *Bacillus licheniformis*: variable hybrid stability and the nature of bacterial species. *Evolution,* **43**, 1585–609.

Dykhuizen, D.E. & Green, L. (1991). Recombination in *Escherichia coli* and the definition of biological species. *Journal of Bacteriology,* **173**, 7257–68.

Ferrari, F.A., Trach, K., LeCoq, D., Spence, J., Ferrari, E. & Hoch, J.A. (1985). Characterization of the *spo0A* locus and its deduced product. *Proceedings of the National Academy of Sciences, USA,* **82**, 2647–51.

Glaser, P., Kunst, F., Arnaud, M., Coudart, M.P., Gonzales, W., Hullo, M.F., Ionescu, M., Lubochinsky, B., Marcelino, L., Moszer, I., Presecan, E., Santana, M., Schneider, E., Schweizer, J., Vertes, A., Rapoport, G. & Danchin, A. (1993). *Bacillus subtilis* genome project: cloning and sequencing of the 97 kb region from 325 (degrees) to 333 (degrees). *Molecular Microbiology,* **10**, 371–84.

Gouy, M. & Gautier, C. (1982). Codon usage in bacteria: correlation with gene expressivity. *Nucleic Acids Research,* **10**, 7055–74.

Hall, B.G. & Sharp, P.M. (1992). Molecular population genetics of *Escherichia coli*: DNA sequence diversity at the *celC, crr* and *gutB* loci of natural isolates. *Molecular Biology and Evolution,* **9**, 654–65.

Harry, E.J., Partridge, S.R., Weiss, A.S. & Wake, R.G. (1994). Conservation of the

168 *divIB* gene in *Bacillus subtilis* W23 and *B. licheniformis*, and evidence for homology to *ftsQ* of *Escherichia coli*. *Gene* (in press).

Harry, E.J. & Wake, R.G. (1989). Cloning and expression of a *Bacillus subtilis* division initiation gene for which a homolog has not been identified in another organism. *Journal of Bacteriology*, 171, 6835–9.

Hastrup, S. (1987). Gene expression system. Patent EP 0242220-A 1.

Henkin, T.M., Moon, S.H., Mattheakis, L.C. & Nomura, M. (1989). Cloning and analysis of the *spc* ribosomal protein operon of *Bacillus subtilis*: comparison with the *spc* operon of *Escherichia coli*. *Nucleic Acids Research*, 17, 7469–86.

Herzer, P.J., Inouye, S., Inouye, M. & Whittam, T.S. (1990). Phylogenetic distribution of branched RNA-linked multicopy single-stranded DNA among natural isolates of *Escherichia coli*. *Journal of Bacteriology*, 172, 6175–81.

Hey, J. & Kliman, R.M. (1993). Population genetics and phylogenetics of DNA sequence variation at multiple loci within the *Drosophila melanogaster* species complex. *Molecular Biology and Evolution*, 10, 804–22.

Istock, C.A., Duncan, K.E., Ferguson, N. & Zhou, X. (1992). Sexuality in a natural population of bacteria – *Bacillus subtilis* challenges the clonal paradigm. *Molecular Ecology*, 1, 95–103.

Itaya, M. & Tanaka, T. (1991). Complete physical map of the *Bacillus subtilis* 168 chromosome constructed by a gene-directed mutagenesis method. *Journal of Molecular Biology*, 220, 631–48.

Iwakura, M., Kawata, M., Tsuda, K. & Tanaka, T. (1988). Nucleotide sequence of the thymidylate synthase B and dihydrofolate reductase genes contained in one *Bacillus subtilis* operon. *Gene*, 64, 9–20.

Jang, J.S., Kang, D.O., Chun, K.J. & Byun, S.M. (1992). Molecular cloning of a subtilisin J gene from *Bacillus stearothermophilus* and its expression in *Bacillus subtilis*. *Biochemical and Biophysical Research Communications*, 184, 277–82.

Kimura, M. (1980). A simple method for estimating evolutionary rates of base substitutions through comparative studies of nucleotide sequences. *Journal of Molecular Evolution*, 16, 111–20.

Kimura, M. (1983). *The Neutral Theory of Molecular Evolution*. Cambridge University Press, Cambridge.

Kreuzer, P., Gaertner, D., Allmansberger, R. & Hillen, W. (1989). Identification and sequence analysis of the *Bacillus subtilis* W23 *xylR* gene and *xyl* operator. *Journal of Bacteriology*, 171, 3840–5.

Kunst, F. & Devine, K.M. (1991). The project of sequencing the entire *Bacillus subtilis* genome. *Research in Microbiology*, 142, 905–12.

Lewis, P.J. & Wake, R.G. (1989). DNA and protein sequence conservation at the replication terminus in *Bacillus subtilis* 168 and W23. *Journal of Bacteriology*, 171, 1402–8.

Li, W.-H. (1993). Unbiased estimation of the rates of synonymous and nonsynonymous substitution. *Journal of Molecular Evolution*, 36, 96–9.

Li, W.-H. & Sadler, L.A. (1991). Low nucleotide diversity in man. *Genetics*, 129, 513–23.

Li, W.-H., Tanimura, M. & Sharp, P.M. (1987). An evaluation of the molecular clock hypothesis using mammalian DNA sequences. *Journal of Molecular Evolution*, 25, 330–42.

Li, W.-H., Wu, C.-I. & Luo, C.-C. (1985). A new method for estimating synonymous and nonsynonymous rates of nucleotide substitution considering the relative likelihood of nucleotide and codon changes. *Molecular Biology and Evolution*, 2, 150–74.

Maynard Smith, J. (1992). Analyzing the mosaic structure of genes. *Journal of Molecular Evolution,* **34**, 126–9.

Maynard Smith, J., Smith, N.H., O'Rourke, M. & Spratt, B.G. (1993). How clonal are bacteria? *Proceedings of the National Academy of Sciences, USA,* **90**, 4384–8.

Milkman, R. (1973). Electrophoretic variation in *Escherichia coli* strains from natural sources. *Science,* **182**, 1024–6.

Milkman, R. & Bridges, M.M. (1990). Molecular evolution of the *Escherichia coli* chromosome. III. Clonal frames. *Genetics,* **126**, 505–17.

Milkman, R. & Bridges, M.M. (1993). Molecular evolution of the *Escherichia coli* chromosome. IV. Sequence comparisons. *Genetics,* **133**, 455–68.

Montorsi, M. & Lorenzetti, R. (1993). Heat-stable and heat-labile thymidylate synthases B of *Bacillus subtilis*: comparison of the nucleotide and amino acid sequences. *Molecular and General Genetics,* **239**, 1–5.

Nakamura, L.K. (1987) Deoxyribonucleic acid relatedness of lactose-positive *Bacillus subtilis* strains and *Bacillus amyloliquefaciens*. *International Journal of Systematic Bacteriology,* **37**, 444–5.

Nelson, K. & Selander, R.K. (1992). Evolutionary genetics of the proline permease gene (*putP*) and the control region of the proline utilization operon in populations of *Salmonella* and *Escherichia coli*. *Journal of Bacteriology,* **174**, 6886–95.

Ochman, H. & Selander, R.K. (1984). Standard reference strains of *Escherichia coli* from natural populations. *Journal of Bacteriology,* **157**, 690–3.

Ochman, H. & Wilson, A.C. (1987). Evolution in bacteria: evidence for a universal substitution rate in cellular genomes. *Journal of Molecular Evolution,* **26**, 74–86.

Ogasawara, N., Nakai, S. & Yoshikawa, H. (1994). Systematic sequencing of the 180 kilobase region of the *Bacillus subtilis* chromosome containing the replication origin. *DNA Research,* **1**, 1–14.

Olsen, G.J., Woese, C.R. & Overbeek, R. (1994). The winds of (evolutionary) change: breathing new life into microbiology. *Journal of Bacteriology,* **176**, 1–6.

Piggot, P.J., Amjad, M., Wu, J.-J., Sandoval, H. & Castro, J. (1990). Genetic and physical maps of *Bacillus subtilis* 168. In *Molecular Biological Methods for Bacillus*, ed. C.R. Harwood & S.M. Cutting, pp. 493–540, John Wiley & Sons, Chichester.

Priest, F.G., Goodfellow, M., Shute, L.A. & Berkeley, R.C.W. (1987). *Bacillus amyloliquefaciens* sp. nov., nom. rev. *International Journal of Systematic Bacteriology,* **37**, 69–71.

Priest, F.G., Goodfellow, M. & Todd, C. (1988). A numerical classification of the genus *Bacillus*. *Journal of General Microbiology,* **134**, 1847–82.

Rayssiguier, C., Thaler, D.S. & Radman, M. (1989). The barrier to recombination between *Escherichia coli* and *Salmonella typhimurium* is disrupted in mismatch-repair mutants. *Nature,* **342**, 396–401.

Roberts, M.S. & Cohan, F.M. (1993). The effect of DNA sequence divergence on sexual isolation in *Bacillus*. *Genetics,* **134**, 401–8.

Rössler, D., Ludwig, W., Schleifer, K.H., Lin, C., McGill, T.J., Wisotzkey, J.D., Jurtshuk, Jr., P. & Fox, G.E. (1991). Phylogenetic diversity in the genus *Bacillus* as seen by 16S rRNA sequencing studies. *Systematics and Applied Microbiology,* **14**, 266–9.

Saitou, N. & Nei, M. (1987). The neighbor-joining method; a new method for reconstructing phylogenetic trees. *Molecular Biology and Evolution,* **4**, 406–25.

Seki, T., Oshima, T. & Oshima, Y. (1975). Taxonomic study of *Bacillus* by deoxyribonucleic acid-deoxyribonucleic acid hybridization and interspecific transformation. *International Journal of Systematic Bacteriology,* **25**, 258–70.

Selander, R.K., Caugant, D.A. & Whittam, T.S. (1987). Genetic structure and variation in natural populations of *Escherichia coli*. In *Escherichia coli and Salmonella typhimurium, Vol. 2, Cellular and Molecular Biology*, ed. F.C. Neidhardt, J.L. Ingraham, K.B. Low, B. Magasanik, M. Schaechter & H.E. Umbarger, pp. 1625–48. Washington DC: American Society for Microbiology.

Selander, R.K. & Levin, B.R. (1980). Genetic diversity and structure in *Escherichia coli* populations. *Science*, **210**, 545–7.

Sharp, P.M. (1991). Determinants of DNA sequence divergence between *Escherichia coli* and *Salmonella typhimurium*: codon usage, map position and concerted evolution. *Journal of Molecular Evolution*, **33**, 23–33.

Sharp, P.M. (1994). Identification of genes encoding ribosomal protein L33 from *Bacillus licheniformis*, *Thermus thermophilus* and *Thermotoga maritima*. *Gene*, **139**, 135–6.

Sharp, P.M., Higgins, D.G., Shields, D.C., Devine, K.M. & Hoch, J.A. (1990). *Bacillus subtilis* gene sequences. In *Genetics and Biotechnology of Bacilli*, ed. M.M. Zukowski, A.T. Ganesan & J.A. Hoch, pp. 89–98. San Diego: Academic Press.

Sharp, P.M., Kelleher, J.E., Daniel, A.S., Cowan, G.M. & Murray, N.E. (1992a). Roles of selection and recombination in the evolution of type I restriction-modification systems in Enterobacteria. *Proceedings of the National Academy of Sciences, USA*, **89**, 9836–40.

Sharp, P.M., Nolan, N.C., Ni Cholmain, N. & Devine, K.M. (1992b). DNA sequence variability at the *rplX* locus of *Bacillus subtilis*. *Journal of General Microbiology*, **138**, 39–45.

Sharp, P.M., Shields, D.C., Wolfe, K.H. & Li, W.-H. (1989). Chromosomal location and evolutionary rate variation in Enterobacterial genes. *Science*, **246**, 808–10.

Sharp, P.M., Stenico, M., Peden, J.F. & Lloyd, A.T. (1993). Codon usage: mutational bias, translational selection, or both? *Biochemical Society Transactions*, **21**, 835–41.

Shields, D.C. & Sharp, P.M. (1987). Synonymous codon usage in *Bacillus subtilis* reflects both translational selection and mutational biases. *Nucleic Acids Research*, **15**, 8023–40.

Sorokin, A.V., Zumstein, E., Azevedo, V., Ehrlich, S.D. & Serror, P. (1993). The organization of the *Bacillus subtilis* 168 chromosome region between *spoVA* and *serA* genetic loci, based on sequence data. *Molecular Microbiology*, **10**, 385–95.

Stahl, M.L. & Ferrari, E. (1984). Replacement of the *Bacillus subtilis* subtilisin structural gene with an *in vitro*-derived deletion mutation. *Journal of Bacteriology*, **158**, 411–18.

Wolfe, K.H. & Sharp, P.M. (1993). Mammalian gene evolution: nucleotide sequence divergence between mouse and rat. *Journal of Molecular Evolution*, **37**, 441–56.

Wood, H., Devine, K.M. & McConnell, D.J. (1990). Characterisation of a repressor gene (*xre*) and a temperature-sensitive allele from the *Bacillus subtilis* prophage, PBSX. *Gene*, **96**, 83–8.

Young, M., Mauël, G., Margot, P. & Karamata, D. (1989). Pseudo-allelic relationship between non-homologous genes concerned with biosynthesis of polyglycerol phosphate and polyribitol phosphate teichoic acids in *Bacillus subtilis* strains 168 and W23. *Molecular Microbiology*, **3**, 1805–12.

DNA SEQUENCE VARIATION AND RECOMBINATION IN *E. COLI*

ROGER MILKMAN AND MELISSA MCKANE

Department of Biological Sciences, The University of Iowa, Iowa City, Iowa 52242-1324, USA

INTRODUCTION

The small size of the bacterial genome permits the study of recombination at the DNA sequence level. Standard wild strains, such as the ECOR (*E. coli* reference) strains (Ochman & Selander, 1984), differ from one another and from strain K12 by as much as 3% and occasionally more, providing a level of resolution some two orders of magnitude better than that obtained with closely spaced genetic markers.

Comparative sequencing of 35–40 ECOR strains and K12 over a continuous 10.4 kb region in and near the *trp* operon reveals a variety of distinct sequence types, which, by definition, differ from one another by more than 1%. These sequence types are often seen to occur in a large set of strains, sometimes throughout the entire region sequenced, but more often in mosaic stretches called clonal segments. In the latter case, there is often one prominent sequence type in which a variety of clonal segments are embedded. This prominent sequence type is called a clonal frame (Milkman & McKane Bridges, 1990).

The sequence patterns observed have led to attempts at reconstructing the recent genomic evolution of the species *E. coli*. The approach embodies estimates of the nucleotide substitution rate, favorable mutation rate, and selection coefficients in the context of a paradigm of clonal descent compromised by recombination. An effort is in progress to explain the mosaic distribution of sequence types in terms of estimated rates and patterns of likely mechanisms of recombination. It is hoped that realistic values will result from this effort, forming part of a reconciliation of parameter values sufficient to produce a useful quantitative model of the chromosomal evolution of the species.

THE PARADIGM: PERIODIC SELECTION

The pre-recombination view of bacterial propagation was incorporated into a scheme for the distribution of variation within a bacterial species based on clonal selection by Atwood, Schneider & Ryan (1951), who observed periodic changes in the frequencies of neutral markers in batch cultures.

They attributed these changes to the now familiar process of hitchhiking, in which an entire genotype is carried to higher frequency, in the absence of recombination, by selection for a newly arisen favourable allele. Over the entire species, this process could result in the spread of clones to prominence, provided that broadly favourable mutations occurred at a very low rate. This rate would have to be far lower than that of a single nucleotide substitution, estimated by Drake (1991; personal communication) at 3×10^{-10} per nucleotide per generation. Frequent favourable mutations would simply produce vast numbers of tiny clones. Locally favourable mutations, for example, host-specific alleles, might result (in the absence of recombination) in associations of particular genome-wide genotypes with particular hosts, but this has not been observed (Selander & Levin, 1980).

Evidence for recombination in nature

In 1986, Dykhuizen and Green (1991) sequenced a set of strains in *gnd* that had previously been sequenced by Milkman and Crawford (1983) in *trp*. The *gnd* region was chosen because it displayed an unusually high degree of variation in the electrophoretic mobility of its product, 6-phosphogluconate dehydrogenase (Milkman, 1973; Selander & Levin, 1980). Sequencing confirmed this high degree of variation, but, far more important, the similarities among the strains were highly discordant between the two loci. This fact argued strongly that recombination had been taking place between the two loci. Analysis of DNA sequence variation, preceded by extensive restriction analysis of ECOR strains (Milkman & McKane Bridges, 1990; unpublished data) confirmed this conclusion, as will be detailed later. The clonality of the periodic selection model thus had to be modified to include recombination.

Meroclones

In general, a group of organisms that share *in their entirety* a common ancestor constitute a clone (Allaby, 1985). All the exclusive descendants of a specific common ancestor (for example, cell A, B, or C) constitute a clone (Fig. 1); clearly, clones can be described that are nested. It turns out, however, that the clones produced by periodic selection are soon compromised by recombination: specifically, segments of the chromosome are replaced by DNA having a different ancestry. These are the clonal segments embedded in the clonal frame, the remainder of the original clonal chromosome. The clonal frame persists in recognizable form until successive replacements eventually reduce it to a minority among the sequence types observed. Figure 2 illustrates the early progression of such replacements, whose size is much exaggerated for visual clarity; the reduction of the clonal frame is thus exaggerated as well. Clonal frames persist long enough to be

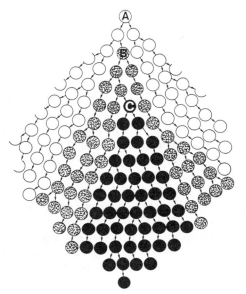

Fig. 1. Clones, defined by arbitrarily chosen common ancestors. In the present context, clones *of particular interest* would likely arise millions of generations apart.

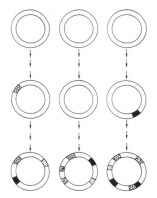

Fig. 2. Recombinational replacements. To the extent that replacement DNA comes from outside the clone, the clonal frame is reduced in extent (as illustrated) and will eventually become unrecognizable.

evident in the restriction analysis of PCR fragments at various chromosomal locations, and to account for the clustering of groups of ECOR strains characterized by Multi-locus enzyme electrophoresis (MLEE), as illustrated by a phenogram based on 38 enzyme loci (Herzer *et al.*, 1990). Each cluster of strains is no longer a strict clone, but its members do share substantial common recent ancestry, and the term *meroclone* has been proposed (Milkman, 1995) to describe them.

Dynamic parameters and their values

Unrecombined portions of a genome should diverge during descent at a rate that is a function of *retained* nucleotide substitution. Aside from favourable substitutions, assumed to occur at negligible frequency, only neutral substitutions will be retained. At present, a simple model is offered (Milkman & McKane Bridges, 1993), which assumes an average of one neutral alternative per codon (mainly in the third position) of translated DNA. Thus, the rate of nucleotide substitution mentioned earlier would be divided by three to recognize only the eligible sites. Next, because only one neutral alternative is posited, only one of the three possible pathways of substitution (for example, $T \rightarrow C$, $T \rightarrow A$ or $T \rightarrow G$) is acceptable, and so the rate is divided by three again. Thus the retained nucleotide substitution rate is $(3 \times 10^{-10})/9 = 1/3 \times 10^{-10}$. Now the divergence rate d, is twice that, or $2/3 \times 10^{-10}$ because substitution can occur in either of two diverging lines. And, finally, time in generations is calculated as

$$t = \frac{1}{2d} \ln \frac{x_e}{x_e - x_o}, \quad \text{or} \quad \frac{1}{4/3 \times 10^{-10}} \ln \frac{x_e}{x_e - x_o} \tag{1}$$

where x_e is the divergence expected at equilibrium and x_o is the observed divergence. Thus, here we would expect that, at equilibrium, 50% of the eligible sites would differ between two lines, since there are two neutral alternatives. This calculation is based on a reversible first-order reaction (Matthews & van Holde, 1990) applied to two states, namely, *same* and *different*, for two homologous nucleotides. Since only one-third of the sites are eligible, the overall sequence divergence observed at equilibrium would be $0.500/3 = 0.167$. So

$$t = 0.75 \times 10^{10} \ln \frac{0.167}{0.167 - x_o}. \tag{2}$$

The number of generations necessary for a clone to rise to prominence (a worldwide frequency of, say 0.10) is approached by the formula

$$N = (1 + s)^t, \tag{3}$$

where N is the number of cells, s is the selection coefficient ($=$ proportionate selective advantage), and t is once again the number of generations, this time the approximate number since the origin of the favourable allele. (It is understood that the *absolute* frequency gain from 1 to about $1/s$ cells is brought about primarily by random genetic drift, which turns out actually to be faster in the lucky few than inexorable selection would have been.) Assuming a world population of 10^{20} *E. coli* cells, a relative frequency of

0.10 would correspond to 10^{19} cells. If $s = 0.00001$, this works out to about 4.4 million generations, or (at 200 generations per year), about 22 000 years. In this number of generations, two lines descended from a common ancestor would diverge by 0.01% of their nucleotides, or 1 per 10 000. This is the order of magnitude of pairwise differences seen in several of the 39 ECOR strains sequenced over 10.4 kb, as will be seen.

It is now of interest to determine the pattern and rate of recombination in nature in *E. coli*, in order to see how this fits in with the mosaic sequence patterns observed, as well as the ages of clones (in generations) so roughly estimated. This turns out to be a fairly complex matter, and it must wait until the DNA sequence variation has been described in detail, and some simple recombination experiments have been described.

DNA SEQUENCE VARIATION

The ECOR strains and their origin

The ECOR strains were assembled by Ochman and Selander (1984) from two sources. Of the strains, 40 were part of a collection of 829 strains (Milkman, 1973) isolated primarily from faecal samples obtained from zoo animals, and domestic and wild animals from various parts of the world. Some clinical isolates from humans also formed part of the collection. The other 32 ECOR strains were of human origin, and the complete set of 72 were chosen on the basis of diversity in host, geographic origin, and multi-locus enzymes electrophoresis (MLEE) phenotype. None of the strains is pathogenic; with that qualification, the set appears to be fairly representative of the species, but isolates from a broader set of hosts (whales, marsupials, monotremes, possibly reptiles and amphibians), and from locales that have been physically isolated for millions of years, might make a valuable addition for comparative studies, particularly in relation to some of the rare fragmentary sequences that have been observed.

Restriction analyses

A set of restriction analyses were performed as a preliminary survey of sequence variation in the ECOR strains (Milkman & McKane Bridges, 1990). For each analysis, 1.5 kb PCR (polymerase chain reaction) fragments were amplified from genomic DNA of 40–60 strains and digested with each of 4–8 restriction enzymes. Of the fragments, eight were contiguous and located in and near the *trp* operon. Several more were at increasing distances from this group, and others were scattered around the chromosome. In all cases, there was considerable restriction fragment length polymorphism (RFLP), and certain groups of strains regularly clustered together. In addition, some strains clustered with one or another of two relatively similar groups, depending on the PCR fragment; and there were other, scattered

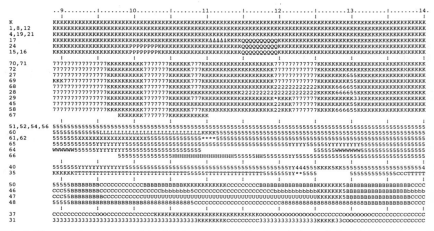

Fig. 3. The mosaic structure of the DNA sequences. Each symbol stands for 50 base pairs and represents a specific sequence type. The scale shows the distance in kilobases from the start of the *trp* operon. ECOR strain numbers are given at the left; sequences identified by more than one strain are essentially identical for all of them. A sequence type is distinguished from all others by a difference of over 1% and at least three nucleotide substitutions. (An indel counts as two substitutions.) The consensus sequence designation (C) is not associated with a particular group of strains. Blank regions have not been sequenced, due to technical difficulties which include likely extreme sequence differences (as in ECOR 64) and large, imprecisely defined deletions (as in ECOR 66). Near position 10 800, ECOR 46 is identical to ECOR 51 at four specific nucleotides and is thus labeled '5' there. This exemplifies a local designation, which requires at least three specific identities. Other symbols: a = small, broadly distributed sequence; * = not assignable. The entire 10.4 kb region is illustrated in Milkman (1995).

changes of affinity as well. Because all the regions analysed had been sequenced for K12 or another standard strain, it was possible to order the restriction site polymorphisms, and this suggested occasional mosaic relationships among strains. This was confirmed in several cases by sequencing. These observations justified an outright comparative sequencing study.

The present 10.4 kb data set

We have now sequenced 39 ECOR strains and (where necessary) K12, with a few short gaps and a very few long ones, in the region of the contiguous PCR fragments. A central 4.4 kb region has been published (Milkman & McKane Bridges, 1993), and a total of 10.4 kb is now complete. The study bears out in great detail the indications of the RFLP analysis: the strains examined fall into three major groups, meroclones, and an assortment of variously related strains. A striking linear mosaicism is observed, in that two sequences may be similar or identical for a considerable distance and then diverge, with one or both showing new similarities. This pattern is illustrated in Fig. 3. The meroclones are clustered with K12, with ECOR 70 and 71, and with ECOR 51, 52, 54 and 56, respectively. Thus there are 11 strains in the K12 *meroclone*, 11 in the *ECOR* 70 *meroclone*, and 10 in the *ECOR* 51 *meroclone*, comprising 31 of the 40 strains in the study. Of the remainder, all share some similarities. Also, in the RFLP studies, ECOR 40 was essentially inseparable from ECOR 38, 39, and 41 (which have not been sequenced) and the same is true for ECOR 50 and ECOR 49 (the latter being sequenced only fragmentarily). These groupings show a striking resemblance to the neighbour-joining phenogram based on extensive MLEE studies of the entire ECOR set by Selander and colleagues, and others (Herzer *et al.*, 1990).

The sequence variation includes the mosaic patterns (segmental clonality) already mentioned, as well as the basic nucleotide substitutions and length polymorphisms. In Fig. 3, seven strains in the K12 meroclone share one sequence type only, as do four strains in the *ECOR* 51 meroclone. The same sequence type is seen as a clonal frame, prominent but not exclusive, in other members of the group, and often as smaller clonal segments in other strains. [The *ECOR* 70 *meroclone's* characteristic sequence type ('7') is often more prominent elsewhere in sequences and restriction patterns.] Long stretches of distinct sequence types are obvious, but short stretches cannot always be resolved. The symbols in Fig. 3, which stand for 50 bp, are assigned on the basis of similarity to a reference strain, or on the basis of difference from all other strains. The sequence types of limited stretches of DNA are distinguished on the basis of the previously noted >1% difference from one another and an absolute number of at least three nucleotide differences. Shifts in the identification of a particular stretch of DNA, first with one reference type and then another, require at least three identities with the new reference type in specific (= non-consensus) nucleotides. These arbitrary criteria do not overcome all ambiguity, much less do they define ancestry precisely, but they do appear to result in a coherent picture that has a logical relationship to the forces at play.

The nucleotide substitutions are predominantly in the third position of codons in translated DNA. Synonymous first-position changes in leucine codons are quite frequent, and amino acid replacements also occur. Non-translated DNA has about 1.2 times as many polymorphic positions as translated DNA, indicating that the non-translated DNA is far from non-specific. Transitions exceed transversions, as expected from a prevalence of synonymous substitutions, but neutrality is more stringent than simple synonymy in that not all synonymous codons are equally plentiful and efficient. Most of the sequence types differ by up to 3%, with some local variation, but ECOR 64 contains a 282-bp stretch that differs from K12 by 23%. Nevertheless, it is clearly homologous to K12 and the other strains; third-position substitutions predominate.

Length polymorphisms range from single-nucleotide indels to a large and varied lambdoid phage called Atlas (Stoltzfus, 1991; Milkman & McKane

Bridges, 1993; Campbell, 1995). In 23 of the 72 ECOR strains, either a functional phage (demonstrated in ECOR 2 and 12) or any of several inactive, reduced versions are inserted between positions 8888 and 8889. Deletions are occasionally shared by several strains. A 319-bp deletion is found in five strains of the ECOR 51 group; elsewhere, a 1-bp deletion occurs in 8 members of that group and two others. Most deletions are in untranslated DNA, but a very few single occurrences cause frame shifts. In *tonB* (Postle & Reznikoff, 1978; Postle & Good, 1983), which produces a protein that acts as a receptor for phages T1 and ϕ80 as well as some colicins, a 3-bp deletion in 11 strains eliminates a glutamine residue and a single 6-bp insertion adds a lysine and a proline. These changes, plus several substitutions involving proline, probably influence recognition specificity. Finally, a non-homologous substitution that does not affect length replaces translated and untranslated DNA adjacent to Atlas with an IS*l*k element.

Other comparative sequence studies have been directed at one or more specific issues, such as evidence for recombination (Dubose, Dykhuizen & Hartl, 1988; Dykhuizen & Green, 1991; Biserčić & Reeves, 1991; Nelson, Whittam & Selander, 1991, 1992; Hall & Sharp, 1992). Analysis at a higher taxonomic level is described by Sharp (this symposium).

RECOMBINATION

The mosaic pattern of sequence similarities raises at once the question of its relationship to recombination. It should be pointed out that there appear to be no local hotspots of exchange, nor any direct functional consequence of a given discontinuity. Recombination must repeatedly pock a chromosome that once shared a genome-wide clonal ancestry with many others. How do the recombinational events add up to the observations?

The major mechanisms of recombination in *E. coli* are likely to be transduction (Robeson, Goldschmidt & Curtiss, 1980) and conjugation (Umeda & Ohtsubo, 1989); and perhaps natural transformation cannot yet be excluded (Stewart, 1989). Most experiments on recombination in *E. coli* have been done using varieties of K12 that differ in genetic markers and infrequent large indels (Perkins *et al.*, 1993), but neither in sequence (except for the few marker genes) nor in restriction-modification systems. The results of many such experiments indicate that the donor DNA is incorporated in one piece in about 80% of the cases observed in K12 (Smith, 1989). Multiple exchanges are certainly known, but they must generally be demonstrated by selection because of their low frequency. These conclusions may well apply to crosses within any given strain. It should be added that physical and genetic analyses have not always been easy to reconcile (Porter, 1987).

Mismatches (Radman, this symposium) and restriction (Murray, this symposium) are known to be barriers to recombination. But can they play other roles? At the levels generally characteristic of the ECOR strains, no

Recipient (-) or Donor (D) DNA in 47→K12 Transductants (Restriction Analysis)

```
PCR       A     G  G B      O       L   A S       C               E             B   S    T     A
Fragment  D     O  L U      N       K   L Q       B               D             S   B    A     C
RE*       H H   R H M S S M H H M M H H S T   *   B B B B B B M M M M   H H   H H M M M   S H
          P P   s3 s3 s3 s3 s3 f 3 q         U U U U U U s s s s   P P   P P s s s   3 P
Change K  T n   C n A n T n C T n n n T n     *   T T C G C G G T A n   n T   n n n n A   n A
       ↓  ↓ ↓   ↓ o ↓ o ↓ o ↓ ↓ o o o o ↓ o       ↓ ↓ ↓ ↓ ↓ ↓ ↓ ↓ ↓ o   o ↓   o o o o ↓   o ↓
      47  C     T   C   G   T C           C       G T A G A C C G         C             G     C
Position -2 2   1 1 1 1 1 1 1 1 1 1                                       -4 5 5 6 6 7 8 8   1 1
Number    6 5   6 5 4 3 3 2 2 2 1 0 0 9 7 7 6   5 4 4 3 3 3 1 1 1         9 0 1 3 4 9 0 8   3 4
          0 5   5 3 3 3 9 4 3 8 2 5 9 1 3     1 6 1 7 7 6 9 2 1 5         5 3 3 5 3 2 1     9 3
          0 0   3 7 3 2 0 2 7 7 7 7 3 4 9 2 5 1 8 5 8 0 9 2 5 6 7         1 3 3 5 3 2 3 1   7 3
          0 0   5 4 7 9 7 4 5 3 8 7 8 3 9 0 0/1 4 6 8 0 3 4 7 5 2 8       7 7 5 6 3 0 6 1   6 8
Strain:
47K-1     - -   - - - - - - - - - - D D D   D   D D D D D D D D D D   - -   - - - - -   D -
47K-2     - D   - - - D D D - - - - D D D   D   D D D D D D - - - -   - -   - - - - -   - -
47K-3     - -   - - - - - - - - - - - - -   D D D D D D - - - -       - -   - - - - -   - -
47K-4     - -   - - - - - - - - - - D D D D D   D D D D D D D D D D   D D   D D D D D   D D
47K-5     - -   - - - - - - - - D D - - - -   D   D D D D D D D D D D   - -   - - - - -   - -
47K-6     D D   D D D D D - - - - - - D D D   D   D D D D D - - - - -   - -   - - - - -   - -
47K-7     - -   - - - - - - - - - - - - -     D   D D D D D D D D D D   D D Δ^b - - - - -   - -
47K-8     - -   - - - - - - - - - - - - -     D   D D D D D D - - - -   D -   - D D D D   D D
47K-9     D D   D D D D D D D D D D D D D D   D   - - - - - - - - - -   - -   - - - - -   - -
47K-10    - -   - - - - - - - - - - - - -   D D   - D D D - - D - -     - -   - - D D D   - -
47K-11    - -   - - - - - - - - - - D D   D   D D - - - - - - - - -     - -   - - - - -   - -
47K-12    - -   D D D D D D D D D D D - D D D   - D   - - - - - - - - -   - -   - - - - -   - -
47K-13    - -   - - - - - - - - - - D   D   D D D D D D D D D D         - -   - - - - -   - -
47K-14    - -   D - - - - - - - D D D D   D   - - - - - - - - - -       - -   - - - - -   - -
47K-15    - -   D D D D D D D D D D D D D D   D   D D D D D D D D D D   - -   - - - - -   - -
47K-16    - -   - - - - - - - - - - - D D   D   D D D D D D - - - -     - -   - - - - -   - -
47K-17    - -   D D D D D D D D D D D D D D   D   - - - - - - - - - -   - -   - - - - -   - -
47K-18    - -   - - - - - - - - - - - - D D   D   - - - - - - - - - -   - -   - - - - -   - -
```

*Restriction Enzymes: HP=HinP I; Rs=Rsa I; H3=Hae III; Hf=Hinf I; Ms=MsP I; S3=Sau3A I; Tq=Taq I; and BU=BstU I.

b Deletion from 5135-5223 that is not present in either recipient or donor.

Fig. 4. Restriction analyses distinguish donor (D) and recipient (−) DNA at the positions listed above (base pairs from the start of the *trp* operon, which runs from right to left in this Figure). Transductant strains are numbered at the left. When a new site is created in ECOR 47, the nucleotide substitution is given. When it is lost, the word 'no' is printed vertically, even though the sequence for ECOR 47 is known in part of the region surveyed.

evidence has appeared that mismatches affect the *pattern* of recombination. Restriction, however, is another story (Pittard, 1964; Harris & Christensen, 1966), and a general role for restriction in natural recombination has been envisioned (DuBose *et al.*, 1988).

McKane (1994) has studied transductants of three ECOR strains into K12, using RFLP analysis over a 40 kb region (Fig. 4), and sequencing (data not shown). Of 18 ECOR 47 → K12 *trp*A33 transductants (selected only for Trp$^+$), 8 had more than one discrete region of donor DNA. Indeed, one transductant had five donor bands. A striking contrast appeared, however, when the two transductants with the longest stretches of donor DNA (47K-4 and 47K-9 in Fig. 4) were back-transduced (backcrossed) to the K12 *trp*A33 recipient. All 15 back-transductants of 47K-9 were identical to 47K-9, and 13 of the 15 47K-4 back-transductants examined were identical to 47K-4. In the other two, the donor DNA of one was abridged on one side, and one on both sides (Fig. 5). While the original transductants averaged donor segments of 6 kb, the only doubly bounded donor segment among the back-transductants was about 10 kb in length. The unbounded ones were of course at least 25 kb in length, and the singly bounded one was about 25 kb. Because we know where the restriction-modification genes are in K12 (Kelleher & Raleigh, 1991), we are reasonably certain that the transductants were identical to the recipient in restriction, though they differ in sequence to varying extents.

Recipient (-) or Donor (D) DNA in 47K-4 Back-transductants (Restriction Analysis)

PCR	L A S	C	E	B	S T	A
Fragment	K L Q	B	D	S	B A	C
RE*	M H H S T *	B B B B B M M M M	H H	H H M M M	S H
	s 3 f 3 q	U U U U U s s s	P P	P P s s s	3 P
Change ↓	n n n T n *	T T C G C G G T A n	n T	n n n n A	n A
↓ 47	o o o ↓ o	↓ ↓ ↓ ↓ ↓ ↓ ↓ ↓ ↓ o	o ↓	o o o o ↓	o ↓
	C	C G T A G A C C G	C	G	C
Position Number	1 1 0 0 9 7 7 6	5 4 4 3 3 3 1 1 1 5	-4 5 5 6 6 7 8 8		1 1 3 4
	8 2 5 9 1 3	1 6 1 7 7 6 9 2 1 5	9 0 1 3 4 9 0 8		9 3
	7 3 4 9 2 5	1 8 5 8 0 9 2 5 6 7	1 3 3 5 3 2 3 1		7 3
	7 8 3 9 0 0/1	4 6 8 0 3 4 7 5 2 8	7 7 5 6 3 0 6 1		6 8

Strain:						
47K-4	- D D D D	D	D D D D D D D D D	D D	D D D D D	D D
4B-1,4-15 ....[-]	D D D D	D	D D D D D D D D D	D D	D D D D D	D D
4B-2 ....[-]	D D D D	D	D D D D D D D D D	D D	D D D D D	- -
4B-3 ....[-]	- D D D	D	D D D D D D D D D	- -	- - - - -	- -

*Restriction Enzymes: Ms=Msp I; H3=Hae III; Hf=Hinf I; S3=Sau3A I; Tq=Taq I; BU=BstU I

[-]: this position is K12 in both donor (47K-4) and recipient.

Fig. 5. Thirteen back-transductants of 47K-4 (second row) are identical to their parent transductant (top row; see also Fig. 4). The other two (third and fourth rows) differ from 47K-4. to 47K-4.

From these and similar experiments, we conclude that restriction can alter the pattern of recombination in transduction, producing small fragments that can be individually incorporated. And, because multiple donor segments were seen in transductions involving three different ECOR strains as parents, it would appear that there is restriction-modification polymorphism among these strains. Thus a set of incorporations may result from a single entrant molecule, making one recombination event at the cell level look like several at the chromosomal level. This has to be reckoned with when we try to estimate recombination rate in the process of reconstructing the micro-evolution of the *E. coli* chromosome. The adaptive significance, if any, of this multiplicity of small incorporations may lie in the refinement of the recombination event, for while the genotype may be reshaped by fairly gross replacements, the recombinational sculpting of proteins may well profit from a finer knife.

An interesting sidelight to McKane's transduction experiment is provided by 30 kb of sequence data, obtained in order to determine with maximal precision the borders between donor and recipient DNA. Not a single nucleotide of the 30 000 proved to be other than either the donor or recipient type; thus the recombination process was not detectably mutagenic.

Now we can compare the mosaic pattern observed in the original transductants with that observed in the sequence comparisons. One difference is obvious at the outset: the transductants could only alternate between the donor and recipient sequences, while the 10.4 kb ECOR comparisons frequently reveal the presence of several sequence types. Also, the segments observed in the transductants have a considerably larger average size than the clonal segments in the ECOR comparison. Successive recombination events are required to explain the observations. This raises the question whether the successive incorporation of large continuous stretches of DNA

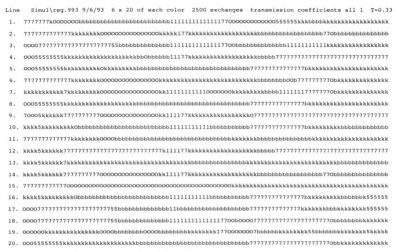

```
Line   Simul\reg.993 9/6/93  6 x 20 of each color  2500 exchanges  transmission coefficients all 1   T=0.33

 1.  7777777k0000000obbbbbbbbbbbbbbbbbbbbbbbbbbb1111111111111177000000000000000555555kkkbbbbbkkkkkkkkkkkkkkkkk
 2.  777777777777kkkkkkkk000000000000000000kkkkk177kkkkkkkkkkkkkkkkkkkkbbbbbbbbbbbbbbbbbbb77obbbbbbbbbbbbbbbb
 3.  000077777777777777777777755bbbbbbbbbbbbbbbb11111111111111177oobbbbbbbbbbbbbbb11111111111kkkkkkkkkkkkkkkkk
 4.  00055555555kkkkkkkkkkkkkkkkkkkkkbb0bbbbbbb11177kkkkkkkkkkkkkkkkkkkkkbbbbb777777777777777777777777777777777
 5.  00055555555kkkkkkkkkk0kkkkkkkkkkbbbbbbbbbbbbbbbbbbbbbbbbbbbbbbbb7777777777777777bkkkkkkkkkkkkkkkkkkkkkkkk
 6.  777777777777kkkkkkkk000000000000000kkkkkkkkkkkkkkkkkkkkkkkkkbbbbbbbbb00b7777777770bkkkkkkkkkkkkkk
 7.  kkkkkkkkkkk7kkkkkkkkk00000000000000000kk11111111110000000kkkkkkkkbbbb11111117777770bkkkkkkkkkkkkkk
 8.  00055555555kkkkkkkkkkkkkkkkkkkkkbbbbbbbbbbbbbbbbbbbbbbbbbbbbbb7777777777777777bkkkkkkkkkkkkkkkkkkkkk
 9.  70005kkkkkk77777777770000000000000000kk111177kkkkkkkkkkkkkkkk0777777777777777777777777777777777777777
10.  kkkk5kkkkkkkk0bbbbbbbbbbbbbbbbbbbbbbbbbbb11111111111bbbbbbbbbb7777777777777777bkkkkkkkkkbbbbbbbbbbbbbb
11.  777777777777kkkkkkkk0000bbbbbbbbbbbbbbbbbbbbbbbbbbbkkkkkkbbbbbbbbbbbbbbbkkkkkkkkkkkkkkkkkkkkkkk
12.  kkkk5kkkkkk7777777777777777777777777777k111177kkkkkkkkkkkkkkkkkkkkkbbbbb777777777777777777777777777
13.  kkkk5kkkkkk7kkkkkkkkkkkkkkkkkkkkkkkkkkkkkbbbbbbbbbbbbbbbbbbbbbbbkkkkkkkkkkkbbbbbbbbbbbbbb
14.  kkkk5kkkkkk777777777700000000000000000kk111177kkkkkkkkkkkkkkkkkkkkbbbbbbbbbbbbbbbbbb770bbbbbbbbbbbbbbb
15.  777777777770000000000000000000000000000000000000000000kkkkkkkkkkkkkkkkkkkkkkkkkkkkkkkkkkkkkkkkkkkk
16.  kkkk5kkkkkkkk0bbbbbbbbbbbbbbbbbbbbbbbbbbb11111111111bbbbbbbbbb7777777777777777bkkkkkkkkkbbbbbbbk555555
17.  000077777777777777777777755bbbbbbbbbbbbbb11bbbbbbbbbbbbbbb7777777777777777kkkkkkkbbbkkkkkkkk555555
18.  000077777777777777777777755bbbbbbbbbbbbbb1111111111111177oobo00007777777777777777777777770bbbbbbbbbbbbb
19.  000000kkkkkkkkkkkkkkk0000bbbbbbbbb0000bbbbbbbbb1770000000077bbbbbkkkkkkkkk55bbbbbbbkkkkkkk5kkkkk
20.  00055555555kkkkkkkkkkkkkkkkkkkkkbbbbbbbbbbbbbbbbbbbbbbbbbbbbbb7777777777777777777777770bbkkkkkkkkkkkkkk
```

Fig. 6. A computer simulation of successive large replacements generates patterns similar to those in Fig. 3. A number of identities are seen the entire set of 120 (only 20 are shown). The 2500 total exchanges are distributed randomly among the 120 strains, which participate both as donors and as recipients. Thus each strain averages about 21 donations and 21 receptions. In the present example, the ratio of the (constant) donor segment length to the length of the observation window is set so that in an expected 0.33 of the cases, the entire observed DNA segment is replaced. The symbols stand for the original six sequences, rather than being associated with sets of identical or nearly identical strains, as has to be done with the actual compared sequences. The transmission coefficients for specific donor-recipient combinations are the probabilities of a replacement occurring. T is the probability of a total replacement over the length observed.

could alone result in the patterns observed. And a very simple computer simulation has shown that it can indeed. The program as used deals with up to six different sequence types identified by colour on the screen and by symbol in a printout. Hundreds of strains with any initial sequence-type frequencies can be mated at random (or with variable success rates). For typographical convenience, 100 characters are used, and these may represent any number of nucleotides in a given simulation. All the events in the experiment are recorded in order, so that the outcome can be confirmed by reconstruction. It is very easy to produce a mosaic pattern similar to those observed in the comparisons (Fig. 6).

RECONCILIATION OF SEVERAL ANALYTIC APPROACHES

Trees

The MLEE approach has been highly successful in clustering strains and isolates by inferred genotype. More recently (Wang *et al.*, 1994), RAPD (random amplified polymorphic DNA) analysis has proven extremely sensitive also. This technique exploits PCR amplification using random primers. Both of these techniques measure the proportion of similar phenotypes produced by compared genomes; neither distinguishes between discontinuous, varied levels of similarity and genome-wide constant levels of similarity. Thus neither MLEE nor RAPD can produce a true phylogeny, since that requires a genome-wide common ancestry.

Sequence comparisons can, and do, reveal discontinuous levels of related-ness, and they illustrate that the proper unit of clonality in a recombining organism is the clonal segment, not the entire genome. Nevertheless, the agreement between sequencing and MLEE in the general clustering of ECOR strains suggest that MLEE can indicate the meroclones, the groups of strains which share clonal relationships over a large part of the genome, presumably the clonal frame. RFLP analysis, which is intermediate in resolution and the practicality of investigating scattered chromosomal regions, is consistent with the other approaches.

Clocks

The divergence time in generations can be estimated from eqn. 2 for a pairwise comparison between homologous sequences. The accuracy of this estimate depends upon the accuracy (and the uniformity over time) of the estimated nucleotide substitution rate, as well as on the accuracy of the estimate of x_e. With respect to translated DNA, the estimate of x_e may be incorrect in discounting neutral non-synonymous substitutions, and the possible frequent existence of more than two equivalent codons in a four-codon set. The numbers observed in the data set so far indicate that the estimate of 0.5 is acceptable.

For clonal frames only, the divergence time should be consistent with the recombination rate, provided once again that such a rate, uniform over time, can be estimated accurately. This estimation is complicated by the likelihood that at least two different recombination processes are important (conjugation and transduction), and that each of these categories contains important variants. Further, restriction-modification polymorphisms in the species seem likely, and differences between donor and recipient can have a large effect on the pattern and perhaps on the frequency of recombination. Moreover, the average number of events (defined as molecular entry into a cell), needed to account for a clonal segment, may vary with process and with restriction differences. Finally, for distributional or other reasons, strains may vary in their accessibility to recombination with different strains. On the other hand, it is possible that all these variables are random with respect to clonal frame, and in this case a standard recombination rate might be estimable. If not, there might be something to learn from different substitution/recombination ratios.

ISSUES AND PROSPECTS

Bastions of polymorphism

According to the current paradigm of periodic clonal selection progressively compromised by recombination, the persistence of the clonal frame should be random with respect to chromosomal position. Furthermore, there is no *general* reason to expect extreme local variation in the degree of recombination observed, i.e. if there is chromosomal localities that evince extremely high levels of recombination, some local explanation is suggested. One such region is the *gnd* locus (Milkman, 1973; Selander, Caugant & Whittam, 1987; Dykhuizen & Green, 1991; Biserčić & Reeves, 1991). Enzyme electrophoresis shows an unusual degree of allelic diversity, and RFLP and sequence analysis indicate wide differences and little linear coherence of these differences. In short, it appears that selection opposes the spread of a single sequence type by hitchhiking and favours the retention of many variants. One focus of such adaptive polymorphism might be the cell surface antigen influenced by *rfb* (Reeves, 1992; Selander *et al.*, 1987), which is located near *gnd*. This mechanism would make the *gnd* locus a passive hitchhiker with a variety of existing *rfb* variants, including some that had been around for a long time, accumulating neutral substitutions. Indeed, one would expect the *his* operon, which ends about 2 kb from *gnd* (Rudd, 1992), to show a comparable level of variation, and RFLP studies on two PCR fragments in the *his* operon show just that (Reeves, personal communication; Milkman, unpublished data). Similar studies of other regions around genes influencing surface antigens have not to my knowledge been done, except for comparative sequencing near and including *tonB*. Since *tonB* codes for a protein which incidentally acts as a phage receptor (as discussed earlier) one might expect a high level of variation nearby, but it is not seen. The forces at play may well differ in these two cases, and there are others to be explored in *E. coli*, including *rfa* and *tonA*.

Whether or not mismatch frequency at, say, the 2–5% level influences the rate and pattern of conjugation, the resolution afforded by such sequence variation would be sufficient to explore other factors, including restriction and chromosomal position. It is also possible that a broader sample of *E. coli* strains might provide extensive regions of greater sequence difference than currently available.

At present, the prospects of a prompt definition of the quantitative roles of transduction and conjugation, and of the calibration of a system of strains with uniformly probable incorporation of one another's DNA, seem unlikely, but steps toward those goals and thus toward the establishment of a recombinational clock seem both feasible and worthwhile.

ACKNOWLEDGEMENTS

We thank Kerri Pohlmann for technical assistance and Philip J. Zee for writing the simulation program. This work was supported by a grant from the National Science Foundation (BRS-9020173).

REFERENCES

Allaby, M., ed. (1985). *The Oxford Dictionary of Natural History*. Oxford University Press, New York.

Atwood, K. C., Schneider, L. K. & Ryan, F. J. (1951). Selective mechanisms in bacteria. *Cold Spring Harbor Symposium of Quantitative Biology*, **16**, 345–55.

Biserčić, M., Feutrier, J. Y. & Reeves, P. R. (1991). Nucleotide sequences of the *gnd* genes from nine natural isolates of *Escherichia coli*: evidence of intragenic recombination as a contributing factor in the evolution of the polymorphic *gnd* locus. *Journal of Bacteriology*, **173**, 3894–900.

Campbell, A. (1995). Cryptic prophages. In Escherichia coli *and* Salmonella *Cellular and Molecular Biology*, 2nd edn. (Neidhardt, F. C. ed.), American Society for Microbiology, Washington, DC, in press.

Drake, J. W. (1991). A constant rate of spontaneous mutation in DNA-based microbes. *Proceedings of the National Academy of Sciences, USA*, **88**, 7160–4.

DuBose, R. F., Dykhuizen, D. E. & Hartl, D. L. (1988). Genetic exchange among natural isolates of bacteria: recombination within the *phoA* gene of *Escherichia coli*. *Proceedings of the National Academy of Sciences, USA*, **85**, 7036–40.

Dykhuizen, D. E. & Green, L. (1991). Recombination in *Escherichia coli* and the definition of biological species. *Journal of Bacteriology*, **173**, 7257–68.

Hall, B. G. & Sharp, P. M. (1992). Molecular population genetics of *Escherichia coli*: DNA sequence diversity at the *celC*, *crr* and *gutB* loci of natural isolates. *Molecular Biology and Evolution*, **9**, 654–65. [Legends to Figs. 1 and 2 were transposed.]

Harris, D. J. & Christensen, J. R. (1966). Pl lysogeny and bacterial conjugation. *Journal of Bacteriology*, **91**, 898.

Herzer, P. J., Inouye, S., Inouye, M. & Whittam, T. S. (1990). Phylogenetic distribution of branched RNA-linked multicopy single-stranded DNA among natural isolates of *Escherichia coli*. *Journal of Bacteriology*, **172**, 6175–81.

Kelleher, J. E. & Raleigh, E. A. (1991). A novel activity in *Escherichia coli* K-12 that directs restriction of DNA modified at CG dinucleotides. *J. Bacteriology*, **173**, 5220–3.

McKane, M. (1994). Transduction experiments suggest that restriction enzymes play an important role in recombination patterns of *E. coli* in nature. MS thesis, The University of Iowa, Iowa City.

Mathews, C. K. & van Holde, K. E. (1990). *Biochemistry*, pp. 343–4. The Benjamin/Cummings Publishing Co. Redwood City, California.

Milkman, R. (1973). Electrophoretic variation in *Escherichia coli* from natural sources. *Science*, **182**, 1024–6.

Milkman, R. (1995). Recombinational exchange among clonal populations. In Escherichia coli *and* Salmonella *Cellular and Molecular Biology*, 2nd edn. (Neidhardt, F. C. ed.), American Society for Microbiology, Washington, DC, in press.

Milkman, R. & Crawford, I. P. (1983). Clustered third-base substitutions among wild strains of *Escherichia coli*. *Science*, **221**, 378–80.

Milkman, R. and McKane Bridges, M. (1990). Molecular evolution of the *E. coli* chromosome. III. Clonal frames. *Genetics*, **126**, 505–17. *Corrigendum Genetics*, **126**, 1139.

Milkman, R. & McKane Bridges, M. (1993). Molecular evolution of the *E. coli* chromosome. IV. Sequence comparisons. *Genetics*, **133**, 455–68.

Nelson, K., Whittam, T. S. & Selander, R. K. (1991). Nucleotide polymorphism and evolution in the glyceraldehyde-3-phosphate dehydrogenase gene (*gapA*) in natural populations of *Salmonella* and *Escherichia coli*. *Proceedings of the National Academy of Sciences, USA*, **88**, 6667–71.

Nelson, K., Whittam, T. S. & Selander, R. K. (1992). Evolutionary genetics of the proline permease gene (*putP*) and the control of the proline utilization operon in populations of *Salmonella* and *Escherichia coli*. *Journal of Bacteriology*, **174**, 6886–95.

Ochman, H. & Selander, R. K. (1984). Standard reference strains of *E. coli* from natural populations. *Journal of Bacteriology*, **157**, 690–3.

Perkins, J. D., Heath, J. D., Sharma, B. R. & Weinstock, G. M. (1993). *XbaT* and *BlnT* genomic cleavage maps of *Escherichia coli* K-12 strain MG1655 and comparative analysis of other strains. *Journal of Molecular Biology*, **232**, 419–45.

Pittard, J. (1964). Effect of phage-controlled restriction on genetic linkage in bacterial crosses. *Journal of Bacteriology*, **87**, 1256–7.

Porter, R. D. (1987). Modes of gene transfer in bacteria. In *Genetic Recombination*. (Kucherlapati, R. and Smith, G. R. eds.), pp. 1–41). American Society for Microbiology, Washington, DC.

Postle, K. & Good, R. F. (1983). DNA sequence of the *Escherichia coli tonB* gene. *Proceedings of the National Academy of Sciences, USA*, **80**, 5235–9.

Postle, K. & Reznikoff, W. S. (1978). *Hind*II and *Hind*III restriction maps of the *attϕ*80-*tonB-trp* region of the *Escherichia coli* genome, and location of the *tonB* gene. *Journal of Bacteriology*, **136**, 1165–73.

Reeves, P. R. (1992). Variation in O-antigens, niche-specific selection and bacterial populations. *FEMS Microbiology Letters*, **79**, 509–16.

Robeson, J. P., Goldschmidt, R. M. & Curtiss III, R. (1980). Potential of *Escherichia coli* isolated from nature to propagate cloning vectors. *Nature, London*, **283**, 104–6.

Rudd, K. (1992). Alignment of *E. coli* DNA sequences to a revised, integrated genomic restriction map. In *A Short Course in Bacterial Genetics*, (Miller, J. H., ed.), pp. 2.3–2.43. Cold Spring Harbor Laboratory Press, Cold Spring Harbor, NY.

Selander, R. K., Caugant, D. A. & Whittam, T. S. (1987). Genetic structure and variation in natural populations of *Escherichia coli*. In: Escherichia coli *and* Salmonella typhimurium *Cellular and Molecular Biology*. (Neidhardt, F. C. ed.), pp. 1625–48. American Society for Microbiology, Washington, DC.

Selander, R. K. & Levin, B. R. (1980). Genetic diversity and structure in *Escherichia coli* populations. *Science*, **210**, 545–7.

Smith, G. (1991). Conjugational recombination in *E. coli*: myths and mechanisms. *Cell*, **64**, 19–27.

Stewart, G. J. (1989). The mechanism of natural transformation. In *Gene Transfer in the Environment*. (Levy, S. B. and Miller, R. V. eds.), pp. 139–64. McGraw-Hill Publishing Co., New York.

Stoltzfus, A. B. (1991). A survey of natural variation in the *trp–tonB* region of the *E. coli* chromosome. PhD thesis, The University of Iowa, Iowa City.

Umeda, M. & Ohtsubo, E. (1989). Mapping of insertion elements IS*1*, IS*2* and IS*3* on the *Escherichia coli* K12 chromosome. Role of the insertion elements in the formation of Hfrs and F factors and in rearrangements of bacterial chromosomes. *Journal of Molecular Biology*, **208**, 601–14.

Wang, G., Whittam, T. S., Berg, C. M. & Berg, D. E. (1994). RAPD (arbitrary primer) PCR is more sensitive than multilocus enzyme electrophoresis for distinguishing related bacterial strains. *Nucleic Acids Research*, **21**, 5930–3.

THE POPULATION GENETICS OF THE PATHOGENIC *NEISSERIA*

BRIAN G. SPRATT, NOEL H. SMITH, JIAJI ZHOU, MARIA O'ROURKE AND EDWARD FEIL

Microbial Genetics Group, School of Biological Sciences, University of Sussex, Falmer, Brighton BN1 9QG, UK

INTRODUCTION

The genus *Neisseria* includes a closely related group of Gram-negative diplococci that are primarily commensals of the mucous membranes of mammals (Knapp, 1988). Several are opportunistic pathogens, but two species, *N. meningitidis* (the meningococcus) and *N. gonorrhoeae* (the gonococcus), are important pathogens of man. Meningococci are divided into 13 subgroups on the basis of the structures and antigenicity of their polysaccharide capsules, but most disease is caused by isolates of serogroups A, B and C (Schwartz, Moore & Broome, 1989). Meningococci of all serogroups appear to be very uniform by DNA–DNA hybridization and are very closely related to gonococci, while the commensal *Neisseria* are more distantly related (Guibourdenche, Popoff & Riou, 1986).

The human commensal *Neisseria*, and *N. meningitidis*, are normally found in the upper respiratory tract, although meningococci may invade the host and cause septicaemia or meningitis (meningococcal disease). *N. gonorrhoeae*, the causative agent of gonorrhoea, are normally found in the urino-genital tract of humans, although sexual activity can also result in their isolation from the rectum and the naso-pharynx. In the latter location they can co-exist, at least transiently, with the other human *Neisseria* species. Serogroup A *N. meningitidis* cause epidemic meningitis, with attack rates as high as 500/100 000, which is now largely confined to parts of sub-Saharan Africa, China and the Middle East (Achtman, 1990, 1994). In Europe, and other western countries, meningococcal disease is mainly due to isolates of serogroups B and C which cause endemic disease with attack rates of about 3/100 000, although approximately 10-fold higher attack rate are found during periods of hyper-endemic disease (Schwartz *et al.*, 1989).

In this chapter we will review recent data on the population genetics of the pathogenic *Neisseria* with particular emphasis on the extent and significance of recombination within these naturally transformable species.

BACTERIAL SEX

Bacteria have three para-sexual processes: transduction, conjugation and transformation. Transducing phage have not been reported in the pathogenic *Neisseria*. Conjugal plasmids are found, but there is no evidence for the mobilization of chromosomal genes (Biswas, Thompson & Sparling, 1989). However, like several other bacterial species, the pathogenic *Neisseria* are naturally competent for transformation, although they are unusual in being fully competent for DNA uptake throughout their life cycle. It will therefore be assumed that all of the recombinational events discussed below are mediated by genetic transformation.

Uptake of DNA during transformation is promoted by a 10-bp uptake sequence that is commonly found downstream of gonococcal and meningococcal genes (Biswas *et al.*, 1989). There is indirect evidence that some commensal *Neisseria* species have similar uptake sequences (Dougherty, Asmus & Tomasz, 1979). Intra-specific transformation using purified DNA is rather efficient, with as many as 15% of the cells being transformed for a marker. Transformation frequencies between *Neisseria* species are believed to decrease according to the sequence divergence between them. In the laboratory, recombination between species that differ by 23% in nucleotide sequence (for example, *N. meningitidis* and *N. flavescens*) can readily be detected (Bowler *et al.*, 1994). The transformation frequency between the two closely related pathogenic *Neisseria* species is only about 10-fold less than the intra-species frequencies, and relatively high recombination rates, both within and between pathogenic the *Neisseria*, have been measured merely by co-cultivation (Frosch & Meyer, 1992).

Transformation can be equated with sex because it results in the generation of recombinant offspring, that is, the chromosome contains genes derived from each parent. However, sex in bacteria is fundamentally different from sex in eukaryotes. Recombination by transformation involves only small segments of DNA although the average replacement size *in vivo* is not known. Because recombinant cells contain a disproportionate amount of DNA from one parent, the effect of recombination on the linkage of bacterial genes is much less severe than it is in eukaryotes (Maynard Smith *et al.*, 1991). Sex and reproduction are not linked in bacteria, which allows the rate of recombination, relative to the rate of asexual spread, to vary between species. Differences in the ratio of recombination to asexual spread will generate differences in bacterial population structures. The neutral mutation rate can be used as a measure of the asexual reproduction rate, but the rate of recombination is harder to measure because observable recombination depends on a variety of factors, including the degree of polymorphism within the population (see below). Transformation is not a passive process but is under the genetic control of several genes (Facius & Meyer, 1993). Selection should, therefore, be able to modify the rate of recombination within a genus or a species.

ELECTROPHORETIC STUDIES OF THE PATHOGENIC *NEISSERIA*

The most definitive data on the extent of recombination between house-keeping genes (assortative recombination) in bacterial species is derived from multi-locus enzyme electrophoresis (MLEE). This technique, applied to bacteria only in the last ten years, measures allelic variation within housekeeping genes, in large population samples (Selander & Musser, 1990; Maynard Smith *et al.*, 1993). Electrophoretic mobility variants of enzymes (electromorphs) identified by starch-gel electrophoresis are equated with alleles and, in combination, can be used to estimate the overall genotypic variation between strains. Extensive MLEE studies on strains of *E. coli* and *Salmonella*, primarily by Selander and co-workers, established that the population structures of these species were basically clonal (Selander & Musser, 1990). That is, in simple terms, electromorphs will change more frequently by mutation than by recombination, leading to strong linkage disequilibrium between alleles in the population. The clonal nature of these populations implies that their evolutionary biology is dominated by periodic selection and stochastic extinction, and the existence of independent non-recombining lineages that are identical by descent (clones) (Levin, 1981).

One of the earliest MLEE studies of the population structure of *Neisseria* was carried out by Caugant *et al.* (1986) on 466 isolates of serogroup B and C meningococci. This study identified a distinctive cluster of isolates with closely related multi-locus genotypes (the ET-5 complex) that had been causing hyper-endemic meningitis on a worldwide scale. A more comprehensive survey of 688 isolates, of all major serogroups, distinguished 331 distinct electrophoretic types and established that *N. meningitidis* was more heterogeneous than any bacterial species previously analysed (Caugant *et al.*, 1987). The authors detected strong linkage disequilibrium between many of the loci examined, implying that assortative recombination in *N. meningitidis* is not sufficiently frequent to randomize associations between genes. Furthermore, the authors suggested that the extensive geographical distribution of the same electrophoretic type (ET) and the recovery of organisms of identical genotype over a period of 15 years, was evidence of a basically clonal population structure (Caugant *et al.*, 1987). However, this study also demonstrated that, with the exception of serogroup A strains, the standard serogrouping/typing schemes for *N. meningitidis* (based on anti-genic differences between capsular polysaccharides and outer membrane proteins) did not accord with multi-locus genotypes. In an extreme case, representatives of five different serogroups were found among a group of 19 strains (ET-47) with identical electrophoretic mobility patterns for all of the 15 enzymes that were studied.

Serogroup A isolates are genotypically less heterogeneous than meningococci of other serogroups and tend to form a single phylogenetic group (Caugant *et al.*, 1987). Detailed analyses of over 500 isolates have identified 84 distinct electrophoretic types, which have been divided into nine major lineages (or subgroups) consisting of isolates of closely related ETs (Wang *et al.*, 1992; Achtman, 1994). Isolates from particular epidemics, or pandemics, tend to belong to the same subgroup and have identical (or almost identical) serosubtypes, even when recovered from different countries at different times (Wang *et al.*, 1992). The observation that each of the major epidemics of serogroup A disease during this century have been caused by isolates from one or other of the nine subgroups suggests that the population may have encountered rounds of periodic selection or bottlenecks (Achtman, 1994).

The application of MLEE to *N. meningitidis* populations has been extremely successful at measuring the heterogeneity of the species, identifying and tracking the progress of epidemics, measuring the carriage rate of particular hyper-endemic or epidemic strains and showing that epidemic and commensal isolates are not distinct genotypes (Caugant, Frøholm & Selander, 1991). These studies have, however, raised questions about the extent of recombination in meningococcal populations. The frequent worldwide recovery of the same multi-locus genotype, and the measurement of significant linkage disequilibrium in meningococcal populations, are assumed to be hallmarks of a clonal population structure and suggest that assortative recombination is rare. However, the disparity between standard typing schemes and multi-locus genotypes suggests that recombination may be common. Similarly, there is a disparity between the large amount of electrophoretic variation in meningococci and the uniformity detected both by DNA–DNA hybridization (Caugant *et al.*, 1987) and the sequencing of housekeeping genes (Zhou & Spratt, 1992). This anomaly could be a consequence of extensive recombination, which can produce greater electrophoretic variability in a population that is rather uniform in sequence than in a population in which recombination is rare.

DOES LINKAGE DISEQUILIBRIUM IMPLY CLONALITY?

The conclusion that meningococcal populations are clonal relies heavily on the demonstration of high coefficients of linkage disequilibrium and the frequent recovery of particular multi-locus genotypes (ETs). However, both these indices of clonality can be generated in a freely recombining (panmictic) population. For example, the samples chosen for MLEE may be biased by the recent spread of a highly successful clone, which introduces linkage disequilibrium into an essentially non-clonal population ('epidemic' clonality). Alternatively, inappropriate sampling may result in the analysis of populations that are genetically or geographically isolated, which will identify the disequilibrium between the populations, and may obscure the occurrence of frequent recombination within each population.

A more critical test of the extent of clonality in bacterial species was applied by Maynard Smith *et al.* (1993) to MLEE data sets of various bacterial species and is known as the index of association (I_A). This statistic measures the variance in the genetic distance between all pairs of strains in a sample, and compares the observed variance with that expected under the hypothesis of frequent mixis. The index of association can be used to compare electrophoretic data sets from any species, and can indicate whether 'epidemic' clonality is obscuring the underlying population structure. This analysis showed that, despite the occurrence of linkage disequilibrium in many bacterial MLEE data sets, clonal bacterial population structures are not as common as previously thought. Relatively high recombination rates within populations are sometimes obscured by 'epidemic' clonality or by the inappropriate lumping together of dissimilar populations. Maynard Smith *et al.* (1993) also pointed out that clonality is a quantitative concept and suggested there is a spectrum of bacterial population structures that range from highly clonal to non-clonal.

HOW CLONAL ARE THE PATHOGENIC *NEISSERIA*?

Analysis by MLEE of 227 isolates of *N. gonorrhoeae*, obtained worldwide over a 26 year period, reached the surprising conclusion that the alleles in the population did not depart significantly from linkage equilibrium ($I_A = 0$), suggesting that recombination must be much more common in the gonococcus than in most other bacterial species (O'Rourke & Stevens, 1993). Furthermore, no correlation existed between serovar, or auxotype, and multi-locus genotype, a result entirely concordant with a panmictic population structure for which electrophoretic types do not mark clones. Linkage equilibrium has also been reported in a population of gonococci from Spain (Vázquez *et al.*, 1993). The non-clonal population structure implies that mixed infections must be common.

Although gonococcal populations appear to be non-clonal, this does not preclude the recovery of isolates of the same ET from geographically and temporally isolated sources (often cited as a hallmark of clonal populations), since these can arise by the recombinational assortment of the most common alleles in the population (Maynard Smith *et al.*, 1993). In the study of O'Rourke and Stevens (1993) the frequencies of recovery of the most common ETs were not significantly different from those predicted by the random association of alleles. As expected, the most commonly encountered ET (ET-1) possessed the most common allele at each locus. Recently, it has been shown that there is as much genetic variation in the ET-1 isolates as there is in a randomly selected group of isolates (O'Rourke & Spratt, 1994). The repeated recovery of indistinguishable isolates in the study of

O'Rourke and Stevens was probably due to the limited number of enzymes that were examined and the uniformity of gonococci (see below) which results in the presence of predominant alleles at most of the loci that were examined.

The structures of *N. meningitidis* populations are more complex. For the data set prepared by Caugant *et al.* (1987) as representative of the entire *N. meningitidis* population (but which are largely serogroup B and C isolates), the value of I_A was high when all the strains were included (I_A = 1.96 ± 0.05), but was greatly reduced when a single representative of each ET was analysed (I_A = 0.21 ± 0.08), and virtually disappeared when clusters of ETs were used (Maynard Smith *et al.*, 1993). This dependence of I_A on the way in which the population is analysed, which is not seen in data sets from highly clonal species, such as *Salmonella*, was interpreted as a reflection of the 'epidemic' population structure of *N. meningitidis*, in which highly success-ful transient hyper-endemic strains occasionally emerge and spread world-wide, so that they represent a considerable proportion of isolates causing meningococcal disease (Maynard Smith *et al.*, 1993). The over-representation of these isolates (for example, the ET-5 complex) introduces linkage disequilibrium into a population which is fundamentally non-clonal, or weakly clonal. Analysing one example of each ET, or one example of each cluster of ETs, reduces the effects of these transient events within the population and shows the underlying population structure. As pointed out by Lenski (1993), the subdivision of bacterial populations in this way should be carried out cautiously, as subdividing the data set may reduce the significance of I_A values when small data sets are analysed.

Serogroup A meningococci appear to be much more clonal than sero-group B and C isolates. Recently, the I_A value for a population of serogroup A isolates (Wang *et al.*, 1992) was shown to be high when all 292 isolates were included (I_A = 2.40 ± 0.08), and, in contrast to the entire *N. meningitidis* data set, the I_A value only reduced slightly when single representatives of each ET were used (I_A = 1.35 ± 0.14) (Maynard Smith, J., personal communication).

We therefore appear to have population structures within the pathogenic *Neisseria*, that range from the non-clonal gonococci to the rather highly clonal serogroup A meningococci. How do these population structures relate to the differences in the ecology, epidemiology, and biology of these organisms? Clonal populations arise because there is insufficient recombi-nation to cause the re-assortment of alleles in the population. Limited recombination can arise for biological reasons (a lack of effective sexual mechanisms), ecological reasons (the bacteria rarely meet each other), or epidemiological reasons (the rate of spread is sufficiently high to reduce the effects of recombination). For example, *E. coli* has a clonal population structure even though there is good evidence that different lineages fre-quently co-exist in the gut (Caugant *et al.*, 1981). Clonality here is presumably biological. Serogroup A meningococci are clonal, but they may be as transformable as serogroup B and C meningococci and gonococci (although this assumption needs to be tested), and a major component in the lack of recombination between lineages may be ecological or epidemiological.

Serogroup A populations consist of multiple successful lineages that cause waves of epidemic, or pandemic, meningococcal disease. As a consequence of the rapid spread of these epidemic clones, even a high background recombination rate may be insufficient to break down linkage disequilibrium. The serogroup B and C strains may represent an intermediate situation, where the bulk of the population exists in a commensal state in the nasopharynx, where mixed colonization and genetic exchange occurs. Occasionally a more virulent hyper-endemic strain emerges and spreads sufficiently rapidly to reduce the homogenizing effect of recombination.

HOW STABLE ARE *NEISSERIA* CLONES?

In a panmictic bacterial population, multi-locus genotypes that mark successful clones should be rapidly broken up by recombination. For gonococci, the rate of assortative recombination seems to be sufficiently high for the recovery, from epidemiologically unrelated individuals, of multi-locus genotypes that are identical by descent to be rare. The frequent recovery of isolates of *N. gonorrhoeae* of distinct phenotype (for example, those from disseminated infections, or from the rectums of homosexual men) may reflect a requirement for common traits adapted to a distinct niche, or epistatic fitness interactions between several genes, rather than overall chromosomal similarity by descent.

For the majority of meningococci, other than serogroup A isolates, recombination appears also to be too frequent to permit the long-term survival of clones. However, occasional hyper-endemic strains, such as ET-5 and ET-37, can arise and flourish for relatively short periods (Caugant *et al.*, 1986; Wang *et al.*, 1993). The latter appear to be unstable, and are broken up by recombination, to produce clusters of isolates that, within a matter of decades, become increasingly variable in multi-locus genotypes, serogroup and serosubtype.

Serogroup A meningococci are unusual in that clones are apparently rather stable over long time periods (Olyhoek, Crowe & Achtman, 1987; Wang *et al.*, 1992). However, sufficient variation is found within epidemics of serogroup A disease, even at alleles distinguishable on starch gels, to require that isolates from an epidemic are described as a subgroup, consisting of closely related multi-locus genotypes, rather than a single multi-locus genotype (Wang *et al.*, 1992). This contrasts with the genotypic stability of a highly clonal pathogen, such as *S. typhi*, for which isolates collected worldwide over long time periods are virtually identical in multi-locus genotype at 24 loci (Selander *et al.*, 1990).

RECOMBINATION IN *NEISSERIA*: EVIDENCE FROM SEQUENCES

The analysis of MLEE data, and the absence of clones, or the relative instability of clones, suggests that there is considerable assortative recombination within the pathogenic *Neisseria*. Can we find molecular evidence for recombination between or within *Neisseria* genes? At the DNA sequence level, assortative recombination can be detected by the non-congruence of the dendrograms derived from the nucleotide sequences of different housekeeping genes from the same set of isolates. Recombination can also be inferred by the presence of localized recombinational events within genes.

Evidence for localized recombinations has been found within several gonococcal or meningococcal genes, including the outer membrane protein genes *opa* (Connell *et al.*, 1988) and *porA* (Feavers *et al.*, 1992), the IgA protease gene (Halter, Pohlner & Meyer, 1989), and at the *pilE* locus (Seifert *et al.*, 1988; Facius & Meyer, 1993). In these examples recombination generates diversity, which in most cases is believed to be maintained in the population by frequency-dependent selection applied by the human immune response (Brunham, Plummer & Stephens, 1993). Recombinational events can be detected in the IgA protease and outer membrane protein genes of serogroup A meningococci but, as expected, they appear to be less common than in serogroup B and C isolates. Thus, there is evidence for a history of recombinational events within these genes from serogroup A isolates, but there is a high degree of uniformity in the outer membrane protein variants, and the IgA protease types, produced by isolates of the same subgroup, whereas variation is found within the transient clones of serogroup B and C isolates (Lomholt *et al.*, 1992; Suker *et al.*, 1994).

Even in highly clonal bacteria, such as *E. coli* and *Salmonella*, recombination has been identified as an important mechanism for generating variation in those genes in which diversity may be advantageous to the bacteria, for example, flagellin (Smith, Beltran & Selander, 1990), lipopolysaccharide (Reeves, 1992), and chromosomal restriction systems (Sharp *et al.*, 1992). In contrast, the evolutionary history of housekeeping genes such as *putP* and *gapA* show no evidence of recombination (Nelson, Whittam & Selander, 1991; Nelson & Selander, 1992), reflecting the clonal nature of *E. coli* and *Salmonella* populations. Indeed, an extensive sequence analysis of 4.4 kb of the chromosome close to the tryptophan biosynthetic operon in natural isolates of *E. coli* has demonstrated that recombination events have not been common enough to obscure the clonal frame of this region (Milkman & Bridges, 1990).

Sequencing of housekeeping genes, which are not believed to be subject to strong directional selection, is therefore required to find evidence of extensive recombination in the pathogenic *Neisseria*. In gonococci, attempts to look directly for recombination have been thwarted by the extreme uniformity of their housekeeping genes. For example, the *recA* genes from

ten different gonococci, isolated over a 30-year period from several continents, were identical in nucleotide sequence (Vázquez *et al.*, 1993).
Attempts to detect recombination require the presence of sufficient sequence diversity to recognize localized recombination or to construct
reasonably robust gene trees. The sequences of the *recA*, *fbp* and *argF* genes
have been sequenced from the same eight meningococcal isolates (Zhou &
Spratt, 1992). Meningococci are certainly more variable than gonococci, but
in most cases there is still not sufficient variation to detect recombination.
For example, the sequences of both the *fbp* and *argF* genes of the eight
meningococcal isolates differed at an average of only 0.4% of nucleotide
sites. The *recA* gene was slightly more variable with an average diversity of
0.9% if one of the isolates (strain P63) was excluded. The *recA* gene of strain
P63 differed from those of other meningococcal isolates at between 2.7 and
3.3% of nucleotide sites. Indeed, this *recA* gene was more distantly related
to the other meningococcal *recA* genes than were the *recA* genes of *N.*
gonorrhoeae and *N. lactamica* (Spratt & Zhou, 1994). As the *argF*, *fbp* and
glnA genes of strain P63 were typical of meningococci, the *recA* gene of P63
may have been obtained from a closely related commensal *Neisseria* species
(see below). Although there is not sufficient sequence diversity to build
robust trees, it is clear that there is no correlation between the extent of
sequence divergence between housekeeping genes and the genetic distances
between the isolates derived from electrophoretic studies (Zhou & Spratt,
1992). Similarly, isolates that appear to be closely related when one gene is
examined do not appear to be closely related when another gene is
examined.

The *glnA* gene of both gonococci and meningococci has been found to be
unusually variable (Zhou, J., Smith, N. H. & Spratt, B. G., unpublished
data). In meningococci there is sufficient variation to look for evidence of
recombination. Figure 1 shows the polymorphic sites within the *glnA* genes
of the eight meningococcal isolates studied by Zhou & Spratt (1992).
Inspection of these sequences shows clear evidence of localized recombinational events.

The levels of sequence diversity within all gonococcal, and most meningococcal, housekeeping genes we have examined are therefore too low to
find statistically significant evidence of recombination. However, in the one
meningococcal gene where there is sufficient variation (*glnA*), there is clear
evidence of a history of recombination. Unfortunately, we do not know the
reason for the unusual level of variation in *glnA*. A likely cause would be
that *glnA* is close to a gene at which there is selection for the maintenance of
variation. However, at present we have been unable to find a likely
candidate for such a gene either upstream or downstream of *glnA*.

```
                      1111111111111112222222222222233333333333344444444444444444444444444445555555555
                      1345671246677778889901344578899000112347900233345566677788889990022334568
                      5687388377814780392816439565814069251185028958943925878903690581718173219
S3446  ATTTATCTTTTCCCATCCCCTTCACATCGACCACGTCGTGATTCCCGTTGCCGTTGCCATTCCCGACCCCCTC
HF130  GC.AGCTCCCGTTGTGTTG.CCT.A.C.CCTT.......TG..A...C.ATT.C...TG..TGG.GTTGG..A
P63    GC..G.T.C................GC.....GTAC..CT.CC..T.C...TTCGCGA.CC..GA.....T..
N94II  GCC.G.T.CCGTTGTGT.TTC..CA..GCCTT....AA..G....GAC.ATTAC...TG..TGGA.TTGG.C.
HF46   GCC.G.T.C...............C..C.....AC..C.GC..T..CC....C...TG..T.GAG....T..
HF116  GCC.G...C................G......GTAC..CT.C...T.C...T.C...TG...GA........
44/76  GCC.G.TCCCGTTGTGT.TTC..CA.CGCCTT....AA..G....GAC.ATTAC...TG..TGGA.TTGG.C.
M470   GC..G.TCCCGTTGTGT.TTC..CA.C.CCTT....AA..G....T.C...TTCGCGA.CC..GA.TTGG.C.
       3333331332333333333333333333333333333333331333333331333333331233331333323333331

                      111111111111111111111111111111
                      5555556666666666667788888899999990000000011111111111111112222
                      9999990033446788924000344012266772223468133334445677777881455
                      01456701092535146911471193814035803922293145703951013492885 14
S3446  GTCGCCCTCTGTCACTGTTGCCCTGGACCCGGGGTATTCCTTATGGGTTGACGGCCCGCTC
HF130  AG.AG.GCTCCCTC...CCAT..C...TTTAAAACGCC.TCCGCCC....C.C......CT
P63    .............................................CCGCCC...AC.C.G.T..CT
N94II  .CT..TT......CTCC..........................................A...
HF46   ....................CAA.T..............TTCCGCCC..C.C.C......TCT
HF116  .C.....TC..CC...C..CCTTTC.ATT.........C.TCCGCCCACCA.T.T.T...CT
44/76  .CT..TT......C.CC..........................................
M470   .CT..TT..................................
       2331233133333333333333333313333333333333333133333133333331313333333
```

Fig. 1. Intragenic recombination within the *glnA* gene of *N. meningitidis*. The *glnA* gene was sequenced from the *N. meningitidis* strains listed at the left (from the reference collection of Caugant *et al.*, 1987). Only the polymorphic sites are shown. Nucleotide sites that are the same as those in the S3446 sequence are marked by dots. The numbers above the sequences show (in vertical format) the nucleotide position, and the numbers below the sequences, the position within the codon. Visual inspection shows several clear examples of intragenic (localised) recombination. For example, the sequences of strains 44/76 and M470 are very different from that of strain S3446 in the front half of the gene, but are identical in the back half. The sequences of the former two strains are almost identical, except between nucleotides 438 and 501, where strain M470 contains a region that is identical to that in strain P63.

INTERSPECIES RECOMBINATION IN *NEISSERIA*

In the laboratory, transformation between meningococci and gonococci is surprisingly efficient and transformation also occurs, at lower frequencies, between the pathogenic *Neisseria* and those commensal *Neisseria* that are related sufficiently closely for homologous recombination to be possible (probably ≤25% sequence divergence). Transformation probably provides the most efficient mechanism by which chromosomal genes can be exchanged between closely related species and several examples of putative interspecies recombinational events have been found in nature (Maynard Smith, Dowson & Spratt, 1991). The clearest examples have occurred during the development of penicillin resistance in the pathogenic *Neisseria*.

Non-β-lactamase-producing penicillin-resistant isolates of *N. gonorrhoeae* are widely encountered and produce altered forms of the two high M_r penicillin-binding proteins (PBPs) that have decreased affinity for penicillin (Spratt, 1994). This type of resistance is also emerging in *N. meningitidis*, but so far isolates appear to produce low affinity forms only of PBP 2 (Spratt, 1994). Analysis of the sequences of the PBP 2 genes (*penA*) of penicillin-resistant and penicillin-susceptible isolates of both *N. gonorrhoeae* and *N. meningitidis* have produced clear evidence for recombination between the pathogenic and commensal species (Spratt *et al.*, 1992). Whereas the *penA* genes of susceptible isolates of each species are very uniform, those of

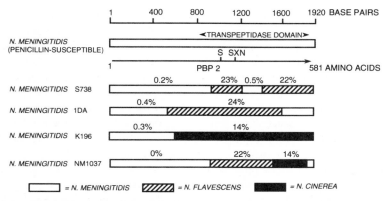

Fig. 2. Interspecies recombination within the *penA* gene of *N. meningitidis*. The upper open rectangle represents the *penA* gene of a penicillin-susceptible meningococcus. The line terminating in an arrow represents PBP 2; the active-site serine residue and the SXN conserved sequence motifs are shown (Spratt, 1994). The four lower rectangles represent the *penA* genes of meningococcal isolates that have reduced susceptibility to penicillin. The percentage sequence divergence between different regions of the *penA* genes and the corresponding regions in the penA gene of the susceptible strain are shown. The highly diverged shaded regions have been introduced by homologous recombination from either *N. flavescens* or *N. cinerea* or, in the *penA* gene of *N. meningitidis* NM1037, from both of these human commensal species. From Spratt (1994).

resistant isolates have a mosaic structure, consisting of regions that are essentially similar to those in the genes of susceptible isolates and regions that are 14–23% diverged in sequence (Fig. 2). The diverged regions have been shown to have been derived from the *penA* genes of the closely related human commensal species, *N. flavescens* and *N. cinerea* (Spratt *et al*., 1992). The affinity for penicillin of PBP 2 from pre-antibiotic era isolates of *N. flavescens* is much lower than that of PBP 2 from the two pathogenic *Neisseria* species. The ability of transformation to promote, at low frequency, the replacement of the genes of the pathogens with their homologues from commensal species has led to the emergence of gonococci and meningococci that contain the 'penicillin-resistant' *penA* gene from *N. flavescens* (Bowler *et al*., 1994). Similar interspecies recombinational events appear to have occurred during the development of sulphonamide-resistant forms of the dihydropteroate synthase of *N. meningitidis*, although, in this example, the donor species have not been identified (Radstrom *et al*., 1992).

The interspecies recombinational events involving the *penA* and dihydropteroate synthase genes have almost certainly been strongly selected by antibiotic usage. Can we find examples of interspecies recombination in genes that are not believed to be under strong directional selection? The increased sequence divergence makes interspecies recombinational events easier to detect than intraspecies events. Putative interspecies recombination can be detected in two ways. First, comparisons of the sequences of housekeeping genes from multiple isolates may identify alleles that are much more diverged than those of other isolates of the same species, and which are even more diverged than some of the closely related commensal species. This situation was found in the case of the *recA* gene of *N. meningitidis* strain P63 (see above). Secondly, comparisions of the sequences of the same gene from gonococci and meningococci may identify mosaic gene structure. Most of a gene may differ by about 2% in sequence (the average divergence between meningococcal and gonococcal housekeeping genes), whereas a part of the gene may be much more diverged. In both cases, the identification of the source of the unusually diverged regions is required to provide the most convincing evidence. The latter method has identified interspecies recombination in the *argF* gene of meningococci (Zhou & Spratt, 1992). In this case, part of the meningococcal gene has been replaced with a region from the *argF* gene of an isolate closely related to *N. cinerea*.

Interspecies recombination can therefore be found in *Neisseria* housekeeping genes. This could mean that recombination is so frequent in *Neisseria* that even quasi-neutral interspecies recombinational exchanges occasionally become fixed in the population. Alternatively, these events may be very rare, but have become fixed as a result of positive selection, acting either on the housekeeping gene, or on a closely linked gene. Surprisingly, the examples of interspecies recombination in the pathogenic *Neisseria* have involved the commensal species, *N. cinerea* and *N. flavescens*, which differ in nucleotide sequence from the pathogens by about 14% and 23%, respectively, and recombination between the two very closely related pathogenic species has not been detected (see below).

·ARE MENINGOCOCCI AND GONOCOCCI A SINGLE POPULATION?

The two pathogenic *Neisseria* species are so closely related (≥98% sequence identity in housekeeping genes) that they would be included in a single species, if it were not for the important differences in the diseases they cause. As extensive recombination occurs within populations of both gonococci and serogroup B and C meningococci, it is of interest to see if their separation into two species is justified on population genetic grounds. Vázquez *et al.* (1993) have used MLEE to examine a combined population of gonococci and serogroup B and C meningococci from Spain. A dendogram based on the electrophoretic data divided all of the 346 gonococcal isolates from all of the 274 meningococcal isolates. The complete separation of the two populations suggests that assortative recombination between gonococci and meningococci is rare. Analysis of the sequences of the *recA*, *fbp*, and *argF* genes also showed no evidence of localized recombination between gonococci and meningococci. Similar results have been obtained with the more variable *glnA* gene (Zhou, J., Smith, N.H. & Spratt, B.G.,

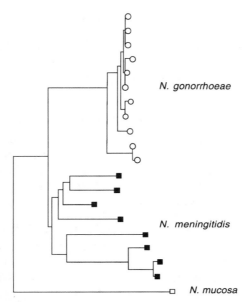

Fig. 3. Relationships between the *glnA* genes of the pathogenic *Neisseria*. A dendrogram representing the sequence relationships between the *glnA* genes of eleven strains of *N. gonorrhoeae* (open circles), eight strains of *N. meningitidis* (closed boxes), and a strain of *N. mucosa* (open box). It should be remembered that this dendrogram is not an evolutionary tree. As a consequence of recombination within the meningococcal and gonococcal populations, the dendrogram reflects only the relatedness between the *glnA* sequences. The dendrogram was constructed by the neighbour-joining method (Saitou & Nei, 1987) applied to a matrix of pairwise sequence differences that had been adjusted for multiple substitutions by the method of Jukes and Cantor (1969). The average pairwise distance between the sequence from *N. mucosa* and the *N. gonorrhoeae* sequences is 9%. The complete separation of all of the gonococcal sequences from all of the meningococcal sequences was obtained in 100% of 500 bootstrap resamplings of the data. Sequence data analysis was carried out within the MEGA program package (Kumar, Tamura & Nei, 1993).

unpublished data). Figure 3 shows a tree, constructed by the neighbour joining method, which shows the clear separation between the gonococcal and meningococcal *glnA* sequences.

These results indicate that gonococci and meningococci are isolated in nature. The isolation of such closely related recombinogenic populations is surprising, and suggests that some barrier must have arisen to allow speciation to occur and be maintained. It has been proposed that gonococci arose as a clone of meningococci that could colonize the genital tract, and that the genetic isolation between them is maintained by the higher frequency of intra- compared to interspecies recombination. The differences in their major niches (genital tract compared to naso-pharynx) probably provides an ecological component of isolation, which may be augmented by a difference in the transformation frequency between, compared to within, populations (Vázquez *et al.*, 1993). Unlike the situation in eukaryotes, where relatively little genetic flow can keep species coherent, in *Neisseria* the localized nature of sex may make separation of a population into two isolated populations (species) possible, even when there is a relatively small difference in the extent of recombination between populations compared to that within populations (Vázquez *et al.*, 1993).

CONCLUDING REMARKS

Studies of most bacterial species suggest that recombination occurs in nature, and indeed may be highly important in generating variation, but that it is infrequent compared to mutation, such that bacterial populations consist largely of independent clonal lineages. In recent years there has been increasing evidence that recombination may be more frequent in naturally transformable species (e.g. *Neisseria* and *Bacillus subtilis*; Istock *et al.*, 1992), although there are clear exceptions as *Haemophilus influenzae* are transformable, but are highly clonal (Maynard Smith *et al.*, 1993).

Some of the best evidence for extensive recombination in nature derives from studies of the pathogenic *Neisseria*, which are particularly interesting as differing population structures, and different ecology/epidemiology, are found within this group of highly related bacteria (Maiden, 1993). The reason for these differences are at present unclear. Certainly, the lifestyle of the clonal serogroup A meningococci is very different from that of the non-clonal gonococci. During epidemic spread the former are essentially 'hit and run' pathogens, where pressures for variation from the host immune system may be much less strong than in gonococci, which (at least before the rapid cure effected by antibiotics) cause chronic infections of the genital tract. The absence of selection from the immune system for variation in cell surface genes is suggested by the surprising finding that isolates from epidemics of serogroup A disease vary more by MLEE than by serological criteria (Achtman, 1994).

Is the relative absence of recombination in serogroup A meningococci, compared to serogroup B and C isolates, and particularly gonococci, due to differences in the efficiency of recombination (transformability), or is it a consequence of their very different epidemiology? At present the answers are not clear, although the efficacy of recombination, *in vitro* or in animal models, in the different *Neisseria* can be approached experimentally.

Gonococci appear to be largely isolated from meningococci and behave as biologically valid species. This does not imply that recombination between these species is absent, but simply that it is not sufficient to cause them to behave as a single population. Indeed, there is evidence of recombination between the pathogenic species and commensal species, but these events are presumably rare, and highly localized, and, although they may be of great importance (for example, in the evolution of antibiotic resistance or the generation of antigenic variation), their effect on population structure is almost certainly insignificant. Housekeeping genes of serogroup A isolates are not clearly distinct from those of other serogroups (Zhou & Spratt, 1992) and there is little evidence that there is isolation between these groups. They do, however, appear to form a phylogenetically distinct grouping within the total meningococcal population, and their behaviour is clearly different from meningococci of other serogroups. Serogroup A meningococci may be some distance along the pathway leading to speciation.

Differences in population structure have profound effects on epidemiology. Clonal populations allow the study of long-term epidemiological trends, for example, the tracking of pandemics of serogroup A meningococcal disease. The non-clonal nature of gonococcal populations implies that long-term epidemiological studies should be impossible, as isolates recovered in one decade are likely to be different from those obtained in the next. Fortunately, most epidemiological studies are concerned with the short-term spread of isolates between sexual contacts, and it is unlikely that recombinations impairs these types of study. However, the non-clonal nature of gonococcal populations makes it difficult to find methods that adequately index the relatedness between isolates. The current methods based on serotype/auxotype combinations are likely to be misleading, not only because serotypic markers are unsatisfactory in principle (because they are subject to positive selection and convergence), but since isolates that show the same serotype/auxotype combination may be genetically diverse isolates that have the same serotype/auxotype as a result of recombination.

Finally, how frequent does assortative recombination have to be compared to mutation to result in bacterial populations like those of *N. gonorrhoeae* that appear to be in linkage equilibrium? Simulations suggest that apparent panmixis can occur if observable alterations in the electrophoretic mobility of housekeeping enzymes occur at least 20 times as frequently by transformation as they do by mutation (Maynard Smith *et al.*, 1993; Maynard Smith, J., in press). This ratio can be compared to estimates for highly clonal species where observable alterations probably occur considerably less frequently by recombination than by mutation (Milkman & Bridges, 1990). Of course, the observable transformational exchanges in gonococcal populations will depend both on the extent of mixed infections and the presence of appropriate polymorphisms between the alleles of the co-infecting isolates. Observable mutational alterations are also only a subset of the non-synonymous mutations, which are themselves a subset of all of the mutations that appear in housekeeping genes. It is therefore difficult to estimate any actual rate of transformation in populations of gonococci, although it appears to be considerable.

REFERENCES

Achtman, M. (1990). Molecular epidemiology of epidemic bacterial meningitis. *Reviews in Medical Microbiology*, **1**, 29–38.

Achtman, M. (1994). Clonal spread of serogroup A meningococci: a paradigm for the analysis of microevolution in bacteria. *Molecular Microbiology*, **11**, 15–22.

Biswas, G. D., Thompson, S. A. & Sparling, P. F. (1989). Gene transfer in *Neisseria gonorrhoeae*. *Clinical Microbiology Reviews*, **2**, S24–8.

Bowler, L. D., Zhang, Q.-Y., Riou, J.-Y. & Spratt, B. G. (1994). Interspecies recombination between the *penA* genes of *Neisseria meningitidis* and commensal *Neisseria* species during the emergence of penicillin resistance in *N. meningitidis*: natural events and laboratory simulation. *Journal of Bacteriology*, **176**, 333–7.

Brunham, R. C., Plummer, F. A. & Stephens, R. S. (1993). Bacterial antigenic variation, host immune response, and pathogen–host coevolution. *Infection and Immunity*, **61**, 2272–6.

Caugant, D. A., Levin, B. R. & Selander, R. K. 1981. Genetic diversity and temporal variation in the *E. coli* population of a human host. *Genetics*, **98**, 467–90.

Caugant, D. A., Frøholm, L. O., Bovre, K., Holten, E., Frasch, C. E., Mocca, L. F., Zollinger, W. D. & Selander, R. K. (1986). Intercontinental spread of a genetically distinctive complex of clones of *Neisseria meningitidis* causing endemic disease. *Proceedings of the National Academy of Sciences, USA*, **83**, 4927–31.

Caugant, D. A., Mocca, L. F., Frasch, C. E., Frøholm, L. O., Zollinger, W. D. & Selander, R. K. (1987). Genetic structure of *Neisseria meningitidis* populations in relation to serogroup, serotype, and outer membrane protein pattern. *Journal of Bacteriology*, **169**, 2781–92.

Caugant, D. A., Frøholm, L. O. & Selander, R. K. (1991). Molecular epidemiology of meningococcal disease. *Medecine et Maladies Infectieuses*, **21**, 191–4.

Connell, T. D., Black, W. J., Kawula, T. H., Barritt, D. S., Dempsey, J. A., Kverneland, K., Stephenson, A., Schepart, B. S., Murphy, G. L. & Cannon, J. G. (1988). Recombination among protein II genes of *Neisseria gonorrhoeae* generates new coding sequences and increases structural variability in the protein II family. *Molecular Microbiology*, **2**, 227–36.

Dougherty, T. J., Asmus, A. & Tomasz, A. (1979). Specificity of DNA uptake in genetic transformation of gonococci. *Biochemical Biophysical Research Communications*, **86**, 97–104.

Facius, D. & Meyer, T. F. (1993). A novel determinant (*comA*) essential for natural transformation competence in *Neisseria gonorrhoeae* and the effect of a ComA defect on pilin variation. *Molecular Microbiology*, **10**, 699–712.

Feavers, I. M., Heath, A. B., Bygraves, J. A. & Maiden, M. C. J. (1992). Role of horizontal genetic exchange in the antigenic variation of the class 1 outer membrane protein of *Neisseria meningitidis*. *Molecular Microbiology*, **6**, 489–95.

Frosch, M. & Meyer, T. F. (1992). Transformation-mediated exchange of virulence determinants by co-cultivation of pathogenic *Neisseria*. *FEMS Microbiology Letters*, **100**, 345–50.

Guibourdenche, M., Popoff, M. Y. & Riou, J.-Y. (1986). Deoxyribonucleic acid relatedness among *Neisseria gonorrhoeae*, *N. meningitidis*, *N. lactamica*, *N. cinerea* and '*Neisseria polysaccharea*'. *Annals of the Institut Pasteur/Microbiology*, **137B**, 177–85.

Halter, R., Pohlner, J. & Meyer, T. F. (1989). Mosaic-like organization of the IgA protease genes in *Neisseria gonorrhoeae* generated by horizontal exchange *in vivo*. *EMBO Journal*, **8**, 2737–44.

Istock, C. A., Duncan, K. E., Ferguson, N. & Zhou, X. (1992). Sexuality in a natural population of bacteria–*Bacillus subtilis* challenges the clonal paradigm. *Molecular Ecology*, **1**, 95–103.

Jukes, T. H. & Cantor, C. R. (1969). Evolution of protein molecules. In *Mammalian Protein Metabolism* (Munro, H. N., ed.), pp. 21–132. Academic Press, New York.

Knapp, J. S. (1988). Historical perspectives and identification of *Neisseria* and related species. *Clinical Microbiology Reviews*, **1**, 415–31.

Kumar, S., Tamura, K. & Nei, M. (1993). MEGA: Molecular Evolutionary Genetics Analysis, version 1.01. The Pennsylvania State University, University Park, PA 16802.

Lenski, R. E. (1993). Addressing the genetic structure of microbial populations. *Proceedings of the National Academy of Sciences, USA*, **90**, 4334–6.

Levin, B. R. (1981). Periodic selection, infectious gene exchange and the genetic structure of *E. coli* populations. *Genetics*, **99**, 1–23.

Lomholt, H., Poulsen, K., Caugant, D. A. & Kilian, M. (1992). Molecular polymorphism and epidemiology of *Neisseria meningitidis* immunoglobulin A1 proteases. *Proceedings of the National Academy of Sciences, USA*, **89**, 2120–4.

Maiden, M. C. J. (1993). Population genetics of a transformable bacterium: the influence of horizontal genetic exchange on the biology of *Neisseria meningitidis*. *FEMS Microbiology Letters*, **112**, 243–50.

Maynard Smith, J., Dowson, C. G. & Spratt, B. G. (1991). Localized sex in bacteria. *Nature, London*, **349**, 29–31.

Maynard Smith, J., Smith, N. H., O'Rourke, M. & Spratt, B. G. (1993). How clonal are bacteria? *Proceedings of the National Academy of Sciences, USA*, **90**, 4384–8.

Milkman, R. & Bridges, M. M. (1990). Molecular evolution of the *Escherichia coli* chromosome. III. Clonal frames. *Genetics*, **126**, 505–17.

Nelson, K. & Selander, R. K. (1992). Evolutionary genetics of the proline permease gene (*putP*) and the control region of the proline utilization operon in populations of *Salmonella* and *Escherichia coli*. *Journal of Bacteriology*, **174**, 6886–95.

Nelson, K., Whittam, T. S. & Selander, R. K. (1991). Nucleotide polymorphism and evolution in the glyceraldehyde-3-phosphate dehydrogenase gene (*gapA*) in natural populations of *Salmonella* and *Escherichia coli*. *Proceedings of the National Academy of Sciences, USA*, **88**, 6667–71.

Olyhoek, T., Crowe, B. A. & Achtman, M. (1987). Clonal population structure of *Neisseria meningitidis* serogroup A isolated from epidemics and pandemics between 1915 and 1983. *Reviews of Infectious Diseases*, **9**, 665–82.

O'Rourke, M. & Spratt, B. G. (1994). Further evidence for the non-clonal population structure of *Neisseria gonorrhoeae*: extensive genetic diversity within isolates of the same electrophoretic type. *Microbiology*, **140**, 1285–90.

O'Rourke, M. & Stevens, E. (1993). Genetic structure of *Neisseria gonorrhoeae* populations: a non-clonal pathogen. *Journal of General Microbiology*, **139**, 2603–11.

Radstrom, P., Fermer, C., Kristiansen, B. E., Jenkins, A., Skold, O. & Swedberg, G. (1992). Transformational exchanges in the dihydropteroate synthase gene of *Neisseria meningitidis*: a novel mechanism for the acquisition of sulphonamide resistance. *Journal of Bacteriology*, **174**: 5961–8.

Reeves, P. R. (1992). Variation in O-antigens, niche-specific selection and bacterial populations. *FEMS Microbiology Letters*, **100**, 509–16.

Saitou, N. & Nei, M. (1987). The neighbor-joining method: A new method for reconstructing phylogenetic trees. *Molecular Biology and Evolution*, **4**, 406–25.

Seifert, H. S., Ajioka, R. S., Marchal, C., Sparling, P. F. & So, M. (1988). DNA transformation leads to pilin antigenic variation in *Neisseria gonorrhoeae*. *Nature, London*, **336**, 392–5.

Selander, R. K., Beltran, P., Smith, N. H., Helmuth, R., Rubin, F. A., Kopecko, D. J., Ferris, K., Tall, B. D., Cravioto, A. & Musser, J. M. (1990). Evolutionary genetic relationships of clones of *Salmonella* serovars that cause human typhoid and other enteric fevers. *Infection and Immunity*, **58**, 2262–75.

Selander, R. K. & Musser, J. M. (1990). In *Molecular Basis of Bacterial Pathogenesis* (Iglewski, B. H., Clark, V. L., eds.), pp. 11–36. Academic Press, San Diego.

Schwartz, B., Moore, P. S. & Broome, C. V. (1989). Global epidemiology of meningococcal disease. *Clinical Microbiology Reviews*, **2**, S118–24.

Sharp, P. M., Kelleher, J. A., Daniel, A. S., Cowan, G. M. & Murray, N. E. (1992). Roles of selection and recombination in the evolution of type I restriction-modification systems. *Proceedings of the National Academy of Sciences, USA*, **89**, 9836–40.

Smith, N. H., Beltran, P. & Selander, R. K. (1990). Recombination of *Salmonella* phase 1 flagellin genes generates new serovars. *Journal of Bacteriology*, **172**, 2209–16.

Spratt, B. G. (1994). Resistance to antibiotics. *Science*, **264**, 388–93.

Spratt, B. G., Bowler, L. D., Zhang, Q.-Y., Zhou, J. & Maynard Smith, J. (1992). Role of interspecies transfer to chromosomal genes in the evolution of penicillin resistance in pathogenic and commensal *Neisseria* species. *Journal of Molecular Evolution*, **34**, 115–25.

Spratt, B. G. & Zhou, J. (1994). Recombination between chromosomal genes of pathogenic and commensal *Neisseria* species. In *Proceedings of the 8th International Pathogenic Neisseria Conference*, in press.

Suker, J., Feavers, I. M., Achtman, M., Morelli, G., Wang, J.-F. & Maiden, M. C. J. (1994). The *porA* gene in serogroup A meningococci: evolutionary stability and mechanism of genetic variation. *Molecular Microbiology*, in press.

Vázquez, J. A., de la Fuente, L., Berron, S., O'Rourke, M., Smith, N. H., Zhou, J. & Spratt, B. G. (1993). Ecological separation and genetic isolation of *Neisseria gonorrhoeae* and *Neisseria meningitidis*. *Current Biology*, **3**, 567–72.

Wang, J.-F., Caugant, D. A., Li, X., Hu, X., Poolman, J. T., Crowe, B. A. & Achtman, M. (1992). Clonal and antigenic analysis of serogroup A *Neisseria meningitis* with particular reference to epidemiological features of epidemic meningitis in the People's Republic of China. *Infection and Immunity*, **60**, 5267–82.

Wang, J.-F., Caugant, D. A., Morelli, G., Koumaré, B. & Achtman, M. (1993). Antigenic and epidemiological properties of the ET-37 complex of *Neisseria meningitidis*. *Journal of Infectious Diseases*, **167**, 1320–9.

Zhou, J. & Spratt, B. G. (1992). Sequence diversity within the *argF*, *fbp* and *recA* genes of natural isolates of *Neisseria meningitidis*: interspecies recombination within the *argF* gene. *Molecular Microbiology*, **6**, 2135–46.

NATURAL SELECTION AND THE SINGLE GENE

DANIEL E. DYKHUIZEN

Department of Ecology and Evolution, SUNY at Stony Brook, Stony Brook, NY 11794, USA

INTRODUCTION

What causes natural selection? The causes of natural selection can be distinguished from the effects. Natural selection changes gene frequency. So the effect of natural selection is to change gene frequency. The models describing the effects of natural selection are extensively discussed in population genetic textbooks (see Chapter 4 of Hartl & Clark (1989)) and are usually thought of as describing natural selection. However, they do not include the organismal or ecological causes of selection. It is these causes that now need to be understood.

In physics, Newton provided the effect laws of force. He postulated that an object remained unchanged in velocity unless affected by a force. The mathematical description of how a force affects the movement of an object is the core of Newtonian physics. A mathematical description of how force arises, the causes of force, has been a major objective of physics since Newton's time, leading to an understanding of electricity, magnetism, thermodynamics, chemical bonds and the various nuclear forces of modern physics. Clearly, the understanding of the causes of force in physics has been much messier and more difficult than describing the effects. But is it this attempt to understand the causes of force that has led to the many practical uses of physics, the many inventions like electric lights and automobiles that make life easier.

Natural selection in biology is analogous to force in physics. The effect laws were understood first. They were simple and elegant. By understanding the effect laws, major intellectual insights were achieved: the circling of the planets or the dynamics of genetic change. In physics, only later, under various guises were the cause laws worked out, providing major practical benefits. The cause laws were more difficult to formulate with a unification of all the causes in one formula still not achieved. I am suggesting, by analogy, that it is important to study the causes of natural selection, that a working out of the cause laws will provide practical benefits from the one part of biology still thought to be of no strategic importance. Strain degeneration in the pharmaceutical industry is an example of natural selection. An understanding of the causes of that natural selection could provide a means to minimize or eliminate the problem.

I am also suggesting, again by analogy, that working out the cause laws will be difficult, data intensive and will not lead to a single simple unifying theory. Different mathematical formulae will be developed for the description of the evolution of life history characteristics than for the evolution of enzymes. However, as with physics, the hope is that a single theory can encompass all of the causes of natural selection.

As in physics, the cause laws can not be worked out without using the experimental method. There are four methods of doing evolutionary biology (and science in general): the descriptive, the comparative, the theoretical, and the experimental. The first two have provided us with most of what we know about adaptation. But they, necessarily, look at patterns and it is difficult to determine the dynamics of the process from observation. The theoretical is based upon what we know or postulate to be true, extending our knowledge through logic. This means that the major theoretical advances have depended upon our understanding of genetics or upon a postulated fitness surface, usually justified by some sort of optimization argument. The environmental causes of natural selection can not be determined using these methods. The experimental method depends upon the testing of alternative hypotheses through controlled experiments. During experiments, the dynamics of a process can be directly observed, and the various factors that may influence these dynamics can be directly controlled and manipulated so that equations describing the causes of natural selection can be devised. As in other sciences, the experimental studies in evolution are not to recreate nature in the laboratory, but to understand evolutionary dynamics.

There are many ways to study evolution experimentally, not just the method to be described in this paper. One can measure the change in the proportions of various phenotypes over a single generation. One can allow a population to evolve in a particular environment, measuring changes in fitness and life history traits. The evolutionary dynamics at the level of the phenotype are observed and then the genetic basis of the change is sought. This is often very difficult. Another disadvantage of these top-down approaches is more subtle. Since phenotypic changes are followed in terms of fitness, the underlying fitness functions for genetic variation can not be discovered.

The approach advocated in this paper is a bottom-up approach. In this approach strains with known genetic differences are allowed to compete and the fitness difference between the strains is measured. These fitness differences are then ascribed to the known genetic differences after all the proper experimental controls have been done. This measurement has to be done before there is any change to the system as, for example, new advantageous strains arising during periodic selection. In other words, the initial fitness differences between two alleles are measured, given a certain environment and a certain starting ratio. These experiments are analogous to experiments in enzymology where one is measuring the initial kinetics of the system. As with enzymology, a single experiment, a single rate measurement or a single selection coefficient, does not tell one much about the properties of the system. The system can only be understood by systematically changing input variables and uniting the resulting selection coefficients in a systematic way. The disadvantage of the bottom-up approach is that one is limited by the current knowledge in designing the experiments. Thus, top-down experiments are also important to develop a complete understanding of the dynamics of natural selection.

OVERVIEW OF THE EXPERIMENTAL APPROACH

Measuring the initial kinetics of selection between alleles of a single locus in bacteria is conceptually easy. An isogenic pair of strains are made in the laboratory, which differ at a known number of loci. The strains are not allowed to evolve over time as in long term experiments as exemplified by the paper of Lenski (1994). The strains are then mixed and allowed to grow in a constant defined environment. The relative frequencies are followed over time. When the log of the ratio of the strains is plotted against time, the points fall on a straight line with the slope of this line equal to the selection coefficient. This selection coefficient is the difference between the Malthusian parameters of growth. Usually time is plotted in terms of clock time so that the selection coefficient is often expressed as a rate per hour. This can be converted to a rate per generation by dividing by the dilution rate, d. The relative fitness is 1 minus this selection coefficient (Dykhuizen & Hartl, 1983b; Lenski et al., 1991).

Tubes, plates and chemostats all create different environments and are expected to select for different traits (Dykhuizen, 1990). While these experiments can be done in tubes or on plates, the ones described below were done in chemostats. Chemostats are continuous culture devices where fresh medium enters the growth chamber at a constant rate through a port and spent medium and cells are removed through an overflow siphon. The culture is constantly mixed by pumping filtered humidified air through the growth chamber, which is immersed in a water bath to maintain a constant temperature. The details of using chemostats as experimental tools for the study of natural selection have been given elsewhere (Dykhuizen, 1993).

Normally the phenotypic differences between the alleles of interest are difficult to assay, as, for example, enzyme alleles that can only be distinguished by protein electrophoresis. Thus strains are marked with alleles at another locus which must have the following properties: 1. They must be easy to score. 2. The scoring must be 'clean', i.e. very few false positives or false negative. 3. The alleles should be selectively neutral in the environments used in the experiment. 4. The strains used must be asexual so the alleles used for scoring remain linked to the alleles of interest. For sugar-limited chemostats, resistance to phage T4 is such a marker and has been used in all the experiments discussed in this paper. Cells are removed from the chemostat, diluted appropriately to obtain between 100 and 10 000 colonies per plate. Cells are plated with excess phage T4 so that only the resistant cells survive and produce colonies. From the dilution factor and the number of cells on the plate, the number of resistant cells (R) per ml in the chemostat can be calculated. Cells are plated also without T4 to obtain an estimate of the total number of cells per ml in the chemostat. The number of sensitive cells (S) is estimated as the difference between the total and the resistant numbers. If the proportion of R is greater than 0.8, then the estimate in ln(R/S) is biased and the sampling variance is large (Hartl & Dykhuizen, 1981).

It is crucial to understand that, in this experimental system, the inter-actions between the various genotypes and the environment are free to vary, even though the genotypes and the environment are determined by the experimenter. This is true even though the environment is artificial, and it is true even if the genetic variation has been created in the laboratory. Since the interactions between genotype and environment are free to vary, fitness is not under direct control of the scientist. Natural selection differs from artificial selection at this point. In artificial selection, the fitnesses of the phenotypes are determined by the investigator, while in natural selection the environment does this. These experiments are examples of natural selection, not artificial selection.

EXPERIMENTAL EXAMPLES

The three studies reviewed here all used related strains of *E. coli* K12 as the genetic background and energy-carbon limited chemostats with a phosphate buffered minimal medium as the environment. The first study measured the fitness of naturally occurring alleles of five loci, all coding for enzymes in central metabolism. When the data from all the experiments is combined, the resulting pattern of selection coefficients suggests that most of the variation is selectively neutral (Hartl & Dykhuizen, 1985). The second study measured the fitness of various alleles of two genes of the lactose operon, the permease and the β-galactosidase, showing that natural selection can be parameterized (Dykhuizen & Dean, 1990). This study was done in a single environment, a lactose-limited chemostat. The third study used the results of the second study to investigate the relationship between enzyme activity and the intensity of selection along an environmental gradient (Dykhuizen & Dean, 1994). Each of these experiments required multiple competition experiments and a theoretical expectation before insight was achieved.

Selective neutrality of electrophoretic alleles

E. coli is highly polymorphic. Virtually every locus is polymorphic (94%) and the estimates for mean allelic diversity per locus (H) range between 0.343 to 0.542 (Selander, Caugant & Whittam, 1987). H can be thought of as the probability that the alleles of an 'average' locus will be different if two random strains of *E. coli* are chosen. Since *E. coli* is haploid, the polymorphism can not be maintained by heterozygote advantage as was postulated for diploids, suggesting that this polymorphism is selectively neutral and these polymorphisms are maintained by a process of mutation and genetic drift (Kimura, 1983; Milkman, 1973).

A series of experiments (Dykhuizen & Hartl, 1980; Dykhuizen, de Framond & Hartl, 1984*a,b*; Dykhuizen & Hartl, 1983*a*; Hartl & Dykhuizen, 1981) were done to experimentally investigate the selective differences between alleles of five genes: *gnd* (6-phosphogluconate dehydrogenase), *zwf* (glucose-6-phosphate dehydrogenase), *pgi* (phosphoglucose isomerase). *edd* (phosphogluconate dehydratase) and *eda* (2-keto-3-deoxygluconate aldolase). Each allele was transduced into a K12 background from a wild-strain by P1 transduction. The isogenicity of the strains was checked by competing the strains in an environment where the amorphic mutation of the locus is selectively equivalent with wild-type (Dykhuizen & Hartl, 1980). For example, in a chemostat limited by a mixture of ribose and succinate, a strain carrying a null mutation of *gnd* is selectively equivalent to a strain carrying the wild-type allele. So the isogenicity of strains carrying various naturally occurring alleles of *gnd* is confirmed by competing the strains in ribose-succinate-limited chemostats and showing they are equivalent in fitness. Another control done was to show that other environments were selective by competing the null mutation against wild-type and showing there was strong selection against the null mutation. For example, *gnd⁻* is strongly selected against in both gluconate and glucose limited chemostats. The various alleles of *gnd* from nature were competed in both gluconate and glucose limited chemostats and the selective difference measured.

Of the 30 alleles tested from all five loci, 9 were selected under some conditions, the others were selectively equivalent to the best strain all the time. For example, three of the seven *gnd* alleles were selected in gluconate-limited chemostats, but none were selected in glucose-limited chemostats. Thus the conclusion seems to be that 70% of the alleles are neutral and 30% are not.

However, one could postulate that none of the alleles are selectively neutral (Gillespie, 1991). The argument is as follows: The limit of detection in this system is a selection coefficient of about 0.002 h^{-1}. Selection coefficients s of less than this are important in nature because the effective population size N_e of *E. coli* is much larger than a few thousand. When the multiplicand of s and N_e is greater than about 10, selection will dominate over drift and the alleles can not be considered selectively equivalent or neutral.

This was answered (Hartl & Dykhuizen, 1985) by noting that the two resources, glucose and the uronic acids, that are commonly found in nature furnished no evidence for selection, while the rare resource, gluconate, provided most of the cases of selection. The pattern of the association of selection with certain resources is important here, since common resources would apply stronger selection in nature than rare ones. This pattern could only be seen when over 200 individual chemostat experiments were looked at as a block.

The selection in the chemostat should be much more than selection in nature, where the environment is more varied. A statistical estimate of the fitness of the *gnd* alleles from DNA sequence data is $s = 1.6 \times 10^{-7}/$ generation (Sawyer, Dykhuizen & Hartl, 1987), compared to an average of about $s = 2 \times 10^{-2}/$generation from the chemostat date. This estimated selection coefficient is not that different from the inverse of an estimate of the effective population size of 5×10^{7} (Maruyama & Kimura, 1980). Thus, even some of the alleles of *gnd* that are selectively different in chemostats are likely to be selectively neutral in nature. This argument has been given in detail for *lacZ* (Dean, Dykhuizen & Hartl, 1988). Thus the distribution of selection coefficients as they are associated with various sugars provides strong evidence for the neutral theory, that all the alleles are selectively neutral in nature.

While these experiments provide insight into what could be happening in nature, as the above arguments show, they do not settle the argument about what proportion of electrophoretic alleles are selectively neutral in nature. This difficulty with interpretation arises because the experiments do not recreate nature in the laboratory. A better use of this experimental method is to understand evolutionary dynamics, as the next two experiments have done.

The genetics of flux and fitness

What is the relationship between enzyme activity and fitness? How does fitness change as the activity of a particular enzyme changes? What is the relationship between the enzymes of a particular pathway and their effects on fitness? A study was done to investigate these questions (Dean, 1989; Dean, Dykhuizen & Hartl, 1986; Dykhuizen, Dean & Hartl, 1987) and the results have been summarized (Dykhuizen & Dean, 1990).

The pathway used in this study is the lactose pathway. Lactose diffuses through the cell wall via porin pores. It is transported across the cell membrane via the lactose permease and then is cleaved into glucose and galactose by the β-galactosidase. The glucose and galactose are metabolized to provide energy and carbon for cell growth. When a series of mutations of *lacZ*, the enzyme that codes for β-galactosidase, were tested, the plot of enzyme activity against relative fitness formed a rectangular hyperbola. This can be explained theoretically if the relationship between enzyme activity and fitness is broken into two parts: the effect of enzyme activity on the flux of lactose through the pathway and the effect of flux on fitness.

The relationship between enzyme activity and flux is described by two closely related theories of metabolic control (Kacser & Burns, 1973) and biochemical systems theory (Savageau, 1976). The relationship between flux J, for the lactose pathway and enzyme activity E, where E is the ratio of the maximal velocity V_{max}, over the saturation constant k_m, can be written as

$$J = \frac{S}{\dfrac{1}{D} + \dfrac{1}{E_2 \, K_{eq\,1.2}} + \dfrac{1}{E_3 \, K_{eq\,1.3}}}$$

where S is the concentration of lactose in the external medium, D is the unsaturable diffusion constant of the porin pores, E_2 and E_3 are enzyme activities of the permease and β-galactosidase, respectively, and $K_{eq\,1.j}$ is the product of the thermodynamic equilibrium constants from the first to the jth substrate. These equilibrium constants are a function of the change in Gibbs free energy of the reaction, the gas constant, and the temperature. If all components of the system are constant except for E_3, the activity of β-galactosidase, then the equation can be written as

$$J = \frac{S\,E_3}{E_3\,C_1 + C_2}$$

where C_1 and C_2 are constants. This is a rectangular hyperbola.

The relationship between flux and fitness is a linear one. The more lactose a cell can acquire and use when lactose is the growth limiting nutrient, the more fit the cell. Consequently, the theoretical relationship between enzyme activity and fitness agrees with the observed relationship.

This can be extended to two enzymes. The theoretical relationship is

$$\frac{1}{w} = \frac{a}{O_1} + \frac{b}{O_2} + c$$

where w is the relative fitness, a, b and c are constants, and the O_1 and O_2 are the relative enzyme activities for the permease and β-galactosidase, respectively. This relationship was confirmed experimentally (Dean, 1989; Dykhuizen *et al.*, 1987).

The importance of this result is that, if a new lactose operon is isolated and the activities of the permease and the β-galactosidase are experimentally measured, the relative fitness of this operon can be predicted (given that the environment remains the same). Thus, through many individual chemostat experiments a predictive equation relating the change in single genes to fitness has been confirmed and this equation can then be used to predict the fitness of other alleles of these genes. This work has achieved the goal of formally describing the causes of natural selection for a simple metabolic system in a simple environment. From this start greater complexity can be investigated, using more complex metabolic systems or more complex environments.

The ecology of flux and fitness

What happens in a more complex environment than a lactose-limited chemostat, say a chemostat limited by a mixture of lactose and glucose? While a mixture of two sugars is not what an ecologist would consider a complex environment, it is more than twice as complex as a single sugar environment. The mixture of glucose and lactose is the complex environment in this experiment, all other components of the environment remain the same. How does fitness change as the proportions of lactose and glucose change? Or how does fitness change across an environmental gradient? The theorectical prediction is that fitness on two limited resources (lactose and glucose) is equal to the fitness on each resource weighted by the proportion of that resource in the fresh medium entering the chemostat (Dykhuizen & Dean, 1994) or

$$w = w_L \frac{S_{Lo}}{S_{Lo} + S_{Go}} + w_G \frac{S_{Go}}{S_{Lo} + S_{Go}}$$

where w_L is the fitness in a lactose limited chemostat, w_G is the fitness in a glucose limited chemostat, and the S_{Lo} and the S_{Go} are the concentrations of lactose and glucose entering the chemostat. This simple linear equation depends upon a number of assumptions (Dykhuizen & Dean, 1994), including the assumption that the phenotypes remain constant over all environments.

When this equation was tested, two out of the three strain combinations gave the predicted straight line. For the third, fitness was a non-linear function of the proportion lactose (Fig. 1). The resulting line can be described equally well by a rectangular hyperbola or a square root function of the percentage lactose. Even though the lactose operon is constitutive in all the strains, there is still a small amount of residual repression in strain TD4. This strain was selected against on lactose because of a mutation in $lacZ$, the gene for β-galactosidase. This mutation lowers the enzyme activity to 1.1% of wild-type. It was postulated that the repression further lowered the amount of enzyme in this strain such that it was selected against more than predicted. This postulate was tested by adding to the chemostat IPTG (isopropyl-β-D-thiogalactopyranoside). IPTG is a powerful non-metabolizable inducer of the lactose operon which causes full expression regardless of the concentration of lactose in the environment. When it is

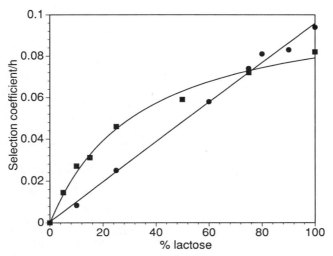

Fig. 1. The relation between the intensity of selection and the percentage of the sugar that is lactose for strains TD2 and TD4. TD4 carries a β-galactosidase with low activity and is selected against on lactose. In the absence of IPTG (Squares), the relation is non-linear. In the presence of 10 μM IPTG (Circles), the relationship is linear.

added to the chemostat medium, any residual amount of regulation will be eliminated. The presence of 10 μM IPTG in the chemostat medium restores the predicted linear relationship. This shows that changing the phenotype, changes the fitness.

There are three determinates of fitness. The first is the genetic determinate. Mutational differences between alleles can affect enzyme activity, the molecular phenotype, thereby affecting flux and fitness. The second is the environmental source of the selection. In this case the relative abundance of lactose and of glucose determines the degree to which a phenotype is targeted by natural selection. The selection changes in a predictable way as the environment changes along the gradient of lactose amounts. The third is the epigenetic determinate. Interactions between the genotype and the environment may modify phenotypes, thereby changing fitness, even though the sources of selection remain unchanged. Lactose as a component of the environment has two separate effects on fitness. It acts as a source of selection while at the same time has an epigenetic effect via gene regulation. The epigenetic effects of lactose are removed by addition of IPTG or by use of constitutive mutations, thus allowing isolation of the selective effects.

DISCUSSION

Clearly, the environment is important and must be considered when the causes of natural selection are studied. Figure 2 shows the interaction

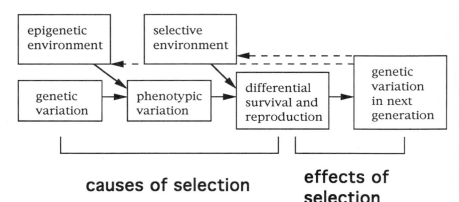

Fig. 2. A cartoon of natural selection, its causes and its effects. The solid arrows show the major direction of cause and effect. The dotted lines are to imply that changes in the genotypes of organisms can change the environment in such a way that these changes effect natural selection. Thus there is a recursive element to evolutionary change which makes predictions of long-term evolutionary trends difficult.

between the three components of natural selection, the genotype, the selective environment and the epigenetic environment. Selection arises as a consequence of the interaction of the phenotype with the selective environment, not the genotype. Why then in evolutionary biology has there been such an emphasis on genetics and so little on the environment?

The history of evolutionary biology has to be considered to understand this emphasis. At the turn of the century, there were many conflicting theories of evolution: Neo-Lamarckian, orthogenetic, saltational, and Neo-Darwinian. This diversity of opinion was created by a paradox pointed out by Fleeming Jenkin (Jenkin, 1867). He showed that natural selection could not be the primary means of evolutionary change if blending inheritance were true. The theories of blending inheritance, such as Darwin's theory of pangenesis (Darwin, 1868), proposed that the intermediate phenotypes of offspring were a result of the blending together the traits of the parents. For example, Darwin proposed that each trait was caused by many gemmules or particles. The children of a white mother and black father would be of intermediate skin colour because the colour of the parents was blended together as paints are blended. As pointed out by Fleeming Jenkin, heritable variation is rapidly lost if blending inheritance is true. The effect of any advantageous mutation is rapidly lost, blended away before selection can act to change the phenotype of the population. Neo-Lamarckian theory was then appealing. Mutations are directed by the environment. Evolution is then driven by mutation, not selection. The saltationists connected large mutational changes with speciation, allowing evolutionary change to happen during speciation with species selection promoting adaptation. The various orthogenetic theories proposed that evolutionary change was directed by an internal working out of a potential. Only the Neo-Darwinians emphasized natural selection as directing evolutionary change and adaptation. Other than August Weismann, not many prominent biologists were Neo-Darwinians.

At the turn of the century, therefore, very few biologists thought that natural selection was an important cause of evolution. But the controversy showed that an understanding of heredity was critical. Mendelian genetics is a particulate, not a blending, theory of heredity. Variation hidden in the children comes out in the grandchildren. There is no loss of variation. A single copy of a gene, an advantageous new mutation, can spread to dominate the entire population and not be blended away. Thus the paradox of Fleeming Jenkin disappears. There is no problem with natural selection being the major cause of adaptive evolution. Now we assume that natural selection drives evolution. The modern synthesis, the evolutionary theory we teach today, rests upon the insights of population genetics. Thus it is not surprising that we biologists, particularly we evolutionary biologists, are fascinated with genetics. It has given us so much. However, genetics is not enough, not even population genetics. A systematic study of the environmental causes of natural selection are now required.

Why is it that population genetics is not enough? To create population genetics, certain simplifying assumptions about the environment had to be made. The causal agents in population genetics are mutation, migration, effective population size and natural selection along with mixis or recombination if multiple loci are considered. These are the agents which cause changes in gene frequency in populations. The change in gene frequency is then the agent of evolutionary change. Mutation, migration, effective population size and natural selection are the primitive terms, the assumed initial agents of evolutionary change in population genetics. How they arise or what causes them is not considered part of population genetics. Thus consideration of the environment and how it effects evolution was not considered part of population genetics.

The study of the causes of mutation and mixis are subfields of genetics: mutation research and recombinational genetics respectively. The study of the causes of migration and effective population size or demography have traditionally been subfields of population ecology. The study of the causes of natural selection, the most important cause of evolutionary change, has been variously called ecological genetics and evolutionary ecology. Much work has been done, particularly related to mimicry and visual predation (Brower, 1988; Kettlewell, 1973) and to life history strategies (Clutton-Brock, 1988), but also on selection for differences in enzyme characteristics (Hochachka & Somero, 1984) and the effects of environmental stress on natural selection (Hoffmann & Parsons, 1991). The research on the causes of natural selection is scattered. There should be a recognizable subfield of evolutionary biology whose goal is to understand the causes of natural selection. To do this, biology will have to focus once again on how the organism responds to its environment, to the ecology of natural selection.

ACKNOWLEDGEMENTS

I thank Rich Lenski and the Center for Microbial Ecology for hospitality while I wrote this paper. I thank all those that have worked on these projects or discussed them with me, especially Daniel L. Hartl, Tony Dean, Pedro Silva, Lin Chao, Rich Lenski, Dave Guttman and Ing-Nang Wang. This is contribution No. 902 from Graduate Studies in Ecology and Evolution, State University of New York at Stony Brook.

REFERENCES

Brower, L. P. (ed.). (1988). *Mimicry and the Evolutionary Process*. University of Chicago Press, Chicago.

Clutton-Brock, T. H. (ed.). (1988). *Reproductive Success*. University of Chicago Press, Chicago.

Darwin, C. (1868). *The Variation of Plants and Animals under Domestication*. Murray, London.

Dean, A. M. (1989). Selection and neutrality in lactose operons of *Escherichia coli*. *Genetics*, **123**, 441–54.

Dean, A. M., Dykhuizen, D. E. & Hartl, D. L. (1986). Fitness as a function of β-galactosidase activity in *Escherichia coli*. *Genetic Research, Cambridge*, **48**, 1–8.

Dean, A. M., Dykhuizen, D. E. & Hartl, D. L. (1988). Fitness effects of amino acid replacements in the β-galactosidase of *Escherichia coli*. *Molecular Biology of Evolution*, **5**, 469–85.

Dykhuizen D. & Hartl, D. L. (1980). Selective neutrality of 6PGD allozymes in *E. coli* and the effects of genetic background. *Genetics*, **96**, 801–17.

Dykhuizen, D. E. (1990). Experimental studies of natural selection in bacteria. *Annual Reviews of Ecology Systems*, **21**, 373–98.

Dykhuizen, D. E. (1993). Chemostats used for studying natural selection and adaptive evolution. *Methods Enzymology*, **224**, 613–31.

Dykhuizen, D. E., de Framond, J. & Hartl, D. L. (1984a). Potential for hitchhiking in the *eda–edd–zwf* gene cluster of *Escherichia coli*. *Genetics Research, Cambridge*, **43**, 229–39.

Dykhuizen, D. E., de Framond, J. & Hartl, D. L. (1984b). Selective neutrality of glucose-6-phosphate dehydrogenase allozymes in *Escherichia coli*. *Molecular Biology Evolution*, **1**, 162–70.

Dykhuizen, D. E. & Dean, A. M. (1990). Enzyme activity and fitness: evolution in solution. *Trends Ecology Evolution*, **5**, 257–62.

Dykhuizen, D. E. & Dean, A. M. (1994). Predicted fitness changes along an environmental gradient. *Evolution Ecology*, **8**, 1–18.

Dykhuizen, D. E., Dean, A. M. & Hartl, D. L. (1987). Metabolic flux and fitness. *Genetics*, **115**, 25–31.

Dykhuizen, D. E. & Hartl, D. L. (1983a). Functional effects of PGI allozymes in *Escherichia coli*. *Genetics*, **105**, 1–18.

Dykhuizen, D. E. & Hartl, D. L. (1983b). Selection in chemostats. *Microbiology Review*, **47**, 150–68.

Gillespie, J. H. (1991). *The Causes of Molecular Evolution*. Oxford University Press, New York.

Hartl, D. L. & Clark, A. G. (1989). *Principles of Population Genetics*, 2nd edn. Sinauer Associates, Sunderland, Massachusetts.

Hartl, D. L. & Dykhuizen, D. E. (1981). Potential for selection among nearly neutral allozymes of 6-phosphogluconate dehydrogenase in *Escherichia coli*. *Proceedings of the National Academy of Sciences, USA*, **78**, 6344–8.

Hartl, D. L. & Dykhuizen, D. E. (1985). The neutral theory and the molecular basis of preadaptation. In *Population Genetics and Molecular Evolution* (T. Okta & K. Aoki eds.), pp. 107–124. Japan Scientific Societies Press, Tokyo.

Hochachka, P. W. & Somero, G. N. (1984). *Biochemical Adaptation*. Princeton University Press, Princeton.

Hoffmann, A. A. & Parsons, P.A. (1991). *Evolutionary Genetics and Environmental Stress*. Oxford Science Publications, Oxford.

Jenkin, F. (1867). The origin of species. *New British Review*, **46**, 277–318.

Kacser, H. & Burns, J. A. (1973). The control of flux. *Symposia of the Society of Experimental Biology*, **27**, 65–104.

Kettlewell, B. (1973). *The Evolution of Melanism: The Study of Recurring Necessity*. Clarendon Press, Oxford.

Kimura, M. (1983). *The Neutral Theory of Molecular Evolution*. Cambridge University Press, Cambridge.

Lenski, R. E. (1994). Evolution in experimental populations of bacteria. In *Population Genetics of Bacteria* (S. Baumberg, P. Young, E. Wellington & J. Saunders eds.), pp. in press. Cambridge University Press, Cambridge.

Lenski, R. E., Rose, M. R., Simpson, S. C. & Tadler, S. C. (1991). Long-term experimental evolution in *Escherichia coli*. I. Adaptation and divergence during 2,000 generations. *American Naturalist*, **138**, 1315–41.

Maruyama, T. & Kimura, M. (1980). Genetic variability and effective population size when local extinction and recolonisation of subpopulations is frequent. *Proceedings of the National Academy of Sciences, USA*, **77**, 6710–14.

Milkman, R. (1973). Electrophoretic variation in *Escherichia coli* from natural sources. *Science*, **182**, 1024–6.

Savageau, M. A. (1976). *Biochemical System Analysis: A Study of Function and Design in Molecular Biology*. Addison-Wesley, Reading, Massachusetts.

Sawyer, S. A., Dykhuizen, D. E. & Hartl, D. L. (1987). Confidence interval for the number of selectively neutral amino acid polymorphisms. *Proceedings of the National Academy of Sciences, USA*, **84**, 6225–8.

Selander, R. K., Caugant, D. Q. & Whittam, T. S. (1987). Genetic structure and variation in natural populations of *Escherichia coli*. In Escherichia coli *and* Salmonella typhimurium: *Cellular and Molecular Biology* (F. C. Neidhardt, J. L. Ingraham, K. B. Low, B. Magasanik, M. Schaechter & H. E. Umbarger, eds.), pp. 1625–1648. American Society for Microbiology, Washington, DC.

CONDITIONS FOR THE EVOLUTION OF MULTIPLE ANTIBIOTIC RESISTANCE PLASMIDS: A THEORETICAL AND EXPERIMENTAL EXCURSION

BRUCE R. LEVIN

Department of Biology, Emory University, 1510 Clifton Road, Atlanta, GA 30322, USA

INTRODUCTION

Antimicrobial chemotherapy has been one of the most significant (if not *the* most significant) achievements of interventive medicine of this century. Unfortunately, and from an evolutionary perspective predictably, for many pathogenic microbes the era of successful intervention with antibiotics and other chemotherapeutic agents may be coming to an end. As a consequence of antibiotic-mediated selection, strains resistant to one, and commonly more than one of these compounds have been replacing their sensitive ancestors. Despite occasional glimmers of hope (Nowak, 1994), there is no compelling evidence that the frequencies of these resistant strains will wane in the near future (Cohen, 1992, Neu, 1992). Moreover, the pace at which new and modified, non-toxic, antimicrobial agents are being discovered and/or developed has been declining (Neu, 1992; Travis, 1994).

A good deal of antibiotic resistance, particularly that for the Gram negative bacteria, is encoded by genes carried on plasmids. These extra-chromosomal elements commonly determine resistance to four or more antibiotics, and are capable of infectious transmission between lineages of bacteria of the same species and, for some plasmids, between phylogenetically quite different species of bacteria (Falkow, 1975). While we cannot say with absolute conviction that the multiple antibiotic resistance, R-plasmids that currently plague us did not exist prior to the human use of antibiotics, there is support for this interpretation. Despite their current ubiquity, multiple antibiotic resistance plasmids had not been found in bacteria isolated before the antibiotic era (Hughes & Datta, 1983; Datta & Hughes, 1983) and the number of resistance genes carried on individual R-plasmids continued to increase as new antibiotics came into common use (Falkow, 1975). On the other hand, it seems almost certain that the ancestors of contemporary R-plasmids and the transposable elements responsible for the movement of resistance genes to and from plasmids existed prior to the human use of antibiotics (Hughes & Datta, 1983; Datta & Hughes, 1983; Bukahari, Shapiro & Adhaya, 1977). Although the evidence is less compelling, it also seems reasonable to believe that the resistance genes were also present before the antibiotic era, see, for example, Beneveniste and Davies (1973) and Davies (1994). In this interpretation, the most plausible mechanism for the evolution of contemporary R-plasmids is via the coalescence of preexisting plasmids, transposons and antibiotic resistance genes with selection for the evolution of these elements due to the human use of antibiotics.

However, even if correct, this interpretation does not provide a sufficient explanation of the nature and extraordinary pace of R-plasmid evolution, or offer proscriptions about what we can do to control this evolution. What are the genetic processes and ecological conditions responsible for the evolution of multiple antibiotic resistance plasmids? Why are these resistance genes on plasmids rather than chromosomes? Why are there relatively few plasmids carrying multiple resistance genes rather many different plasmids carrying single resistance genes? Is there anything we can do to prevent the evolution of new R-plasmids, or control the rate of spread and the persistence times of existing ones?

I review, here, the results of studies we have done to address some of these questions. A good deal of the material I present has already been published. However, a substantial fraction is new and much of what follows is a report of work in progress. In reference to these endeavours, I use the pronouns 'we' and 'our'. Most of this work has been in collaboration with other investigators, particularly, Richard Condit, Peter Sykora and Carol Laursen.

I divided this presentation into three major sections. The first is theoretical descriptions and diagrammatic presentations of the mathematical and computer simulation models we have developed for R-plasmid evolution and the results of our analyses of the properties these models. The second is experimental, a report of the results of the studies we have done (and are currently doing) on R-plasmid evolution with *E. coli* and its plasmids and transposons. The final section is a discussion of the virtues, implications and limitations of our theoretical experimental excursions into R-plasmid evolution.

THEORY

So far, we have considered three processes, scenarios, by which plasmids carrying multiple antibiotic resistance genes may be generated and evolve in bacterial populations. First, when there is selection for the infectious transfer of resistance genes (transposons) carried on the chromosome. Secondly, when selected, complementary resistance genes are on separate plasmids that segregate at relatively high rates when infecting the same cell. Thirdly, when there is selection for the transfer of complementary resistance genes that are initially on separate plasmids.

Scenario 1. Selection for the infectious transfer of chromosomal resistance genes

When the resistance genes are originally on the chromosome, a plausible course for the evolution of multiple antibiotic resistance is analogous to that of a 'mating-out' protocol. To 'move' transposons carrying antibiotic resistance genes from the chromosome to a conjugative plasmid *in vivo*; (i) one infects the transposons-bearing donor with the self-transmissible plasmid to receive the transposon, (ii) mates those donors with a recipient not carrying that plasmid, and (iii) using the appropriate media, selects for cells of the recipient type with plasmids carrying the resistance transposon, transconjugants. This 'mating out' procedure can be used to simultaneously or sequentially move any number of chromosome-borne resistance transposons (or, in some cases, chromosomal genes) to a plasmid.

In Fig. 1, we present a diagram of this model for the evolution of multiple antibiotic resistance plasmids by this mating out process. The donor population is of two cell types, those with the transposon on the chromosome, but not on the plasmid, D1, and those where a copy of the transposon has moved to the plasmid, D*. Transposon and plasmid-free recipients, R, are of a genotype different from the donor that can receive the plasmid with or without the transposon, transconjugants populations, T1 and T*, respectively. Here, and in the models that follow, the names of the different cell

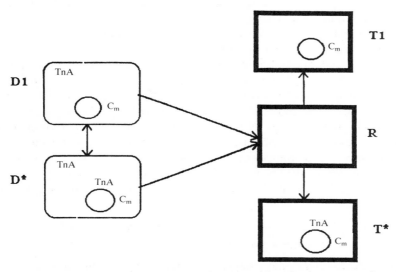

Fig. 1. Model of Scenario 1. R-Plasmid evolution by transposition of resistance genes from the chromosome to a conjugative plasmid and selection of the infectious transfer of those genes. D, donor; R, recipient; T, transconjugant, C_m, chloramphenicol-resistant plasmid; TnA, transposon; *denotes presence of plasmid-borne transposon.

lines (states) are also used to specify the densities (bacteria per ml) of these populations.

The mathematical incarnation of this diagram, and that of the models that follow, is a series of coupled, ordinary differential equations, one for the rates of change in the densities of each of the component populations, D1, D*, R, T1, and T*, in this model, and one for the changes in the concentration of a limiting resource r µg/ml. The parameters of this model are: (i) the resource concentration dependent growth rate, $\Phi(r)$, (a mono-tonic increasing function of r), (ii) the relative growth rates (fitnesses) of the component bacterial populations, a_is $(0 < \alpha_i < 1)$, (iii) the rate of transpo-sition (recombination) λ and, (iv) the conjugative transfer rate parameter γ (Levin & Stewart, 1977, Simonsen et al., 1990). For this chromosome to plasmid scenario, we neglect loss of the plasmids by vegetative segregation.

In this model and those that follow, we assume an 'equable' (chemostat) habitat (Stewart & Levin, 1973) where resources from a reservoir where they are maintained at a concentration C µg/ml enter and are removed at a constant rate, ρ per hour, which is equal to the rate at which organisms and wastes are removed. For the growth function, we employ a Monod (1950) equation, $\Phi(r) = Vr/(K+r)$, where V is the maximum growth rate and K the resource concentration when the growth rate is half the maximum. We assume a constant amount, e µg/ml of the resources, 'conversion efficiency' is required to produce a single new cell (Stewart & Levin, 1973). Interested readers are welcome to write to me for copies of the equations for this and the following models and the computer programs used for the numerical analysis of their properties.

To illustrate the properties of this model, we use numerical solution to the equations for this model generated by SOLVER (a PASCAL Template, Blythe et al., 1990). In Fig. 2, we consider two cases for the evolution of a resistance plasmid in this way; (i) when there is intense selection for transconjugant with the evolved plasmid, T*, Fig. 2(a), and (ii) when the intensity of this selection is weaker and the rates of plasmid transfer and transposition lower, Fig. 2(b). Each of these simulations was initiated with only the donor and recipient, D1 and R, populations. With both sets of parameters, transconjugants carrying the evolved plasmid, T*, ascend and eventually dominate. The rate of this ascent is greatest when the intensity of selection for T* and the rates of transposition and plasmid transfer are high.

Scenario 2. Selection for complementary resistance genes and high rates of segregation in cells carrying multiple plasmids

The model for this scenario depicted in Figure 3 is a simplified version of that in Condit and Levin (1990). We consider two plasmids and a single host genotype which can be in five different states with respect to their carriage of plasmids: (i) plasmid-free, R, (ii) carrying one plasmid, D1, (iii) carrying the other plasmid, D2, (iv) carrying both plasmids, D12, and (v) carrying a recombinant (or cointegrate) plasmid with both resistance genes, D*. For simplicity, we neglect cells carrying other combinations of multiple plasmids, 1-*, 2-*, and 1-2-*. The D1, D2 and D* populations are converted into recipients, R, when they lose their plasmid by vegetative segregation. Segregants of the D12 population enter the D1 or D2 populations at equal

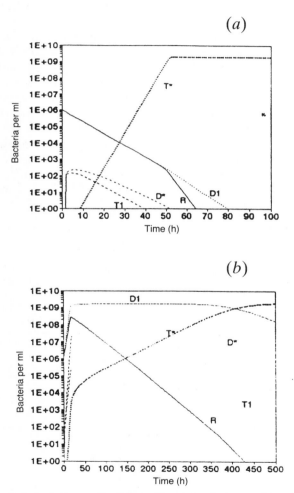

Fig. 2. Scenario 1 simulation results. R-Plasmid evolution by transposition of resistance genes from the chromosome to a conjugative plasmid and selection of the infectious transfer of those genes. In this and the following simulations: $V = 0.7$, $K = 4$, $e = 5 \times 10^{-7}$, $c = 1000$, $\rho = 0.2$. (a) $\alpha_R = \alpha_{D1} = \alpha_{D*} = 0.95$, $\alpha_{T*} = 0.00$, $\gamma = 10\text{–}10$, $\lambda = 10^{-4}$, $\tau = 0$. (b) $\alpha_R = 0.125$, $\alpha_{D1} = \alpha_{D*} = 0.25$, $\alpha_{T*} = 0.00$, $\gamma = 10\text{–}11$, $\lambda = 10^{-5}$, $\tau = 0$.

rates. We assume two segregation rates, τ and τ_{12}, for cells carrying single plasmids and those carrying two plasmids, respectively. We allow for conjugative transfer of the plasmid from the D1, D2, and D* to the R population and from D1 to D2, to form D12 cells.

From our numerical analysis of the properties of this model, we used SOLVER, and initiated all simulations with only the D1 and D2 cell populations. In Fig. 4(a), we illustrate what occurs when the rates of segregation from cells carrying two plasmids, D12, is much greater than that from cells carrying just one, D1, D2 and D, $\tau_{12}, \gg \tau$ and where there is intense selection favouring cells with both plasmid-borne resistance markers, $\alpha_{12} = \alpha_* < \alpha_1 = \alpha_2 = \alpha_R$. In a relatively short time cells with the recombinant plasmid, D*, dominate. The rate of ascent of this R* population depends on the segregation rate, τ_{12}. This can be seen in Fig. 4(b), where we use the same selection regime but vary τ_{12}. When τ_{12}, is on the same order as τ, as it would be for compatible plasmids, the D* population, like D12, increases initially due to selection. However, following its initial ascent the D*/D12 ratio continues to increase, but only slowly. The increase in this ratio is an artifact of our model allowing the formation of the recombinant, *, plasmid from cells carrying separate incompatible plasmids, D12, but not the inverse, D* → D12. If we also allowed for that dissociation, there would be a stable equilibrium with a constant D*/D12 ratio.

Scenario 3. Selection for the infectious transfer of complementary resistance genes carried on separate conjugative plasmids

From the preceeding work it seems apparent that, if all else were equal and the plasmids carrying the reciprocal resistance genes did not segregate at a

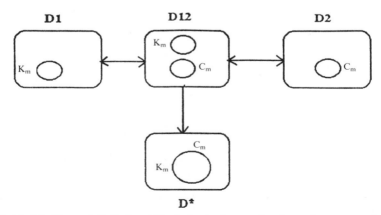

Fig. 3. Model of Scenario 2. R-plasmid evolution by the transfer of complementary resistance genes between plasmids and relatively high rates of segregation in cells carrying multiple plasmids.

higher rate than those carrying single plasmids, it would take some time before the multiply resistant plasmids would evolve to dominance. Moreover, if the rate at which D* dissociated to D1 and D2, was greater than the rate they are formed. λ, then the D*/D12 ratio would decline and the multiply resistant plasmid, *, would not evolve. On first consideration, it

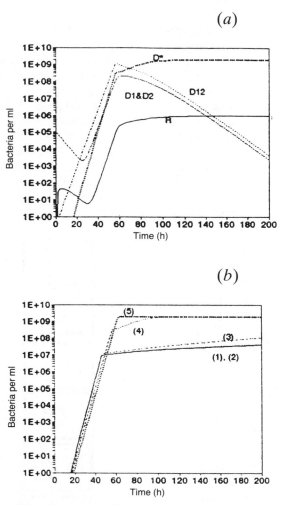

Fig. 4. Scenario 2, simulation results. R-Plasmid evolution by the transfer of complementary resistance genes between plasmids and relatively high rates of segregation in cells carrying multiple plasmids. (a) $\alpha_R = \alpha_{D1} = \alpha_{D2} = 0.95$, $\alpha_{D12} = \alpha_{T*} = 0.00$, $\tau = 0.0001$, $\tau_{12} = 0.01$, $\gamma = 10^{-11}$, $\lambda = 10^{-4}$ (b) $\alpha_R = 0.125$, $\alpha_{D1} = \alpha_{D*} = 0.25$, $\alpha_{T*} = 0.00$, $\gamma = 10^{-11}$, $\lambda = 10^{-5}$, $\tau = 0.0001$:
Lines (1) $\tau_{12} = 0.0001$, (2) $\tau_{12} = 0.001$, (3) $\tau_{12} = 0.01$, (4) $\tau_{12} = 0.10$, (5) $\tau_{12} = 0.40$.

may seem that one way to get around this is to allow for the transfer of the plasmids to a recipient population, the idea being that a single evolved, multiple resistant plasmid would transfer to recipients at a higher rate than two separate plasmids carrying those resistance genes. If, in addition, transconjugants expressing both resistance genes have the highest fitness, then the combination of higher rates of infectious transfer and antibiotic-mediated selection should favour the evolution of the recombinant, *, plasmid. To explore this scenario in a more quantitative way, Peter Sykora and I developed and analysed the properties of a mathematical model of this process (Sykora & Levin, unpublished observations).

In the minimal form of our model, there are two genetically distinct bacterial populations, a donor, D and a recipient, R, and three plasmids, the original plasmids carrying single resistance genes, 1 and 2, and a recombinant or cointegrate plasmid carrying both resistance genes, *. Even though we neglect cells carrying the 1-*, 2-* and 1-2-*, plasmid combinations, there are a total of 13 bacterial populations (Fig. 5). We assume that the evolved plasmid, *, is generated in two ways: first, at a rate λ, in the course of cell division by bacteria carrying both plasmids, D12 and T12, and secondly, in the course of plasmid transfer between cells carrying two plasmids, D12 or T12, and recipients that are plasmid-free. In these matings, fractions of recipients, x_1, x_2, x_{12}, and x_* receive, plasmid 1, plasmid 2, both plasmids

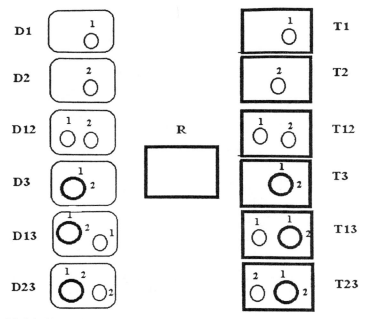

Fig. 5. Model of Scenario 3. R-plasmid evolution by the transfer of complementary resistance genes between pairs of plasmids and selection for the infectious transfer of those genes.

(1 and 2) and the recombinant (cointegrate), *, plasmid respectively, where $x_1 + x_2, + x_{12}, + x_* = 1$. The rates of plasmid transfer are governed by rate constants, γ for the transfer of the plasmids from the cells carrying the original plasmids, 1, 2, and 12 and γ_* for transfer from cells carrying the recombinant or cointegrate plasmid, D^* and T^*.

For the numerical solution to these equations for this model, we used a single step Euler method and a FORTRAN 77 program. In Fig. 6(a), we demonstrate that, by itself, the infectious transfer-selection scheme described above is unlikely to account for the evolution of multiple resistance plasmids. In these runs, the T12 and T* populations are equally fit, have a

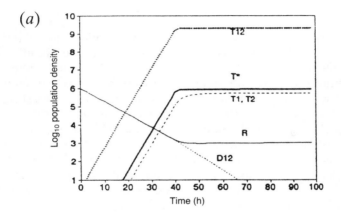

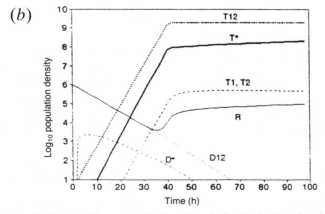

Fig. 6. Scenario 3 simulation results. R-plasmid evolution by the transfer of complementary resistance genes between pairs of plasmids and selection for the infectious transfer of those genes. In both sets of runs $a_R = a_{D1} = a_{D2} = a_{D12} = a_{D^*} = a_T = a_{T1} = a_{T2} = 0.95$. $a_{T12} = a_{T^*} = 0.00$, $\tau = 0.0001$. (a) $x_1 = x_2 = x_{12} = 0.333333$, $x_* = 0.000001$, $\lambda = 0.000001$, $\gamma_D = 10^{-11}$, $\gamma_* = 10^{-9}$ (b) $x_1 = x_2 = x_{12} = 333$, $x_* = 001$, $\lambda = 0.001$, $\gamma_D = \gamma_* = 10^{-11}$.

marked selective advantage over remaining 11 cell lines, and the rate constant of transfer from T* and D*, γ_*, is two orders of magnitude greater than γ. Nevertheless, the T* population remains a small minority. The reason for this is that infectious transfer plays a relatively small role in this system because of the low densities of potential host due to selection against the R cell line. The only way we have been able to achieve substantial frequencies of the T* relative to T12 in this model, is to assume that recombinant or cointegrate plasmids are generated at relatively high rates, $\lambda = 10^{-3}$, and $x_* = 10^{-3}$ (Fig. 6(b)).

EXPERIMENTS

Overview

The experiments we performed were designed to explore the validity and limitations of the above models ('scenarios') of R-plasmid evolution considered in the preceeding section. Using *E. coli* K-12 and some of its plasmids and transposons, we independently estimated the parameters of these models. In chemostat and/or transfer culture, we established conditions similar to those specified in the models and, by selective plating, followed the changes in the densities of bacteria of different plasmid-bearing states. We tested the stability of the 'evolved', *, plasmids by mating cells carrying them with a plasmid-free recipient and by culturing the resulting transconjugants for a number of generations in the absence of selection. Following this excursion into old-fashioned, *in vivo*, genetics we examined the nature of the evolved plasmids *in vitro* with restriction gels.

To date, all of our experiments on the population dynamics and evolution of multiple antibiotic resistance plasmids have been devoted to the second and third scenarios considered in the preceding section. The first scenario, where the R-plasmids are generated from resistance transposons carried on the chromosome seemed too straight forward to bother to actually do the population dynamic-evolution experiments. We have successfully employed the analogous mating-out protocol to move transposons to plasmids. On the other hand, there may well be surprises (things to learn) when we actually perform the experiments exploring this scenario; there almost always are, and I may have experiments of this type to report by the time of the symposium.

Multiple antibiotic resistance from pairs of incompatible plasmids

Our research on this aspect of the problem has already been published (Condit & Levin, 1990). In the experimental portion of this endeavour, we used *E. coli* K-12 and variants of R100-1drd (C$_m$), and Inc FII plasmid that is permanently derepressed for conjugative pili synthesis. Using strains of *E. coli* carrying Tn3 or Tn5 on the chromosome with a mating-out procedure, we constructed three different R100-1s: R5A (C_m K_m), R5H (K_m), and R3 (A_m). By mating and selective plating, we constructed strains carrying pairs of R100 plasmids with complementary resistance markers, the D12 populations. For example, on agar containing Chloramphenicol and Ampicillin, we selected for transconjugants carrying both R100 (C_m K_m) R5A and R100(A_m).

For each of these plasmids, individually, and for most pairs, we estimated the growth rates V, segregation rates τ, and the rate constants of transfer γ to plasmid-free recipients and to recipients carrying one or another R100 variant. These parameter estimation and experimental population cultures were maintained in 50 ml Erlenmeyer flasks containing 10 ml of Luria broth and maintained at 37°C with shaking at approximately 200 rpm.

We initiated the population experiments with bacteria carrying pairs of incompatible plasmids, the D12 cell lines taken directly from the agar used to select for them. The population experiments were performed in medium containing antibiotics selecting for both plasmid-borne resistance markers, e.g. the R5A(Kan Cam)- R3(Amp) cells in medium with Chloramphenicol and Ampicillin. These cultures were allowed to achieve stationary phase at which time a fraction, 2.5×10^{-5}, of the culture was transferred to fresh medium. At each transfer, samples were taken and plated on non-selecting agar and agar containing both antibiotics.

For all pairs of R100 variants examined in our population studies, the ratio of cells growing on the agar containing the two antibiotics to total cell density increased rapidly with time (Fig. 7). We attributed the higher rate of increase of the frequency of stable Amp-Kan (the R5H-R3 pair) resistant cells, relative to that of the Amp-Cam (from the R100-1-R3 combination) to detoxification of both the ampicillin and chloramphenicol by the β-lactamase and Chloramphenicol acetyltransferase produced by cells carrying the latter two plasmids. This inconvenience of biological reality aside, the rate of increase in the frequency of cells with stable inheritance of the resistance markers was gratifyingly within the range anticipated from our models with the estimated segregation rate, $\tau \sim 0.4$.

Following the ascent of the stable multiply resistant cell lines, we isolated cells from the plates containing two antibiotics and tested them for the stable transfer of both resistance genes. Three results obtained; (i) in the majority of cases, both resistance genes were transferred, (ii) less commonly, only single resistance genes transferred, and (iii) in one case, both markers were stably inherited, but we could not transfer the plasmid by conjugation.

To ascertain the nature of the evolved plasmids, we extracted plasmid DNA from the potentially evolved plasmids, digested that DNA with the restriction endonuclease BAMH1, and electrophoresed it on 0.7% agarose gels. The restriction fragment patterns of the DNA extracted from plasmid that transferred both markers were different from those of the DNA extracted from the parental plasmids. Moreover, almost all of 'evolved'

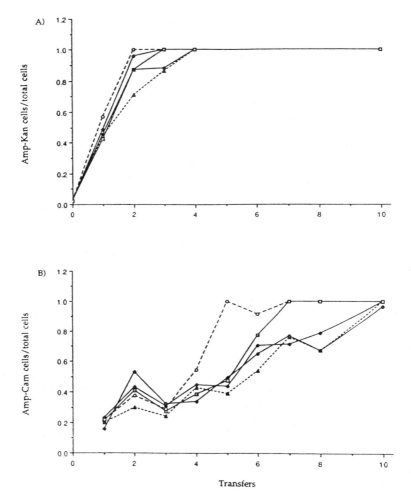

Fig. 7. Experimental results for Scenario 2: R-plasmid evolution by the transfer of complementary resistance genes between plasmids and relatively high rates of segregation in cells
carrying multiple plasmids. Changes in the population density of doubly resistant cells as a
fraction of total cell density. Simultaneous selection for resistance to two antibiotics determined
by genes initially carried on the separate plasmids. Each line represents the results of a single
experiment. Five independent experiments were performed for each case.
(a) Rise of Amp Kan resistant cells in an experiment initiated with bacteria carrying the R100-1
variants, R5H(Kan) and R3(Amp) in a medium containing both Ampicillin and Kanamycin.
(b) Rise of Amp Cam resistant cells in an experiment initiated with bacteria carrying the R100-1
variants, R100-1(Cam) and R3(Amp). For aesthetic reasons, when the estimated density of
doubly resistant bacterial exceeded the total cell density, that ratio was rounded off to 1.0
(Figure from Condit & Levin, 1990).

plasmid generated in independent experiments initiated with the same combinations of parental plasmids differed from each other. We interpreted this as indicating that there are many ways by which multiply resistant plasmids could be generated by the movement of transposons from one plasmid to another. In those cases where the plasmid was isolated at the end of the experiment carried only one resistance gene, the restriction fragment pattern of the DNA of the plasmid isolated at the end of the experiment was identical to that of one of the parental plasmids. We interpreted this to indicate that the stable inheritance of both resistance genes was achieved by the movement of one plasmid-borne transposon to the host chromosome. In the case where the reciprocal resistance genes were stably inherited but the plasmid did not transfer, the restriction fragment pattern of the plasmid DNA extracted differed from that of both parental plasmids. We believe this evolved plasmid was generated by the movement of a resistance transposon into the Tra (conjugative transfer) region of the plasmid.

In these experiments, we used combinations of R100-1 variants because they provided the extreme case of high rates of segregation in cells carrying multiple plasmids, i.e. replicons that are identical save for the resistance transposon. In this study, Rick Condit and I also examined this scenario using pairs of compatible plasmids, combinations of R100-1, R390 (Amp Cam) and Sa(Cam Kan), Inc FII, IncN and IncW plasmids, respectively. After ten transfers with the experimental protocol described above, we found no evidence for the evolution of stable, multiply resistant plasmids from any of the three combinations of compatible plasmids tested, R5H-R390, Sa-R390 and Sa-R3. This negative result is essentially what would be anticipated from the segregation model considered in the preceeding theory section (Scenario 2).

Multiple resistance by infectious transfer from pairs of compatible plasmids

In an effort to evaluate the Scenario 3 model for the evolution of multiple resistance from compatible plasmids, Peter Sykora and I performed a series of parameter estimation and population dynamic experiments with *E. coli* K-12 carrying both the IncFII plasmid R100-1 (Tet Cam) and a variant of the IncW plasmid, Sa, denoted SaH (Kan-Neo) (Sykora & Levin, unpublished observations). In our population experiments, we established two protocols by which selection would favour transconjugants that carry complementary genes borne on these two replicons: (i) by using a Nal^r recipient with a Nal^S donor in glucose-limited minimal medium containing Naladixic acid (NAL), kanamycin (KAN) and chloramphenicol (CAM), and (ii) by using a Lac+ recipient and a Lac$^-$ donor in minimal medium that contained both glucose and lactose as limiting carbon sources as well as KAN and CAM. The antibiotics we used in the experimental media were at concentrations that were relatively low but inhibitory for sensitive cells; 6.25, 12.5 and 5 μgm per ml, respectively for KAN, CAM and NAL. The concentrations we employed in the selecting agar were 12.5, 25 and 20 μgm per ml, respectively.

We inoculated overnight cultures of donors into flasks containing the selecting minimal medium and, at periodic intervals, plated samples on to agar to estimate the total cell density and the density of transconjugants carrying the complementary resistance markers. In the NAL, KAN, CAM medium, only transconjugants carrying both the R100 and SaH resistant markers were able to grow well, and it took approximately 50 hours before the culture reached stationary phase. (Under these culture conditions in the absence of antibiotics, stationary phase is achieved within 8 hours.) In Fig. 8, we present the results of three independent experiments of this type for the NAL, KAN, CAM medium. To distinguish between Kan Cam transconjugants carrying the evolved, *, plasmid and transconjugants of this phenotype carrying these markers on separate plasmids, T12, we purified the cells from the selecting agar, extracted their plasmid DNA, digested that DNA with EcoRI and electrophoresed it on 0.7% agarose gels.

The restriction patterns of the plasmids isolated from these independent cultures were of only two types. The first was that of the original pair of

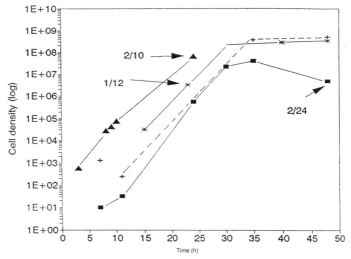

Fig. 8. Experimental results for Scenario 3. R-plasmid evolution by the transfer of complementary resistance genes between pairs of plasmids and selection for the infectious transfer of those genes. Changes in the density of Cam Kan resistant transconjugants for four independent experiments. Each of these experiments was initiated with Nal[S] cells carrying two plasmids, R100-1(Cam) and SaH(Kan) and a Nal[r] recipient. Bacteria were maintained in glucose limited minimal medium containing Chloramphenicol, Kanamycin and Nalidixic acid. The ratios, 2/10, 1/2, and 2/24, are the number of cointegrates detected over the total number of colonies examined for cointegrate formation in three experiments. The separate lines indicate the results of these different experiments. (Sykora & Levin, unpublished observations).

R100–SaH plasmid, the second was what we interpret to be a cointegrate between these two plasmids. The ratios on Fig. 8, 2/10, 1/12, and 2/24, are the number of cointegrate plasmids detected over the total screened at that point in these experiments. Substantially higher frequencies of cointegrate plasmids were observed in the experiments performed in glucose–lactose KAN CAM medium. The restriction fragment profiles of the evolved Kan Cam plasmids in these experiments were identical to those in the NAL KAN CAM medium. Thus, unlike the situation that occurred in the experiments with pairs of incompatible plasmids where there were many different evolved plasmid configurations, a seemingly unique plasmid evolved in these experiments with pairs of compatible plasmids.

How did the 'cointegrate', Kan Cam plasmid evolve as rapidly as observed in these experimental cultures? The simulation Peter Sykora and I used for this study, like the compatible plasmid model in Fig. 6, predicted that a multiply resistant plasmid of this type, T*, would only be anticipated to ascend to detectable frequencies in these, if the rate of formation of the cointegrates, either by the cell carrying two plasmids generating them in the course of cell division, $D12 \rightarrow T^*$ and $T12 \rightarrow T^*$, or during infectious transfer from D12 or T12 to R, was in the order of 10^{-3}, a seemingly very high value.

To obtain independent estimates of this cointegration rate, we performed 'conduction' experiments; (i) *in vitro*, we cut out the Tra region from the SaH plasmid, (ii) by transformation, 'put' the ligated, Tra-less SaH into a plasmid-free *E. coli* K-12 and, (iii) by mating put an R100 (Tet Cam) plasmid into the same transformed cell line. We then mated the resulting R100(Tet Cam)-SaH(Tra⁻ Kan) cells with a K-12 recipient with markers that would enable us to select for transconjugants carrying the R100 (Tet Cam) alone and R100(Tet Cam) with the SaH (Tra⁻ Kan) plasmid. Since the SaH plasmid is Tra⁻, the only way the Kan marker can be transferred to the recipient is by some form of recombination or cointegration with R100. Thus, the ratio of the densities of cells growing on CAM–KAN to those growing on CAM alone provides an estimate of the fraction of cells receiving R100–SaH cointegrates. In the minimal medium in which we performed the experiments, we estimated this fraction to be 4.5×10^{-4} (mean of six experiments). This value is within the range that could account for the seemingly rapid increase in the frequency of recombinant (cointegrate) plasmids observed in our population experiments. Eureka!!

DISCUSSION

We interpret the results of this jointly theoretical and experimental study as support for the view that all three of the mechanisms considered for the origin and evolution of multiple resistance plasmids are plausable. *A priori* each of these processes can account for the evolution of R-plasmids and the conditions (parameter values) for this evolution to occur are biologically realistic. In our experiments testing two of these scenarios with experimental populations of *E. coli*, stable multiple resistant plasmids evolved and the dynamics of this evolution was roughly ('gratifyingly') similar to that anticipated by our models. While we did not evaluate the chromosomal transposon-infectious transfer scenario with population experiments, we can readily generate multiple resistance plasmid by the mating-out protocol on which this model is based.

I do not consider any of our theoretical predictions or experimental results to be unanticipated. To me, the most surprising observation was the high rate at which the R100–SaH conintegrate was formed. On the other hand, the results of these experiments do make predictions about the processes responsible for the evolution of R-plasmid that can be evaluated by retrospective procedures with R-plasmids isolated from natural populations. If R-plasmids evolved by a plasmid to plasmid transposition–segregation model similar to that in scenario 2, and this evolution occurred multiple times, there should be considerable variation in the positions of the resistance genes in otherwise identical R-plasmids. I expect this would also be the case if R-plasmids evolved by a chromosome to plasmid transposition–selection for infectious transfer model similar to that presented here as scenario 1. On the other hand, if R-plasmids evolved by a mechanism similar to the cointegration of separate plasmids – infectious transfer mechanism considered here as scenario 3, then we would anticipate contemporary species of R-plasmids to be nearly invariant chimeras derived from plasmids of different incompatibility groups. Moreover, R-plasmids that have recently evolved by this mechanism may have more than one origin of replication or vestiges of multiple origins. There is, in fact, some evidence for contemporary R-plasmids having multiple ancestries and multiple origins, see Sykora (1992) for a review of these data as well as consideration of the implication of cointegration for plasmid evolution at large.

The three scenarios for R-plasmid evolution considered here are certainly not exhaustive. We have not formally considered non-conjugative plasmids or the role of bacteriophage as vectors for the movement of genes and transposons to and from plasmids. Moreover, by assuming that contemporary R-plasmids evolved from existing plasmids, resistance genes and transposons, we have begged the interesting general question of how and why these accessory genetic elements and resistance genes evolved in the first place. Finally, we have only touched on the question of why these resistance genes and transposons are on plasmids rather than chromosomes. I believe the mating-out, scenario 1, process is at least part of the reason these genes are on infectiously transmitted accessory elements rather than chromosomes.

While plasmid-borne genes upon which we focused here may be the main source of antibiotic resistance for some species of bacteria, for others, including major pathogens like *Mycobacterium tuberculosis* and *Streptococcus pneumoniae*, resistance to multiple antibiotics can be attributed to mutation and possibly recombination of chromosomal genes. To date, little or no theoretical or experimental considerations have been made of the evolution of multiple resistance by mutation and recombination of chromosomal genes. From one perspective, it may seem that the development of this kind of theory and performance of these experiments would not be a particularly unique or worthwhile endeavour. To a large extent, traditional population genetic theory for mutation and selection in haploid organisms (Crow & Kimura, 1970) can be applied directly to the problem of chromosomal resistance. And, in accord with this theory, sequential bouts of mutation and selection-mediated by a single antibiotic would be the most parsimonious explanation for the evolution of multiple chromosomal gene resistance. On the other hand, this may not be the unique explanation and other processes may be involved. A common procedure for treating tuberculosis, short-course chemotherapy (Coombs, O'Brien & Geiter, 1990) involves the simultaneous use of multiple antibiotics that require separate mutations for resistance. Nevertheless, virtually untreatable, multiply resistant strains of *M. tuberculosis* have evolved and are increasing in frequency (Frieden *et al.*, 1993, Goble *et al.*, 1993).

ACKNOWLEDGEMENTS

I wish to thank Richard Condit, Peter Sykora and Carol Laursen for their part in producing the fodder for this review. This endeavour has been, and continues to be, supported by a grant from the US National Institutes of Health GM 33782.

REFERENCES

Beneveniste, R. & Davies, J. (1973). Aminoglycoside antibiotic-inactivating enzymes in actinomycetes similar to those present in clinical isolates of antibiotic resistant bacteria. *Proceedings of the National Academy of Sciences, USA,* **7**, 2276–80.

Bloom, B.R. & J.L. Murray. (1992). Tuberculosis: commentary on a reemergent killer. *Science,* **257**, 1055–61.

Blyth, S.P., Gurney, W.S.C., Maas, P. & Nisbet, R.M. (1990). Solver: Turbo Pascal program templates for integrated coupled sets of ordinary differential equations with possible delays and nonlinearities. University of Strathclyde, 107 Rottenrow, Glasgow G4 0NG, Scotland.

Bukhari, A.I., Shapiro, J.A. & Adhaya, S.L., eds. (1977). *DNA Insertion Elements, Plasmids and Episomes.* Cold Spring Harbor Laboratory, Cold Spring Harbor, NY.

Cohen, M.L. (1992). Epidemiology of drug resistance: implications for a post-antibicrobial era. *Science,* **257**, 1050–5.

Condit, R. & Levin, B.R. (1990). The evolution of antibiotic resistance plasmids: the role of segregation, transposition and homologous recombination. *American Naturalist,* **135**; 573–96.

Coombs, D.L., O'Brien, R.J. & Geiter, L. (1990). USPHS tuberculosis short-course chemotherapy trial 21: effectiveness, toxicity and acceptability. *Annals of Internal Medicine,* **112**, 397–406.

Crow, J. & Kimura, M. (1970). Introduction to Population Genetic Theory. Harper Row, New York.

Datta, N. & Hughes, V.M. (1983). Plasmids of the same Inc groups in enterobacteria before and after the medical use of antibiotics. *Nature, London,* **306**, 616–17.

Davies, J. (1994). Inactivation of antibiotics and the dissemination of resistance genes. *Science,* **264**, 375–88.

Falkow, S. (1975). *Infectious Multiple Drug Resistance.* Pion, London.

Frieden, M.D., Sterling, M.P.H., Pablos-Mendez, A., Kilburn, J.O., Cauthen, G.M. & Dooley, S.W. (1993). The emergence of drug-resistant tuberculosis in New York City. *New England Journal of Medicine,* **328**, 521–6.

Goble, M., Iseman, M.D., Lorie, L.A., Waite, D., Ackerson, L. & Horsburgh, R. (1993). Treatment of 171 patients with pulmonary tuberculosis resistant to isoniazid and rifampin. *New England Journal of Medicine,* **328**, 527–32.

Hughes, V. & Datta, N. (1983). Conjugative plasmids in bacteria of the 'pre-antibiotic' era. *Nature, London,* **302**, 725–6.

Levin, B.R., Stewart, F.M. & Rice, V.A. (1979). The kinetics of conjugative plasmid transmission: fit of a simple mass action model. *Plasmid,* **2**, 247–60.

Monod, J. (1949). The growth of bacterial cultures. *Annual Reviews in Microbiology,* **2**, 371–94.

Neu, H.C. (1992). The crisis in antibiotic resistance. *Science,* **257**, 1064–73.

Nowak, R. (1994). Hungary sees an improvement in penicillin resistance. *Science,* **264**, 364.

Simonsen, L.D., Gordon, D.M., Stewart, F.M. & Levin, B. R. (1990). Estimating the rate of plasmid transfer: an endpoint method. *Journal of General Microbiology,* **136**, 2319–25.

Stewart, F.M. & Levin, B.R. (1973). Partitioning resources and the outcome of interspecific competition: a model and some general considerations. *American Naturalist,* **107**, 171–98.

Sykora, P. (1992). Macroevolution of plasmids: a model for plasmid speciation. *Journal of Theoretical Biology,* **159**, 53–65.

Travis, J. (1994). Reviving the antibiotic miracle? *Science,* **264**, 360–2.

EVOLUTION IN EXPERIMENTAL POPULATIONS OF BACTERIA

RICHARD E. LENSKI

*Center for Microbial Ecology, Michigan State University,
East Lansing, Michigan 48824, USA*

Studies of evolution and of evolutionary processes can readily be carried out in controlled microbial populations.

H. E. Kubitschek, 1974

INTRODUCTION

The basic forces responsible for the evolution of biological populations are now well known. Natural selection and genetic drift cause systematic and random changes, respectively, in the frequency of alleles and genotypes. Mutation, recombination and migration provide the genetic variation on which selection and drift act. Together, all of these processes have produced the tremendous diversity of life on this planet, including the striking adaptations of organisms to their environments. But, although these basic evolutionary forces are well known, it is usually quite difficult to determine their contributions to any particular case of organic evolution or to disentangle their roles more generally.

In science, it is generally much more difficult to determine *a posteriori* the processes that produced some pattern than merely to describe that pattern. Thus, for example, the widespread availability of molecular data has done much less to settle the debate over the roles of genetic drift and natural selection in maintaining genetic variation (Lewontin, 1974; Kimura, 1983; Gillespie, 1991) than it has done to resolve phylogenetic relationships. The analysis of complex dynamical systems is greatly enhanced by being able, at the very least, to observe directly the dynamics and, preferably, to control and manipulate various factors that may influence those dynamics.

To investigate rigorously the dynamics of the evolutionary process, it would be desirable to employ an organism with the following properties: (i) ease of propagation, (ii) rapid generations, (iii) substantial control over environmental variables and over the initial genetic composition of populations, (iv) simple life history, (v) capacity to store organisms in a state of suspended animation, and (vi) potential for replication at all appropriate levels of an experiment. In fact, bacteria such as *Escherichia coli* possess all of these properties. Their short generation times allow evolutionary dynamics to be observed over hundreds and even thousands of generations, and the capacity to freeze their cells permits direct comparison between derived and ancestral types. Their large population sizes ensure a considerable supply of new mutations, so that studies can address the origin as well as the fate of genetic variation and phenotypic novelties. Moreover, their environments and genomes can be readily manipulated in the laboratory, so that rigorous experiments are possible in addition to direct observation.

It should be emphasized that most experimental studies of bacterial evolution are intended to address general issues in evolutionary biology (Dykhuizen, 1990; Lenski, 1992), rather than to recreate natural environments in the laboratory. This research is not concerned with how, for example, *E. coli* has adapted to living in the mammalian gut, but rather it seeks to examine more generally the process by which any population becomes adapted to its environment. How quickly can populations adapt genetically to environmental change? How specific are adaptations with respect to the environment and with respect to the prevailing genetic background? How repeatable is adaptive evolution? Of course, an investigator must always be mindful of the special features of any particular organism, which is the unique product of several billion years of evolution. However, this concern does not alter the fundamental value of experimental studies in a field of research that is otherwise dependent on historical, comparative and theoretical methods of inference.

It is also important to emphasize that these experimental studies of evolution encompass *natural* selection, rather than *artificial* selection as performed by plant and animal breeders. That is, although these experiments employ rather simple and artificial environments, the investigator does not select organisms for breeding on the basis of any particular phenotypic property. Rather, any properties of an organism that enhance its reproductive success and are heritable can respond to natural selection.

This paper reviews a series of studies in experimental evolution that have used bacteria as subject populations. The next section provides an overview of the experimental approach and describes features of evolutionary dynamics that apply especially to asexual populations. The major section that follows then summarizes the results of four evolutionary experiments with bacteria performed by myself and my collaborators, which have addressed a variety of hypotheses. Finally, the specific findings from these studies are pulled together into several more general conclusions.

OVERVIEW OF EXPERIMENTAL EVOLUTION WITH BACTERIA

Experimental approach

Experimental studies of bacterial evolution are conceptually very simple. Populations of cells are propagated, in the laboratory, for many generations in one or more experimentally controlled environments. These environments may be simple or complex, but they must be controlled and reproducible. At the outset of the experiment, a sample of the ancestral bacteria used to found the populations is stored indefinitely, for example as a clone at $-80\,°C$. (Microbiologists sometimes store bacteria rather carelessly. If the ancestral stock was left on a slant at room temperature or maintained by passage in some standard medium, then it might itself evolve, thereby confounding interpretation of the changes that occurred in the experimental populations.) At intervals, samples from the experimental bacterial populations are similarly stored.

In these evolutionary studies, it is often desirable to employ a strain that is strictly asexual (lacks conjugative plasmids and functional prophages) and to found each population with a single cell (achieved operationally using a colony that is the out-growth of a single cell). Barring contamination, any heritable changes from the ancestral state that are observed must then have occurred *de novo* within the experimental population. Contamination from external sources can be excluded as a source of variation by employing a founding strain that carries some distinctive combination of genetic markers. Cross-contamination between populations within an experiment can be ruled out by interspersing additional markers.

After a population has been propagated for some time in a particular environment, the ancestral and derived genotypes may then be compared with respect to any ecological, physiological or genetic properties of interest. Many phenotypic properties are influenced not only by an organism's genotype but also directly by its environment (or by the interaction between genotype and environment). Some environmental effects may be rather persistent; for example, proteins synthesized in response to induction of the *mal* operon may persist for several generations after cells are grown in medium that does not contain maltose (Ryter, Shuman & Schwartz, 1975). Therefore, when comparing phenotypic properties of derived and ancestral genotypes, it is important that both types be allowed to acclimate for several generations to the experimental environment of interest. In that way, it is possible to demonstrate heritable differences between derived and ancestral types, even if the precise genetic basis of those differences are not known.

From an evolutionary standpoint, the most important property of any organism is its Darwinian fitness, that is its propensity to leave descendants. A remarkable feature of bacteria for experimental studies of evolution is that it is possible to estimate directly the fitness of a derived genotype (or the mean fitness of an evolving population) relative to its ancestor. To do so, derived and ancestral types are separately acclimated to the environment of interest, and then they are mixed together and their abundances monitored over time. From these data, each competitor's rate of population change (Malthusian parameter) can be estimated. The relative fitness of the two competitors is then expressed as $w = m_d/m_a$, where m_d and m_a are the Malthusian parameters estimated for the derived and ancestral types, respectively. Some of the studies reviewed here report selection rate constants, given by $r = m_d - m_a$, instead of relative fitnesses. To standardize the values given in this review, relative fitnesses have been computed as $w = 1 + r/d$, where d is the dilution rate of the culture medium (Lenski *et al.*, 1991). It must be kept in mind that fitness depends not only on an organism's genotype but on its environment as well. Thus, it does not make sense to say that genotype A is more fit than B, unless one also specifies in which environment that comparison holds. In this paper, it will be understood that the relevant environment is that in which the evolving populations were propagated, unless explicitly stated otherwise. The derived and ancestral types can be distinguished after they have been mixed together by incorporating into the experimental design a genetic marker than allows strains to be easily scored (for example, by colony colouration on agar medium containing an appropriately reactive dye). Of course, it is necessary to use a balanced experimental design and to perform appropriate controls to establish that the marker does not confound inferences concerning the relative fitness of derived and ancestral types.

Finally, it should be emphasized that evolution experiments with bacteria typically incorporate (at least) two distinct kinds of replication. The first involves repetition of any phenotypic assay for every genotype or population of interest. The second kind of replication is that several independently evolving populations are propagated under each environmental regime. Thus, statistical tests can be performed to determine not only whether any given population has changed relative to its ancestor but also whether a certain pattern of evolutionary change is systematically associated with a particular environment. Given these two levels of replication (plus, as appropriate, different experimental treatments, such as environmental regimes or founding strains), the mathematical machinery embodied in the analysis of variance can be employed to partition the overall variation among a set of phenotypic assays into components due to (i) measurement error, (ii) differences between replicate populations in the same experimental treatment, and (iii) differences between the derived populations and their ancestor or between various treatments.

Dynamical considerations

Evolutionary dynamics in experimental populations of bacteria are expected to differ somewhat from those seen in other model systems, such as fruit flies, that are more commonly employed in population genetics research. The two major differences are that bacteria reproduce asexually and genetic variation is initially absent (by design) from the founding population, and must therefore be generated *de novo* during the course of the experiment.

As a consequence of asexuality, two or more mutations that have

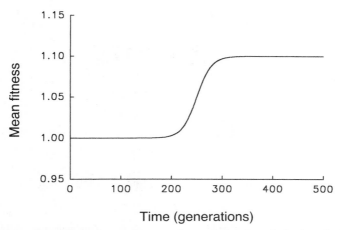

Time (generations)

Fig. 1. Expected step-like trajectory for mean fitness during the substitution of a favourable mutation. The mutation confers a 10% fitness advantage and is initially present at a frequency of 3×10^{-8}. Although increasing in frequency from the outset, the advantageous mutation is effectively hidden from view until it becomes quite common. Modified from Lenski *et al.* (1991).

beneficial effects can be incorporated into an evolving population only if they occur within the same lineage. By contrast, in a population of sexual organisms, beneficial mutations that occur in separate lineages may be recombined on their way to fixation. Moreover, each selectively advantageous mutation that goes to fixation in an asexual population concomitantly sweeps out all other mutations (deleterious, neutral or beneficial) that occurred in every other lineage, although secondary mutations will eventually accumulate in the successful clone. This purging of genetic variation in an asexual population, caused by the substitution of a favourable allele, was described by Atwood, Schneider & Ryan (1951) and is often called periodic selection.

Because of the initial genetic homogeneity of populations and subsequent periodic selection events, adaptation by natural selection depends upon the continual production of new variants by spontaneous mutation. In a large population, any favourable mutation is by definition very rare when it first occurs, having a frequency of only $1/N$. Natural selection is essentially the differential growth of populations and so, like population growth itself, it is an intrinsically exponential (non-linear) process. For example, assuming constant selection, it takes as long for a favourable mutant allele to increase from a frequency of 10^{-7} to 10^{-6} in a population as it takes that same allele to increase from 10% to 90%. Yet, only after the allele has reached a high frequency does it noticeably affect the mean fitness or any other average property of a population. Therefore, the dynamics of adaptation in an asexual population are expected to have an almost step-like quality (Fig. 1),

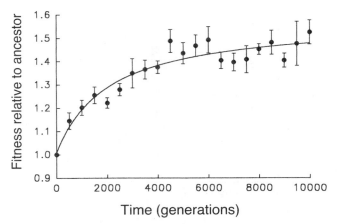

Fig. 2. Trajectory for the grand mean fitness, relative to the common ancestor, during 10 000 generations of experimental evolution with *E. coli*. Error bars give 95% confidence intervals. The curve is the best fit of a hyperbolic model to these data.

as successive advantageous alleles are driven to high frequency by natural selection, each one effectively hidden from view until it becomes common. However, many advantageous alleles are lost by random genetic drift soon after they appear.

SOME RESULTS OF EXPERIMENTAL EVOLUTIONARY STUDIES

The four studies reviewed here all used related strains of *E. coli* B and similar environmental conditions, which consisted of daily serial transfer in a glucose-limited minimal salts medium. Substantive differences in founding genotypes or environments, which gave each study its distinctive character and allowed a range of questions to be addressed, are described below.

Dynamics of adaptation and diversification during long-term experimental evolution

This study (Lenski *et al.*, 1991; Travisano, 1993; Lenski & Travisano, 1994; Vasi, Travisano & Lenski, 1994; Travisano, Vasi & Lenski, in press) has examined the dynamics of adaptation by natural selection, the demographic and ecophysiological bases of adaptation, and the repeatability (across replicate populations) of both the dynamics and the phenotypic bases of adaptation. Twelve populations were founded from the same progenitor strain and propagated in identical environments.

During 10 000 generations, the grand mean fitness relative to the ancestor increased by ~45%, with most of that improvement occurring in the early part of the experiment (Fig. 2). The trajectory for fitness is described

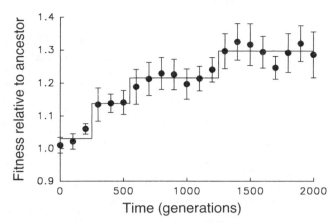

Fig. 3. Step-like changes in mean fitness of a single population of *E. coli* relative to its ancestor. Error bars are 95% confidence intervals. Modified from Lenski & Travisano (1994).

reasonably well by a hyperbolic model, which predicts that the asymptotic value (as $t \to \infty$) for the grand mean fitness will be ~1.57, with half of the eventual improvement attained in ~2000 generations. From these data, it is apparent that there was substantial room for improvement of the ancestral strain in this environment and that most of the improvement occurred fairly quickly. The fact that further improvements become increasingly hard to come by suggests that functional constraints prevent fitness from increasing indefinitely.

As a consequence of the initial genetic homogeneity of the founding population and the purging of genetic variation by subsequent periodic selection events, it was expected that the dynamics of genetic adaptation would exhibit nearly discrete steps (Fig. 1). The data in Fig. 2, however, seem to indicate evolutionary dynamics that are smoothly continuous. In fact, this smooth continuity is an artefact of (i) the 500-generation sample interval, which tends to obscure discrete jumps; and (ii) the pooling of trajectories for the replicate populations, which are out-of-phase with one another as a consequence of the stochastic occurrence of the advantageous mutations that eventually go to fixation. As shown in Fig. 3, the trajectory for mean fitness of a single population, sampled at 100-generation intervals, reveals the step-like dynamics expected for an asexual organism that depends entirely on new mutations for its capacity to adapt genetically by natural selection.

This long-term experiment affords a unique opportunity to examine the repeatability of adaptive evolution. Although the replicate populations were founded from the same ancestral genotype and propagated in identical environments, they may have diverged from one another as a consequence of random mutation and genetic drift. Evolutionary biologists have tended

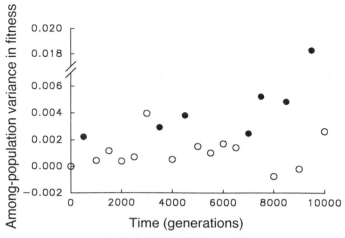

Fig. 4. Trajectory for the among-population variance in mean fitness during 10 000 generations of experimental evolution with *E. coli*. Filled symbols indicate the estimated variance was significantly different from zero ($P < 0.05$). Modified from Lenski & Travisano (1994).

to focus on the consequences of these stochastic processes for genetic changes that are selectively neutral (Kimura, 1983). In this study, however, it was possible to evaluate whether the replicate populations diverged from one another in mean fitness (Lenski *et al.*, 1991; Lenski & Travisano, 1994). Analyses of variance were performed to estimate the trajectory for the among-population variance component for mean fitness, which indicates the variation *above and beyond* the experimental uncertainty in the estimation of mean fitness. Although the variance components have rather large uncertainties (as is typical), these analyses yielded a substantial excess of positive and statistically significant values (Fig. 4), indicating that the populations diverged from one another in their mean fitnesses (Lenski & Travisano, 1994). Moreover, their divergence was sustained for thousands of generations (Fig. 4), even after fitness gains had come to a virtual standstill (Fig. 2). This sustained divergence in mean fitness implies that the replicate populations achieved different adaptive solutions to the experimental environment and, moreover, that these solutions were of unequal selective value to the organism (Lenski & Travisano, 1994).

The fitness improvements show that the derived populations are better adapted to the experimental environment than their ancestor, but they give no indication of the phenotypic bases of these improvements. To that end, an analysis of phenotypic differences between the ancestral and derived genotypes (using isolates of the latter obtained at 2000 generations) was undertaken. The first level of analysis focused on heritable changes in the demographic parameters that govern population growth (Vasi *et al.*, 1994). Recall that the experimental populations were propagated under a serial

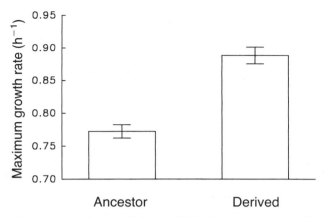

Fig. 5. Change in average maximum growth rate at high glucose concentration for *E. coli*, after 2000 generations under a serial transfer regime in a glucose-limiting minimal salts medium. Error bars are 95% confidence intervals. Data from Vasi *et al.* (1994).

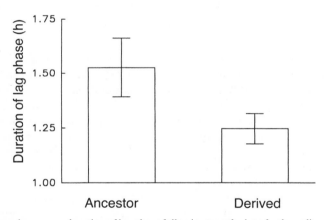

Fig. 6. Change in average duration of lag phase following transfer into fresh medium for *E. coli*, after 2000 generations under a serial transfer regime in a glucose-limiting minimal salts medium. Error bars are 95% confidence intervals. Data from Vasi *et al.* (1994).

transfer regime, in which populations fluctuated between feast and famine conditions. Therefore, a derived genotype could, in principle, be more fit than its ancestor under this regime because of (i) shorter lag phase prior to the commencement of growth following transfer into fresh medium, (ii) higher rate of exponential growth, (iii) greater affinity for the limiting glucose as it becomes depleted, and/or (iv) reduced mortality during stationary phase. The first two changes would indicate adaptation to the feast aspect of the fluctuating environment, whereas the last two would indicate adaptation to the famine aspect. Both exponential growth rate (Fig. 5) and duration of lag phase (Fig. 6) showed substantial improvement (the

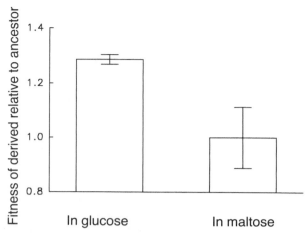

Fig. 7. Comparison of grand mean fitness in glucose versus maltose of *E. coli* propagated for 2000 generations in a glucose-limiting minimal salts medium. Error bars are 95% confidence intervals. Modified from Travisano *et al.* (in press), after converting selection rate constants to relative fitnesses as described in the text.

former increasing and the latter becoming shorter), accounting for most of the gains in fitness (Vasi *et al.*, 1994). By contrast, neither affinity for limiting glucose (expressed as $1/K_s$ in the Monod model of growth) nor death rate during stationary phase (measured from 11–24 h after transfer into fresh medium) showed any improvement. The lack of evolutionary improvement with respect to famine conditions was partly a consequence of the fact that the ancestral strain was already very well adapted to this aspect of the experimental environment, having a high glucose affinity (a low K_s) and a low death rate (Vasi *et al.*, 1994).

The next level of analysis sought to identify the physiological basis of improved fitness (Travisano *et al.*, in press). As a first approximation, cellular metabolism can be divided into two facets: uptake of the limiting nutrient (glucose) and its subsequent utilization (catabolism, biosynthesis, etc.). To discern evolutionary improvements in uptake, derived genotypes were placed in competition with their common ancestor in the same environment, except that maltose replaced glucose in the medium. Maltose is a disaccharide of glucose, and its catabolism once inside a cell is almost identical to that of glucose. However, glucose diffuses across the outer membrane via the porin OmpC (*E. coli* B lacks OmpF) and is actively transported across the inner membrane by the phosphotransferase system (PTS), whereas maltose induces synthesis of a larger porin (LamB) and of a permease (Mal) that is not part of PTS (Nikaido & Vaara, 1987; Postma, 1987; Schwartz, 1987). Figure 7 compares the average fitness of the derived populations in glucose versus maltose. Figure 8 contrasts the extent of the

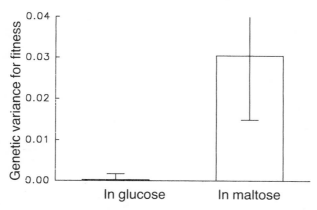

Fig. 8. Comparison of the genetic variance for fitness in glucose versus maltose of *E. coli* propagated for 2000 generations in a glucose-limiting minimal salts medium. Error bars are 95% confidence intervals. Modified from Travisano *et al.* (in press), after converting selection rate constants to relative fitnesses as described in the text.

among-population genetic variance for fitness in these two sugars. Evidently, the replicate populations showed much more improvement in glucose than in maltose, and they were also much less variable in glucose than in maltose. Both of these outcomes support the inference that glucose transport (rather than catabolism, biosynthesis, etc.) was the primary focus of adaptation. Examination of ten additional carbon sources (which vary in their primary modes of transport across the two cell membranes) further supported the inference that the derived genotypes adapted primarily by improvements in their ability to obtain glucose from the environment (Travisano, 1993).

Evolutionary adaptation to changing temperatures

This study (Bennett, Lenski & Mittler, 1992; Bennett & Lenski, 1993; Lenski & Bennett, 1993; Leroi, Lenski & Bennett, in press) has investigated the rate of adaptation to environmental change, the specificity of adaptation with respect to an environmental gradient, and the evolution of specialists and generalists. Six replicate populations in each of the four treatment groups were propagated for 2000 generations at constant 32, 37, or 42 °C, or under a daily alternation between 32 and 42 °C. The common ancestor was a genotype sampled from one of the populations described in the previous section, which had already been propagated for 2000 generations at constant 37 °C in the same medium. The 37 °C treatment thus served as a control for continued adaptation to the ancestral environment, the 32 and 42 °C treatments provided symmetric changes in temperature, and the alternating 32–42 °C treatment provided a change in thermal variability.

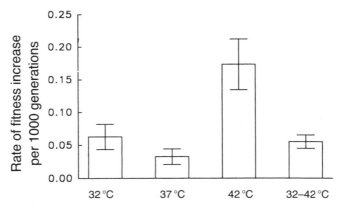

Fig. 9. Rates of increase in mean fitness for *E. coli* populations during 2000 generations at constant 32, 37, or 42 °C or under a daily alternation between 32 and 42 °C. Fitnesses were measured under the same temperature regimes and are expressed relative to a common ancestor that had already been propagated for 2000 generations at 37 °C. Error bars show 95% confidence intervals. Data from Bennett *et al.* (1992).

Adaptation to all three novel regimes was significantly faster than to the ancestral regime (Fig. 9). The extent of adaptation to higher and lower temperatures was strongly asymmetric, with more improvement in the warmer environment (Bennett *et al.*, 1992). This asymmetry was manifest in both the constant and variable regimes, when fitness was assayed at each component temperature of the latter (Leroi *et al.*, in press). The more rapid evolution at the higher constant temperature might, in principle, have been due to an elevated mutation rate (Hoffmann & Parsons, 1991); however, this explanation cannot adequately explain the asymmetry in the extent of adaptation to the two components of the variable environment, because mutations that occurred at either temperature were subject to selection at both (Lenski & Bennett, 1993).

The mean fitnesses of the populations in all four treatment groups were also measured across a range of temperatures in order to evaluate the specificity of thermal adaptation and the generality of tradeoffs in performance across environments (Bennett & Lenski, 1993). Figure 10 shows the range of temperatures over which the populations in each treatment group significantly improved in fitness relative to their common ancestor. Several results are noteworthy from this analysis. (i) Among the three groups of thermal specialists, the temperatures at which mean fitness relative to the ancestor were highest correspond closely to the constant temperatures at which they were propagated for 2000 generations. (ii) Among these thermal specialists, the specificity of temperature adaptation was quite pronounced, with each group's improvement relative to the ancestor extending for only a few degrees in either direction. (iii) In contrast to the three groups of thermal specialists, the thermal generalists (which evolved under a daily

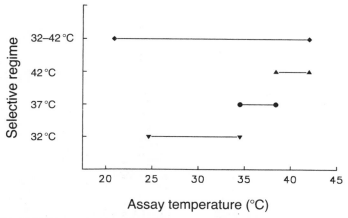

Fig. 10. Specificity of genetic adaptation with respect to environmental temperature in groups of *E. coli* populations that were propagated for 2000 generations at constant 32, 37, or 42 °C or under a daily alternation between 32 and 42 °C. Each solid line indicates the approximate range of temperatures over which the mean fitness of a group was improved significantly relative to the common ancestor ($P < 0.05$). Data from Bennett & Lenski (1993).

alternation between 32 and 42 °C) showed small but significant improvements across a broad range of temperatures, extending even outside the range at which they had been propagated. Thus, the populations propagated under the varying thermal regime evidently became 'jacks of all temperatures but masters of none' (Huey & Hertz, 1984). Also, although each group of thermal specialists was best adapted to its own temperature, reductions in fitness at dissimilar temperatures were the exception rather than the norm. The 42 °C-adapted populations were, on average, as fit as their ancestor at much lower temperatures. And while a few of the populations propagated at 32 and 37 °C were severely disadvantaged above ~40 °C, most were still as fit at these temperatures as the ancestor (Bennett & Lenski, 1993).

Evolutionary compensation for maladaptive pleiotropic effects of chromosomal mutations

Mutations that produce some major phenotypic change, such as resistance to bacteriophage, frequently engender reduced fitness in environments where that phenotype is not specifically favoured (Lenski & Nguyen, 1988). This study (Lenski, 1988*a,b*) sought to test whether the reduction in competitive fitness associated with mutations conferring resistance to bacteriophage T4 was inherent in that phenotype or depended on the underlying genotype. Twenty independent mutations conferring T4-resistance were isolated from the same progenitor, and their fitnesses relative to the progenitor were measured in the absence of phage (Lenski, 1988*a*). Five T4-resistant populations and six populations of the T4-sensitive progenitor were

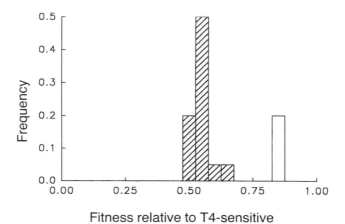

Fitness relative to T4-sensitive

Fig. 11. Frequency distribution of T4-resistant mutants showing their fitness relative to the T4-sensitive progenitor. The shaded portion indicates mutants that were cross-resistant to T7 (even though they had not been exposed to T7), whereas the unshaded portion indicates mutants that retained sensitivity to T7. Modified from Lenski (1988a).

then propagated for 400 generations, in the absence of phage, to determine how the resistance mutations affected the subsequent course of evolution (Lenski, 1988b).

All 20 T4-resistant genotypes were much less fit competitively than their progenitor (Fig. 11). However, the magnitude of the disadvantage varied considerably, with two distinct classes of resistant mutants having mean fitnesses of ~0.55 and ~0.85 relative to their progenitor (Lenski, 1988a). All of the mutants in the former group were cross-resistant to phage T7 (to which the bacteria had not previously been exposed), whereas all those in the latter group remained sensitive to T7. Phages T4 and T7 are not closely related, but they both adsorb to the LPS core in the cell envelope of *E. coli* B. However, T7 requires for adsorption heptose moieties that are more basal in the LPS core than the glucose moiety to which T4 adsorbs (Wright, McConnell & Kanegasaki, 1980). Disruptions to the peripheral structure of the LPS core evidently are less disruptive than those that are more basal. These results show that the fitness cost associated with T4-resistance depends on the genetic basis of the mutation in relation to the physiological architecture of the organism.

After 400 generations, the T4-resistant populations achieved a grand mean fitness that was statistically indistinguishable from that of the sensitive progenitor (Fig. 12). The mean fitness of the evolving sensitive populations meanwhile increased only slightly, so that the average gap in relative fitness between the resistant and sensitive genotypes was quickly reduced from ~35% to <10% (Lenski, 1988b). This convergence in fitness was achieved without any concomitant loss of resistance to either phage T4 or T7; instead,

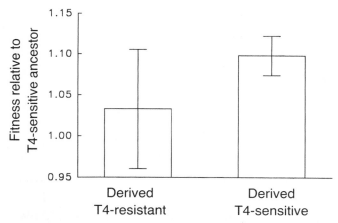

Fig. 12. Convergence in fitness of T4-resistant populations and T4-sensitive populations, during 400 generations of experimental evolution in the absence of bacteriophage T4. Fitnesses were estimated relative to the T4-sensitive ancestor. Error bars show 95% confidence intervals. The derived resistant populations largely overcame their initial competitive disadvantage, but they remained resistant to T4 (and to T7 in those populations that were initially cross-resistant). Data from Lenski (1988*b*).

the evolutionary compensation for the cost of resistance was a consequence of epistatic (non-additive) interactions for fitness between alleles determining resistance and other traits (Lenski, 1988*b*).

Evolutionary adaptation of a bacterium to carriage of an extrachromosomal element

Numerous studies have shown that plasmid carriage is deleterious to bacteria in the absence of antibiotics or other selective agents that specifically select for plasmid-encoded gene functions (Lenski & Nguyen, 1988). These experiments have typically been performed by testing arbitrary combinations of plasmids and hosts, and hence they do not address the possibility that the two genomes may adapt to one another by natural selection so as to minimize the burdensome effects of plasmid carriage on host fitness. Therefore, this study (Bouma & Lenski, 1988; Lenski *et al.*, 1994) examined the evolutionary changes that took place in the interaction between a nonconjugative plasmid and its bacterial host. First, the effect of carriage of pACY184 (which encodes resistance to chloramphenicol and tetracycline) on the fitness of *E. coli* was measured, in the absence of antibiotic. The plasmid-bearing host was then propagated for 500 generations (in a medium containing chloramphenicol, so that plasmid-free segregants would not take over the population), and the change in the effect of plasmid carriage on host fitness that occurred during this period was analysed (Bouma & Lenski, 1988).

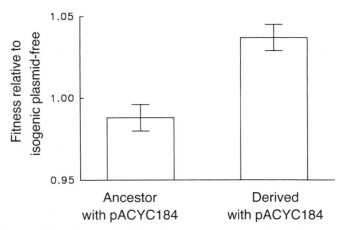

Fig. 13. Change in the effect of plasmid carriage on bacterial host fitness during experimental evolution. Plasmid pACYC184 initially reduced the fitness of *E. coli* (in the absence of antibiotic), relative to its isogenic plasmid-free counterpart. But a derived genotype that evolved with pACYC184 for 500 generations out-competed its own isogenic plasmid-free counterpart, indicating that plasmid carriage conferred some advantage on the derived genotype that it did not confer on the ancestral genotype. Error bars are 95% confidence intervals. Data from Bouma & Lenski (1988), with selection rate constants converted to relative fitnesses as described in the text.

In the absence of antibiotic, the ancestor carrying pACYC184 was significantly less fit than its isogenic plasmid-free counterpart (Fig. 13). After 500 generations, the derived plasmid-bearing cells out-competed the ancestral plasmid-bearing genotype in medium without antibiotic (Bouma & Lenski, 1988). To discern whether genetic changes in the host, the plasmid, or both were responsible for this adaptation, the four combinations of ancestral and derived, host and plasmid genomes were constructed. An analysis of their fitnesses indicated that all of the improvement during the experimental evolution had resulted from changes in the bacterial chromosome and not the plasmid (Bouma & Lenski, 1988). This result indicates that the bacteria adapted by natural selection to their environment, but it does not indicate whether the host adapted specifically to plasmid carriage (as opposed to the culture medium, etc.). To test whether the host had adapted to plasmid carriage, the derived host genotype with the plasmid was placed in competition against the same derived host genotype without the plasmid (the latter obtained as a spontaneous plasmid-free segregant), again in antibiotic-free medium (Bouma & Lenski, 1988). The plasmid-bearing genotype was more fit, indicating that plasmid carriage was not only less burdensome but had actually become beneficial to the host (Fig. 13).

Although these results indicate that the host genome had evolved, they also imply the existence of a plasmid-encoded function that improves the

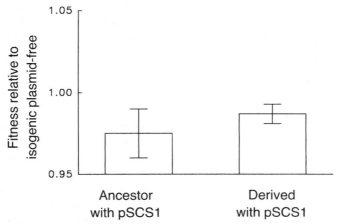

Fig. 14. Benefit of plasmid carriage for the derived host genotype requires tetracycline-resistance function. Plasmid pSCS1 was constructed from pACYC184 by deleting a tetracycline-resistance gene. Fitnesses were measured in the absence of antibiotic and are expressed relative to corresponding (ancestral and derived) isogenic plasmid-free genotypes. A comparison of these data with Fig. 13 indicates that deletion of the tetracycline-resistance gene did not alter the cost of plasmid carriage for the ancestral host, but it eliminated the benefit of plasmid carriage for the derived host. Error bars are 95% confidence intervals. Data from Lenski *et al.* (1994), with selection rate constants converted to relative fitnesses as described in the text.

fitness of the derived (but not the ancestral) host genotype. And, because the plasmid genome is much smaller and more amenable to genetic analysis than the host genome, it was feasible to identify what functions encoded by pACYC184 were responsible for its detrimental and beneficial effects on the ancestral and derived host genotypes, respectively (Lenski *et al.*, 1994). To that end, a number of plasmid derivatives that lacked various parts of the pACYC184 genome were constructed and transformed into the two host backgrounds. The chloramphenicol-resistance function was largely responsible for the detrimental effect of plasmid carriage on the ancestral host genotype (Lenski *et al.*, 1994). This function remained detrimental on the derived host background; thus, deletion of the chloramphenicol-resistance gene increased the advantage of plasmid carriage for the derived host genotype. But the tetracycline-resistance function went from being essentially neutral on the ancestral background to being highly beneficial for the derived host. Thus, deletion of the tetracycline-resistance gene did not affect the cost of plasmid carriage for the ancestral host, but it eliminated the benefit of plasmid carriage for the derived host (Fig. 13 versus Fig. 14). It was further demonstrated that the beneficial effect of this tetracycline-resistance gene was manifest on the derived host background even when it was expressed from a heterologous plasmid background (Lenski *et al.*,

1994). Thus, the host genotype that had evolved with pACYC184 for 500 generations somehow took advantage of this plasmid-encoded resistance gene, so that its expression was beneficial even in the absence of antibiotic.

EVOLUTIONARY ISSUES AND INFERENCES

Adaptation by natural selection

A problem often faced by evolutionary biologists is the inherent difficulty in drawing inferences from comparative and historical data. It is especially difficult to quantify the effects of the various fundamental population genetic *processes* – natural selection, drift, mutation, recombination, and migration – from patterns of phenotypic and genetic similarities and differences. Even deciding what features or organisms are adaptations *per se* (as opposed to products of selection on correlated characters, drift, or derived characters that presently are of no importance) is a difficult and controversial subject (Gould & Lewontin, 1979; Harvey & Pagel, 1991; Bennett, in press).

All four experimental studies reviewed here unequivocally demonstrate not only the outcome but the very process of genetic adaptation by natural selection. Moreover, in these experiments with bacteria, the dynamics of evolutionary change depended on new mutations that arose during the course of the experiment, in contrast to experiments performed with other model systems in population genetics (for example, fruit flies), where selection is presumed to operate largely on standing variation already present in a base population. Thus, bacteria provide a powerful system for studying the origin, as well as the fate, of genetic variation and phenotypic novelties, both of which must ultimately determine the course of evolution in the real world.

Environmental change and adaptive evolution

The experiments reviewed here demonstrate the role of environmental change in accelerating the pace of adaptive evolution. During 10 000 generations under a particular environmental regime, bacteria adapted much faster initially than they did subsequently (Lenski *et al.*, 1991; Lenski & Travisano, 1994). In contrast to selection experiments with higher organisms, this deceleration cannot be attributed to depletion of genetic variation that was present initially in the population (Falconer, 1983), because all of the genetic variation in the experiment with bacteria was generated *de novo* by mutation, which is an on-going process. Instead, the initially rapid evolution must have been a consequence of intense selection triggered by placing the bacteria in an arbitrary and essentially new environment. The subsequent deceleration implies that the bacteria achieved a set of genetic solutions to this environment that became increasingly difficult to improve upon.

The decreasing pace of adaptive evolution in a constant environment was reinvigorated by environmental change. When bacteria taken from this experiment were subjected to either novel thermal regimes or to a continuation of the same environment, more rapid evolution ensued under the novel conditions (Bennett et al., 1992). Changes in genetic composition that throw an organism out of kilter can also accelerate adaptive evolution, as when evolving T4-resistant populations almost caught up with their initially much fitter T4-sensitive counterparts (Lenski, 1988b). Thus, both environmental and genetic perturbations may tend to promote evolutionary change, provided that they are not so severe that a population is driven extinct and also that genetic solutions are accessible (as opposed to requiring several mutations, each of which individually is deleterious).

The phenotypic targets of natural selection and bases of adaptation

Bacteria offer several advantages for studying the dynamics of evolutionary change, including their rapid generations and large populations, the ability to store ancestral and derived genotypes indefinitely in suspended animation, and the general ease of manipulating their environments and genomes. It is interesting, however, that the *modus operandi* in studying adaptation by natural selection in experiments with bacteria is almost the inverse of that employed by evolutionary biologists working in the field. Researchers studying Darwin's finches, for example, begin by observing conspicuous variation in phenotypic traits, such as beak size and shape (Grant, 1986). To support the inference that phenotypic differences are important for the process of adaptation by natural selection, the researchers face a two-fold challenge: (i) to show that these differences are, at least partially, heritable; and (ii) to demonstrate that these differences affect the survival and reproductive success of individuals (Endler, 1986). If either (i) or (ii) does not hold, then the phenotypic variation is unimportant for genetic adaptation by natural selection. By contrast, it is fairly simple to demonstrate that a population of bacteria has adapted to a particular environment (by showing a heritable improvement in fitness relative to its ancestor); but it is more difficult to identify the genetically encoded phenotypic changes that are responsible for the improvement. Thus, the substantial challenge facing this approach is to open the 'black box' that is the experimental organism and determine what has changed and why it yields an improvement.

A 'top-down' approach may be pursued to study the phenotypic bases of adaptation by examining how the ancestral and derived genotypes differ in their reactions to various environments. For example, in the long-term selection experiment with *E. coli* (Lenski et al., 1991), the supply of limiting glucose was altered to show that the bacteria had adapted to periods of resource abundance rather than becoming more tolerant of famine (Vasi *et al.*, 1994), and maltose and other nutrients were substituted for glucose to implicate the importance of adaptations involved in resource uptake (Travisano, 1993; Travisano *et al.*, in press). Temperature and the presence of mutations, viruses, and plasmids were similarly manipulated in the other studies to examine the specificity of adaptations in the derived genotypes (Bennett & Lenski, 1993; Lenski, 1988*b*; Lenski *et al.*, 1994). This specificity, or the lack thereof, may give important clues as to the nature of the underlying changes.

Specificity of adaptation with respect to the environment

One of the most intriguing general findings from these studies is the high degree of specificity of the adaptations. During thousands of generations in an environment in which glucose was the sole carbon and energy source, bacteria dramatically improved their fitness in this environment. Yet their advantage relative to the common ancestor largely disappeared when a glucose disaccharide (maltose) was substituted for glucose in the medium (Travisano *et al.*, in press) and even when glucose itself was reduced to a very low concentration (Vasi *et al.*, 1994). Similarly, when populations adapted to the glucose medium were then subjected to various thermal regimes, their adaptive responses were quite specific with respect to temperature. The populations adapted to constant 32, 37 or 42 °C had maximal fitness (relative to their common ancestor) very near the temperature at which they had been propagated, and they showed little or no improvement only a few degrees away. And the populations propagated under alternating 32 and 42 °C had adapted to this thermal variation as well, by becoming more fit over a wider range of temperatures than the populations propagated under constant thermal regimes (Bennett & Lenski, 1993).

Indeed, this specificity of adaptation extended not only to the external environment, but also to the internal (genetic) environment. Thus, in evolving populations of T4-resistant bacteria, alleles were selected that conferred much larger advantages on that background than on an otherwise isogenic T4-sensitive background (Lenski, 1988*b*). And in evolving populations of plasmid-bearing cells, chromosomal mutants were selected that somehow took advantage of an antibiotic-resistance gene, so that carriage of the plasmid went from being burdensome to beneficial in the absence of antibiotic (Lenski *et al.*, 1994).

This specificity shows clearly the power of bacterial populations for studying evolution by demonstrating the value of being able to control and manipulate the relevant ecological and genetic factors. There is no compelling reason to think that, in the real world, adaptations are any less specific to particular features than those observed in these experiments. In the real world, however, the myriad complexities and changes in ecological and genetic context make it difficult to discern the specificity of adaptations.

How repeatable is adaptive evolution?

Evolution depends on a mix of stochastic and deterministic forces. Evolutionary biologists sometimes fall into the trap of regarding these two types of forces as somehow isolated or separate from one another, typically appealing to mutation and drift to explain variation at the molecular level while invoking natural selection to explain conspicuous phenotypic differences. The demonstrable improvements of the derived genotypes relative to their progenitors in the experiments reviewed here clearly indicate the important role of natural selection, but these adaptive changes also depended on mutations (and on the chance avoidance of loss due to drift when favorable mutations first appeared). Despite the striking specificity of adaptation to the imposed environment, replicate populations (founded from the same ancestor and propagated under identical regimes) none the less diverged from one another in certain respects that, while not important in their immediate environment, could have profound ramifications should the environment subsequently change (Cohan & Hoffmann, 1986; Silva & Dykhuizen, 1993). The populations adapted to growth on glucose were ~100-fold more variable in their fitness in maltose than in glucose (Travisano et al., in press). Among the populations adapted to 32 and 37 °C, some but not all had become much worse at higher temperatures (Bennett & Lenski, 1993). Among mutants resistant to phage T4, some also became T7-resistant but were very poor competitors for glucose, whereas others were fair competitors but remained vulnerable to T7 (Lenski, 1988a).

Thus, the specificity of adaptation with respect to a given selective challenge can often be achieved by a diversity of underlying genetic changes, which may have different consequences for the subsequent ecological and evolutionary success of a population. Some populations may be fortuitously preadapted to forthcoming challenges, while others (just as well adapted to their present circumstances) may have been doomed to future extinction by the accidental genetic details of their present adaptations. This combination of deterministic and stochastic forces gives organic evolution its seeming predictability (the match between organisms and their environments) and its capriciousness (the lack of correspondence between past and future success), which together give rise to the uniqueness of evolutionary history.

ACKNOWLEDGEMENTS

I am grateful to all those who have worked on these projects or discussed them with me (or both), especially Al Bennett, Judy Bouma, Dan Dykhuizen, Armand Leroi, John Mittler, Toai Nguyen, Michael Rose, Sue Simpson, Mike Travisano, and Farida Vasi. This research was supported by grants from the National Science Foundation (BSR-8858820, DEB-9249916, IBN-9208662) and by the National Science Foundation Center for Microbial Ecology (BIR-9120006).

REFERENCES

Atwood, K. C., Schneider, L. K. & Ryan, F. J. (1951). Periodic selection in *Escherichia coli*. *Proceedings of the National Academy of Sciences, USA*, **37**, 146–55.

Bennett, A. F. (In press). Adaptation and the evolution of physiological characters. In *Handbook of Comparative Physiology* (Dantzler, W. H., ed.). Oxford University Press, Oxford.

Bennett, A. F. & Lenski, R. E. (1993). Evolutionary adaptation to temperature. II. Thermal niches of experimental lines of *Escherichia coli*. *Evolution*, **47**, 1–12.

Bennett, A. F., Lenski, R. E. & Mittler, J. E. (1992). Evolutionary adaptation to temperature. I. Fitness responses of *Escherichia coli* to changes in its thermal environment. *Evolution*, **46**, 16–30.

Bouma, J. E. & Lenski, R. E. (1988). Evolution of a bacteria/plasmid association. *Nature, London*, **335**, 351–2.

Cohan, F. M. & Hoffmann, A. A. (1986). Genetic divergence under uniform selection. II. Different responses to selection for knockdown resistance to ethanol among *Drosophila melanogaster* populations and their replicate lines. *Genetics*, **114**, 145–63.

Dykhuizen, D. E. (1990). Experimental studies of natural selection in bacteria. *Annual Review of Ecology & Systematics*, **21**, 378–98.

Endler, J. A. (1986). *Natural Selection in the Wild*. Princeton University Press, Princeton.

Falconer, D. S. (1983). *Introduction to Quantitative Genetics*, 2nd edition. Longman, London.

Gillespie, J. H. (1991). *The Causes of Molecular Evolution*. Oxford University Press, New York.

Gould, S. J. & Lewontin, R. C. (1979). The spandrels of San Marco and the Panglossian paradigm: a critique of the adaptationist programme. *Proceedings of the Royal Society of London, Series B*, **205**, 581–98.

Grant, P. R. (1986). *Ecology and Evolution of Darwin's Finches*. Princeton University Press, Princeton.

Harvey, P. H. & Pagel, M. D. (1991). *The Comparative Method in Evolutionary Biology*. Oxford University Press, Oxford.

Hoffmann, A. A. & Parsons, P. A. (1991). *Evolutionary Genetics and Environmental Stress*. Oxford University Press, Oxford.

Huey, R. B. & Hertz, P. E. (1984). Is a jack-of-all-temperatures a master of none? *Evolution*, **38**, 441–4.

Kimura, M. (1983). *The Neutral Theory of Molecular Evolution*. Cambridge University Press, Cambridge.

Kubitschek, H. E. (1974). Operation of selection pressure on microbial populations. *Symposia of the Society for General Microbiology*, **24**, 105–30.

Lenski, R. E. (1988*a*). Experimental studies of pleiotropy and epistasis in *Escherichia coli*. I. Variation in competitive fitness among mutants resistant to virus T4. *Evolution*, **42**, 425–32.

Lenski, R. E. (1988*b*) Experimental studies of pleiotropy and epistasis in *Escherichia* *coli*. II. Compensation for maladaptive pleiotropic effects associated with resistance to virus T4. *Evolution*, **42**, 433–40.

Lenski, R. E. (1992). Experimental evolution. In *Encyclopedia of Microbiology*, Vol. 2 (Lederberg, J., ed.), pp. 125–140. Academic Press, San Diego.

Lenski, R. E. & Bennett, A. F. (1993). Evolutionary response of *Escherichia coli* to thermal stress. *American Naturalist*, **142**, S47–64.

Lenski, R. E. & T. T. Nguyen (1988). Stability of recombinant DNA and its effects on fitness. *Trends in Ecology & Evolution*, **3**, S18–20.

Lenski, R. E., Rose, M. R., Simpson, S. C. & Tadler, S. C. (1991). Long-term experimental evolution in *Escherichia coli*. I. Adaptation and divergence during 2,000 generations. *American Naturalist*, **138**, 1315–41.

Lenski, R. E., Simpson, S. C. & Nguyen, T. T. (1994). Genetic analysis of a plasmid-encoded, host genotype-specific enhancement of bacterial fitness. *Journal of Bacteriology*, **176**, 3140–7.

Lenski, R. E. & Travisano, M. (1994). Dynamics of adaptation and diversification: a 10,000-generation experiment with bacterial populations. *Proceedings of the National Academy of Sciences, USA*, **91**, 6808–14.

Leroi, A. M., Lenski, R. E. & Bennett, A. F. (In press). Evolutionary adaptation to temperature. III. Adaptation of *Escherichia coli* to a temporally varying environment. *Evolution*.

Lewontin, R. C. (1974). *The Genetic Basis of Evolutionary Change*. Columbia University Press, New York.

Nikaido, H. & Vaara, M. (1987). Outer membrane. In Escherichia coli *and* Salmonella typhimurium: *Cellular and Molecular Biology* (Neidhardt, F. C., ed.), pp. 7–22. American Society for Microbiology, Washington, DC.

Postma, P. W. (1987). Phosphotransferase system for glucose and other sugars. In Escherichia coli *and* Salmonella typhimurium: *Cellular and Molecular Biology* (Neidhardt, F. C., ed.), pp. 127–141. American Society for Microbiology, Washington, DC.

Ryter, A., Shuman, H. & Schwartz, M. (1975). Integration of the receptor for bacteriophage Lambda in the outer membrane of *Escherichia coli*: coupling with cell division. *Journal of Bacteriology*, **122**, 295–301.

Schwartz, M. (1987). The maltose regulon. In Escherichia coli *and* Salmonella typhimurium: *Cellular and Molecular Biology* (Neidhardt, F. C., ed.), pp. 1482–1502. American Society for Microbiology, Washington, DC.

Silva, P. J. N. & Dykhuizen, D. E. (1993). The increased potential for selection of the *lac* operon of *Escherichia coli*. *Evolution*, **47**, 741–9.

Travisano, M. (1993). Adaptation and divergence in experimental populations of the bacterium *Escherichia coli*: the roles of environment, phylogeny and chance (PhD dissertation, Michigan State University, East Lansing, Michigan, USA).

Travisano, M., Vasi, F. & Lenski, R. E. (In press). Long-term experimental evolution in *Escherichia coli*. III. Variation among replicate populations in correlated responses to novel environments. *Evolution*.

Vasi, F., Travisano, M. & Lenski, R. E. (1994). Long-term experimental evolution in *Escherichia coli*. II. Changes in life history traits during adaptation to a seasonal environment. *American Naturalist*, **144**, 432–56.

Wright, A. M., McConnell, M. & Kanegasaki, S. (1980). Lipopolysaccharide as a bacteriophage receptor. In *Virus Receptors, Part 1, Bacterial Viruses*. Chapman and Hall, London.

GENETIC POPULATION STRUCTURE AND PATHOGENICITY IN ENTERIC BACTERIA

THOMAS S. WHITTAM

Institute of Molecular Evolutionary Genetics, Department of Biology, The Pennsylvania State University, University Park, Pennsylvania 16802, USA

INTRODUCTION

Population genetics is the study of evolutionary change in the genetic composition of populations with two complementary avenues of inquiry. On the one hand, there are general questions about the mechanisms of evolution. How do the forces of mutation, migration, natural selection, and genetic drift influence the nature and rate of evolution change? On the other hand, there are specific questions about the history of particular species and populations. In the recently developing field of bacterial population genetics, both avenues are being actively pursued. Fundamental questions are being addressed about the types of population structure, the nature of allelic variation and its organization in genomes, and the roles of different modes of recombination in generating genotypic variation. In the evolutionary study of pathogenic strains, historical questions are of great interest. When did particular pathogenic lineages evolve? What were the major events in the emergence of a novel pathogen?

The purpose of this chapter is to highlight some of the recent findings of the study of genetic variation in natural populations of bacteria with special emphasis on the evolution of pathogenic clones in enteric organisms.

CLONAL ANALYSIS OF ENTERIC PATHOGENS

Our work on population genetics of pathogenic bacteria begins with the characterization of a large collection of isolates for numerous genetic markers for the purpose of identifying the common genotypes in the sample. Of the order of several hundred bacterial isolates are necessary to determine the frequency of genotypes in populations over broad geographic areas or across a variety of disease classes and host species. The ideal genetic markers reveal large amounts of polymorphism which can be interpreted in terms of genetic variation at specific chromosomal loci. In addition, variation that is selectively neutral and irrelevant to disease processes is best for assessing the genetic relatedness of isolates. Finally, we can examine the distribution of phenotypic traits implicated in pathogenicity and virulence, in this way, attempting to elucidate the specialization of pathogenic bacteria.

Definition of a clone

In a rigorous sense, a bacterial clone consists of a single cell and all its descendants representing a monophyletic branch on an evolutionary tree. In this strict sense, a clonal lineage is a closed genetic system that accumulates differences only through genetic processes that occur within a single cell, such as point mutations, inversions, duplications and deletions, and transpositions. Clearly, because of these processes, members of a clone are not necessarily genetically or phenotypically homogeneous. In practice, the term clone is used in a looser sense to refer to an extant subgroup of bacteria within a species that have many similarities derived from a common ancestor not shared by other organisms in the species. Thus *clone* is used in bacterial population genetics in the same sense that *clade* is used to refer to species that belong to a monophyletic branch of an evolutionary tree (Li & Graur, 1991). We refer to closely related groups of clones as clone complexes.

Although the term clone implies strictly asexual reproduction, populations of bacteria have evolved mechanisms for exchanging genetic information between cell lines and comparative nucleotide sequencing of strains of *Escherichia coli*, *Salmonella enterica* and other bacterial species has revealed many cases of mosaic segments of DNA (Milkman & Stoltzfus, 1988; Milkman & Bridges, 1990, 1993; Achtman & Hakenbeck, 1992; Selander *et al.*, 1994). From these studies, it has become evident that, as cell lines grow and spread, mutations accumulate and sections of the genome are replaced through gene transfer. As recombination occurs, the history of genomic divergence becomes obscured. Thus, in one sense, reconstructing the history of clonal divergence involves detecting an ancestral signal against a background of recombinational noise.

Multilocus enzyme electrophoresis

One of the primary tools for studying the amount and organization of genetic variation in natural populations of bacteria is multilocus enzyme electrophoresis (Selander *et al.*, 1986). This technique was established more than 25 years ago in eukaryotic population genetics (Lewontin, 1991) and detects protein polymorphisms caused by amino acid replacements that alter rates of electrophoretic migration. Multilocus enzyme electrophoresis is an efficient method for rapidly screening allelic variants at many structural loci in the hundreds of individuals required for population genetic analysis (Selander *et al.*, 1986).

Studies of protein polymorphisms in *E. coli* (Achtman & Pluschke, 1986; Selander, Caugant & Whittam, 1987) and a variety of agents of infectious diseases in humans (Selander *et al.*, 1987; Musser *et al.*, 1990), including, for example, Legionnaire's disease (*Legionella pneumophila*), meningitis (*Neisseria meningitidis* and *Haemophilus influenzae*), enteric and typhoid fevers (*S. enterica*), and toxic shock syndrome (*Staphylococcus aureus*), have yielded two fundamental discoveries about the nature of genetic variation in natural populations of bacteria. First, bacterial populations are highly variable with multiple alleles detected for most enzyme-encoding loci. The overall levels of genetic diversity are several times greater than those typical of eukaryotic species. Secondly, the structure of bacterial populations is predominantly clonal, featuring a small number of ubiquitous multilocus genotypes. Thus, despite the prevalence of the three parasexual mechanisms of gene transfer in bacteria, transformation, transduction, and conjugation, the rate at which recombination assorts genes into new combinations in nature appears to be very low.

Although protein bands on gels are what are observed in enzyme electrophoresis, these phenotypes are the primary products of known genes and the phenotypic variation observed can be interpreted as allelic variation at these structural loci. This fact opens the door to population genetic analysis and allows one to draw inferences about the overall genetic relatedness of strains. It also makes multilocus enzyme electrophoresis distinct from serotyping, biotyping, phage-typing and other traditional techniques in which differences between phenotypes are not ascribed to allelic differences at specific genes.

MULTILOCUS LINKAGE DISEQUILIBRIUM

A hallmark of the restricted recombination in a clonally structured population is the presence of linkage disequilibrium: the non-random association of alleles in haploid genotypes (Whittam, Ochman & Selander, 1983a,b; Hartl & Dykhuizen, 1984). Multilocus associations at the population level mean that certain combinations of alleles at different genes are found at greater frequency than expected, whereas other combinations are rare or absent. Extensive linkage disequilibrium reflects the genotypic structure of a population which, in principle, decays as a function of the rate of recombination (Levin, 1981). In the absence of natural selection favouring specific gene combinations, frequent recombination will tend to randomize the allele combinations found in genotypes and the genotypic structure of the population will approach linkage equilibrium (Hedrick & Thomson, 1986).

The contrast in genotypic structure between a clonal and a freely recombining population is illustrated in Fig. 1 (Whittam, 1992). The dendrograms show the genetic relationships for 30 haploid genotypes depicted in the patterned bars as distinct allele combinations for 20 loci on a chromosome. In the top panel, cell lineages have diverged in multilocus genotype as mutations create new alleles without recombination. With the accumulation of allele substitutions, linkage disequilibrium builds up, as reflected in the deep branch in the dendrogram and the large variance and bimodal shape of

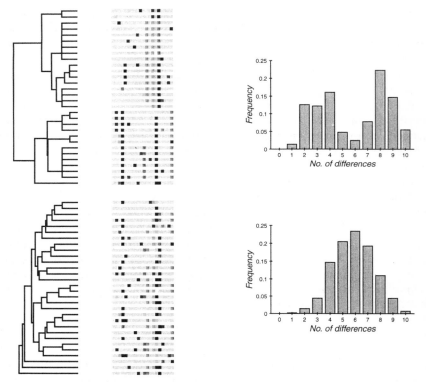

Fig. 1. Extremes of population structure. The top panel shows the dendogram illustrating the genetic relatedness among chromosomes under clonal reproduction (no recombination). Chromosomes comprise 20-locus genotypes (patterned bars) with alleles designated by patterns. The allele-mismatch distribution at the right is bimodal, and has a inflated variance owing to linkage desequilibrium. With free recombination (bottom panel), alleles are randomized on chromosomes, the dendogram is star-like, and the mismatch distribution is unimodal with a smaller variance.

the frequency distribution of the number of allele differences between chromosomes. This distribution is found by comparing all pairs of chromosomal genotypes and tallying the number of allele differences between each pair. With frequent recombination, however, alleles become randomized in genotypes as shown in the bottom panel of the Figure. In this case, there is little linkage disequilibrium. As a result, the dendrogram is bush-like, and there is a reduction in the variance as the mismatch distribution becomes unimodal in shape. Thus, in a population near linkage equilibrium, most genotypes lie near the mode of the distribution and recombination becomes a 'cohesive force' that counters mutational divergence of bacterial lineages.

    The connection between the effect of recombination on variation of the allele-mismatch distribution forms the basis of a statistical method for assessing multilocus linkage disequilibrium. Brown, Feldman & Nevo

Table 1. *Multilocus linkage disequilibrium in serogroups of* E. coli *and* S. enterica

Species Group	Number of isolates	Number of ETs	Number of loci	All isolates		ETs only	
				D	I_A	D	I_A
S. enterica[a]							
typhimurium	340	17	24	0.035	1.03 ± 0.10	0.117	1.22 ± 0.33
paraptyphi B	118	14	24	0.074	2.68 ± 0.13	0.168	0.36 ± 0.37
choleraesius	174	11	24	0.036	2.57 ± 0.12	0.178	1.36 ± 0.42
E. coli[b]							
O55	300	57	20	0.755	4.48 ± 0.08	0.938	1.45 ± 0.18
O111	391	58	20	0.554	2.67 ± 0.07	0.738	1.23 ± 0.19
O157	863	59	20	0.466	7.01 ± 0.05	0.892	1.16 ± 0.18
ECOR	72	68	38	0.173	1.85 ± 0.17	0.133	1.70 ± 0.17

[a]Adapted from Maynard Smith *et al.* (1993) with electrophoretic data from Selander *et al.* (1990).
[b]*E. coli* reference collection (Ochman & Selander, 1984).

(1980) devised an index of multilocus association, designated I_A which is equal to one minus the ratio of the observed variance of the mismatch distribution (V_o) to the expected variance (V_E) in the absence of linkage disequilibrium. I_A is near zero as a population approaches linkage equilibrium, and increases in value with increasing amounts of multilocus linkage disequilibrium. For the example in Fig. 1, the expected variance under linkage equilibrium is $V_E = 2.9$ and is the same for both the top and bottom distributions because the expected variance is a simple function of single-locus diversity (Brown *et al.*, 1980). The observed variance, however, decreases from $V_o = 7.7$ ($I_A = 1.6$) under the strictly clonal model to 3.8 ($I_A = 0.3$) for a population at linkage equilibrium. This variance ratio method was first applied to bacterial populations in studies of natural isolates of *E. coli*, and demonstrated significant and widespread multilocus linkage disequilibrium both globally and in separate geographic populations (Whittam, Ochman & Selander, 1983*a,b*).

   Maynard Smith and colleagues (1993) have recently extended this analysis to eight species of bacteria including *S. enterica*, the pathogenic neisseria, and *Rhizobium* species. They found that strains of several serovars of *Salmonella* were similar to *E. coli* populations with significant inflation of the variance of the mismatch distribution, as measured by I_A (Table 1). Such inflation of the variance represents substantial linkage disequilibrium, and is incompatible with a high rate of assortative recombination of chromosomal genes. These findings support the hypothesis that serogroups of *Salmonella* have predominantly clonal population structures. In contrast, multilocus enzyme data from *Neisseria gonorrhoeae* and *Rhizobium meliloti* were close to linkage equilibrium, supporting the hypothesis that in these populations, recombination occurs frequently (Maynard Smith *et al.*, 1993).

In addition to commensal *E. coli* populations, strains representing serotypes associated with enteric diseases also exhibit evidence of linkage disequilibrium. Table 1 summarizes the genetic diversity and linkage disequilibrium found among strains of three O serogroups (O55, O111, and O157) commonly associated with diarrhoeal diseases in infants, adults, or domesticated animals. In contrast to the major *S. enterica* serovars, strains of the O55, O111, and O157 serogroups of *E. coli* are genetically more variable, with two to three times the number of multilocus genotypes (electrophoretic types or ETs) and more than an order of magnitude greater levels of single-locus diversity than found among *S. enterica* serovars (Table 1). In fact, the amount of genetic diversity among strains within these O-serogroups is several times greater than that found in the *E. coli* reference collection (ECOR), a sample originally established to represent the extent of genetic variation among the global *E. coli* population (Ochman & Selander, 1984). Secondly, with the exception of cholerasuis strains, all of these groups of enteric bacteria show strong multilocus associations as reflected in the significant I_A values when either isolates or ETs are the unit of analysis. The degree of nonrandom association is more extreme when all isolates are examined because the bulk of the strains within a serogroup belong to a limited number of common ETs.

The results in Table 1 demonstrate the general finding that strains of *E. coli* and *S. enterica* associated with enteric diseases have extensive multilocus structuring at the population level. This structure reflects the fact that under natural conditions, assortative recombination produces new genotypes at a relatively low rate; in *E. coli* populations, for example, this rate may be of the order of as 10^{-10}–10^{-9} per generation (Whittam & Ake, 1993).

INTRAGENIC RECOMBINATION AND MOSAIC ALLELES

The amount of linkage disequilibrium estimated from multilocus enzyme electrophoresis data provides information about one type of recombination in bacteria, assortative recombination, in which whole alleles are transferred and give rise to new multilocus genotypes. Intragenic recombination, on the other hand, produces novel mosaic alleles that cannot be distinguished from new alleles that arise by point mutation without knowledge of the nucleotide sequences.

To obtain a more complete understanding of the genetic basis of allelic variation and elucidate the role of recombination, several laboratories have been involved in large-scale sequencing of multiple wild strains of *E. coli* and *S. enterica* (Milkman & Crawford, 1983; DuBose, Dykhuizen & Hartl, 1988; Dykhuizen & Green, 1991; Nelson, Whittam & Selander, 1991;

Hall & Sharp, 1992; Nelson & Selander, 1992; Boyd *et al.*, 1994). For several enzyme-encoding genes, including alkaline phosphatase (*phoA*), glyceraldehyde-3-phosphate dehydrogenase (*gapA*), proline permease (*putP*), and malate dehydrogenase (*mdh*), the general picture that emerges is that substitutions resulting in amino acid replacements are highly constrained relative to silent substitutions, and that intragenic recombination generates mosaic alleles at only a modest rate that is of the order of magnitude of the mutation rate per locus (Whittam & Ake, 1993). Thus, for proteins involved in housekeeping functions (with the exception of *gnd*, see below), recombinant alleles are either neutral, nearly neutral, or deleterious, so that they are usually quickly eliminated by selection or genetic drift.

A completely different picture emerges for proteins on the cell surface that may be under direct positive selection. For example, the *fliC* gene that encodes flagellin, is highly polymorphic among *S. enterica* strains with most alleles exhibiting a mosaic structure comprised of segments of diverse origin (Li *et al.*, 1994). In these cases, natural selection may favour new recombinant alleles as adaptation to escape host immune defence or to evade recognition by flagellatropic bacteriophages (Li *et al.*, 1994). The *rfb* region, that specifies the composition of the surface polysaccharides underlying O antigens, is highly polymorphic in *S. enterica* with mosaic structures composed of segments that differ strongly in GC content (Reeves, 1993). The spread and maintenance of O antigenic variants may again be an adaptive response to immune system selection by different host species.

The influence of diversifying selection on the *rfb gene cluster* may account for the unusually high variability of the *gnd* locus that encoded 6-phosphogluconate dehydrogenase, one of the most polymorphic enzymes in *E. coli* populations (Whittam & Ake, 1993). Biserčić, Feutrier and Reeves (1991) have suggested that the close proximity of *gnd* and *rfb* inhibits genetic drift at the *gnd* locus, presumably because of the action of selection on surface antigenic variation. One possibility is that recombinants involving the *gnd–rfb* region have selective advantages, and increase in frequency under certain conditions that favour strains with specific O antigens. By attaining higher frequencies, these variants are less likely to be lost by random drift or through turnovers caused by periodic selection.

Support for the idea that different types of selection dominate the evolution of certain classes of proteins is seen in a comparison of the degree of amino acid divergence of homologous proteins in *E. coli* and *S. enterica* (Fig. 2). The 179 proteins are classified into six families based roughly on the following functional categories: ribosomal proteins include elongation, peptide-release, and transcription termination factors; DNA replication proteins include the subunits of DNA polymerases and structural histone-like proteins; regulators include repressors and activators that regulate transcription through a variety of DNA-binding sites; signal and transport proteins include membrane-associated proteins that function primarily in

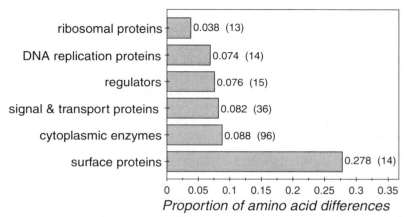

Fig. 2. Divergence of different classes of proteins. The mean percentage amino acid difference for homologous proteins from *E. coli* and *S. enterica*. Most proteins, including housekeeping enzymes, are relatively conserved, diverging by about 7–9% in primary sequence. In contrast, surface proteins, such as flagellin, porin, and pilins, have diverged at more than three times the rate of most cytoplasmic proteins.

sensory transduction or transmembrane transport; cytoplasmic enzymes; and outer surface proteins such as the components of flagella, porins, and pilins.

For most of the bacterial proteins, the mean divergence in amino acid sequences falls into the narrow range of 7.4–8.8%. The bulk of the proteins were on average twice as variable as the most conserved sequences seen among the ribosomal proteins and elongation factors. The proteins that function in the structure and replication of DNA, for example, the histone-like proteins, DNA polymerases, and DNA repair enzymes are the most conserved of these families with an average of 7.4% divergence among 14 pairs of homologous proteins. In contrast, proteins that function primarily on the cell surface show the greatest mean divergence in amino acid sequence, evolving at nearly three times the rate of the cytoplasmic and other proteins from within cells. Presumably the rapid rate of evolution reflects selective pressures in addition to relaxed constraints on surface components. The diversity of surface proteins could result, for example, from selection favouring antigenic variants to evade the immune system or from selection favouring binding to specific variable cell receptors.

A rapid rate of divergence of cell surface components has also been described for homologous proteins from rodents and humans (Murphy, 1993). Murphy classified more than 600 proteins into 14 protein families, and found that the group composed of host defence ligands and receptors (interleukins, interferons, colony-stimulating factors, chemokines, etc.) had a mean amino acid divergence of 35% compared to a range of 1–12% for the 11 of the 12 other families. Thus proteins that play a role in host defence are approximately three-fold more divergent than average cellular proteins, the mirror image of the bacterial surface proteins. Murphy hypothesizes that selective pressure to evade mimicry at the molecular level by pathogenic microorganisms has accelerated the rate of the amino acid replacement in proinflammatory ligands and receptors.

CLONAL NATURE OF ENTERIC PATHOGENS

The variation detected by multilocus enzyme electrophoresis has provided useful systems of genetic markers for analysing the population structure of bacterial species that cause human infectious diseases. A key generalization derived from these studies is that many bacterial pathogens have a clonal population structure with a limited number of geographically widespread, pathogenic clones accounting for most cases of disease (Selander & Musser, 1990). For example, *Shigella sonnei*, a pathogen that causes dysentery, represents a homogeneous, geographically widespread clone, that is in reality a single clonal lineage of *E. coli* (Ochman *et al.*, 1983; Whittam, Ochman & Selander, 1983*a*, *b*). Karaolis, Lan & Reeves (1994) have shown that strains of *S. sonnei* collected from unassociated hosts on different continents over a 40-year period are virtually homogeneous in serotype and electrophoretic type, and are similar in DNA sequence for two housekeeping genes. However, despite the overall low level of variation among sonnei strains, restriction site analysis of ribosomal genes shows a clear change in the frequency of alleles and suggesting a recent clonal replacement (Karaolis *et al.*, 1994).

Selander and coworkers have used multilocus enzyme electrophoresis to analyse the population structure and clonal relationships among common serovars of *S. enterica* associated with enteric disease in human and domesticated animals (Beltran *et al.*, 1988; Selander *et al.*, 1990). Eight serovars (typhimurium, heidelberg, cholerasuis, typhi, paratyphi C, panama, and saintpaul) were each found to be monophyletic and composed of a single predominant electrophoretic type. Because the strains were originally recovered at different times from separate geographic areas, the authors infer that these multilocus genotypes distinguish pathogenic clones with cosmopolitan distributions. In contrast, other serotypes comprised a diversity of enzyme genotypes suggesting that, in these cases, the somatic and flagellar antigens underlying the serotyping scheme have repeatedly converged by recombination and mutation (Selander *et al.*, 1990).

Enteropathogenic E. coli *(EPEC)*

The concept that *E. coli* populations are composed of widespread clones was introduced by the Ørskovs (Ørskov *et al.*, 1976; Ørskov & Ørskov, 1983) to account for the observation of identical phenotypes, in such variable traits as

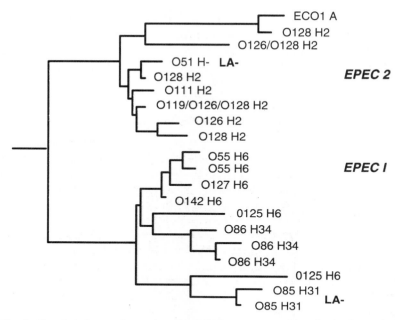

Fig. 3. Clonal phylogeny for strains with EPEC serotypes and localized adherence. Shaded boxes highlight groups on two main branches: EPEC 1 includes strains with either H6 or H34 flagellar antigen whereas EPEC 2 includes strains with H2 antigens. ECO1 A represents the K-12 clone complex.

serotype and biotype, among *E. coli* strains recovered from separate outbreaks of disease. The first group of strains that were epidemiologically linked to enteric disease were the enteropathogenic *E. coli* (EPEC) strains that caused severe outbreaks of infantile diarrhoea in hospitals in Great Britain more than 40 years ago (Levine & Edelman, 1984). Only specific O : H serotypes of *E. coli* were incriminated in outbreaks, and these strains were later shown to have a distinct phenotype defined by the pattern of adhesion to tissue culture cells, called localized adherence (LA) (Cravioto *et al.*, 1979). The LA+ phenotype corresponds with the special ability of bacterial cells to form distinct microcolonies on eukaryotic cells. The full expression of the LA phenotype is mediated by a plasmid that carries an adherence factor (Nataro *et al.*, 1985; Gomes *et al.*, 1989).

Characterization of the electrophoretic types of 50 LA+ strains representing nine EPEC serotypes revealed that most serotypes represent homogeneous bacterial clones (Ørskov *et al.*, 1990). Strains of the same serotype from diverse sources were identical in ET and indistinguishable in outer membrane protein profiles and other phenotypic properties. More importantly, LA+ strains fell into two genetically related groups, clustering that was not evident in the distribution of O serotypes (Fig. 3). Within each cluster, a variety of O antigens were present; however, flagellar antigens (H types) tended to be conserved within a lineage, with EPEC 1 strains typically expressing H6 antigen and EPEC 2 strains having H6 antigens (Fig. 3). In two cases, clones had lost the LA phenotype which was accompanied by a change in the adherence character of the strains from a localized to a perinuclear pattern (Ørskov *et al.*, 1990).

These results indicate that classical EPEC strains represent widespread clones that are organized into two distinct clone complexes. Thus a majority of cases of EPEC disease are caused by infection by members of these clone complexes. Strains of both clusters express the LA phenotype, which has been shown to be plasmid mediated. These two EPEC lineages are only distantly related to other pathogenic *E. coli* (Whittam *et al.*, 1993), suggesting that the LA ability has evolved in divergent chromosomal backgrounds, presumably through the horizontal spread of plasmid-borne genes. The maintenance of LA phenotypes in separate lineages also suggests that this ability confers a selective advantage to enteropathogenic strains.

Clonal diversity and disease

In contrast to the situation in which bacteria of a single clone or clone complex cause most cases of infectious disease (Selander & Musser, 1990), the analysis of genetic variation in *E. coli* strains from certain populations has disclosed surprisingly high levels of clonal diversity. For example, Woodward *et al.* (1993) characterized the multilocus enzyme genotypes of 87 *E. coli* isolates collected from neonatal pigs with diarrhoea from Australia. Although the isolates represented three common O serogroups (O9, O20, and O101) and were collected from one pathological condition in a single country, the genetic diversity was extensive with 73 multilocus genotypes resolved for 19 enzyme loci. There was no single predominant clonal type or common genetic background associated with porcine diarrhoeal disease. Similar results have been obtained from diseased birds in domesticated flocks where three or four diverse clone complexes are typically found among isolates recovered from cases of colibacillosis, airsacculitis, or pericarditis (Whittam & Wilson, 1988*a,b*; White *et al.*, 1990).

There are several hypotheses that may account for the genetically heterogeneous nature of strains recovered from certain disease populations. First, the clinical signs defining a specific condition may be the manifestation of more than one type of infection so that bacteria collected from affected hosts represent a mixture of divergent strains with different virulence factors and modes of pathogenesis. Secondly, strains implicated in disease might share a powerful virulence factor that has been spread horizontally among many lines in the bacterial population. Under this hypothesis, the transmission of a factor by plasmids or phage could promote the virulence of a large number of strains with different chromosomal genomes. Thirdly, a large fraction of the *E. coli* isolates recovered from affected individuals may represent opportunistic infections by normally non-pathogenic strains. In this case, the frequency of opportunistic clones obtained from diseased individuals would reflect the prevalence of those bacterial genotypes in the normal flora. Evidence for this type of mixture of clones is found for strains implicated in urinary tract infections in humans and animals, in which population genetic studies have shown that the variety of ETs and serotypes recovered from affected individuals represent both uropathogenic clones and opportunistic infections (Caugant *et al.*, 1983; Whittam, Wolfe & Wilson, 1989).

EMERGENCE OF A NEW BACTERIAL PATHOGEN

In 1982, several outbreaks of an unusual form of bloody diarrhoea, called haemorrhagic colitis, drew attention to a new pathogenic *E. coli* strain, serotype O157 : H7, with a novel mode of pathogenesis that previously had not been associated with enteric disease (Riley *et al.*, 1983). The strains recovered from these outbreaks were unusual in that they did not possess the virulence determinants typical of other *E. coli* that cause infectious enteric disease: they failed to produce the classical heat-labile or heat stable enterotoxins, lacked invasive abilities, and were serotypically distinct from EPEC strains that have long been associated with worldwide outbreaks of infantile diarrhoea (Riley *et al.*, 1983). Since these initial outbreaks, strains of the O157 : H7 serotype have been incriminated in serious outbreaks of haemorrhagic colitis (HC) and haemolytic uraemic syndrome (HUS), and have been linked to hundreds of sporadic cases of gastrointestinal illness in the United States and Canada (Karmali, 1989). For example, in January 1993, O157 : H7 strains caused a regional outbreak of severe diarrhoea in Washington that resulted in three deaths and more than 600 reported cases of illness (O'Brien *et al.*, 1993). Illnesses associated with *E. coli* O157 : H7 infections as well as diseases caused by other cytotoxin-producing strains (also referred to as Verocytotoxigenic *E. coli* or VTEC), have emerged as a major health problem in North America and Europe (Smith & Scotland, 1988; Karmali, 1989; Cryan, 1990).

At present, the mechanism of virulence and pathogenesis involved in infections of O157 : H7 strains is not fully understood, but it differs from other mechanisms described in pathogenic *E. coli* that cause diarrhoeal disease in humans and animals. Several factors have been implicated in the virulence of O157 : H7 strains including a high level of expression of potent Shiga-like cytotoxins (O'Brien *et al.*, 1984; Newland *et al.*, 1985), carriage of plasmids that encode adhesins mediating bacterial adherence to intestinal cells (Karch *et al.*, 1987; Levine *et al.*, 1987), and production of intimin, a protein encoded by the chromosomal *eaeA* gene which is involved in the intimate attachment of bacteria to enterocytes and subsequent effacement of the microvilli (Jerse *et al.*, 1990; Donnenberg & Kaper, 1992; Yu & Kaper, 1992).

To determine the genetic relationships of O157 : H7 strains to other pathogenic forms of *E. coli*, we used multilocus enzyme electrophoresis to study the genetic diversity and clonal relationships among O157 : H7 isolates and strains of other serotypes implicated in diarrhoeal disease (Whittam *et al.*, 1993). By detecting allelic variation at polymorphic enzyme loci, the multilocus approach provides a sensitive system of genetic markers for characterizing the chromosomal genotypes of strains, and for estimating the overall genomic relatedness of isolates (Selander *et al.*, 1986). Population genetic analysis demonstrated that O157 : H7 isolates from recent epidemics of HC and HUS in North America were a single clone that was not closely allied to Shiga-like cytotoxin-producing strains of other *E. coli* serotypes (Whittam, Wachsmuth & Wilson, 1988), many of which produce a clinically similar form of bloody diarrhoea (Tzipori *et al.*, 1986, 1987). The O157 : H7 clone was also found to be only distantly related to other ETs of the O157 group associated with enteric infections in animals (Whittam & Wilson, 1988*a,b*).

Further comparisons of O157 : H7 strains to a diverse collection of isolates of serotypes associated with infectious diarrhoeal disease revealed that 72% of the isolates belong to 15 major ETs each of which marks a bacterial clone with a widespread geographic distribution (Whittam *et al.*, 1993). Genetically, the O157 : H7 clone is most closely related to a clone of O55 : H7 strains that has long been associated with worldwide outbreaks of infantile diarrhoea. Both O55 : H7 and O157 : H7 strains attach intimately to the surfaces of intestinal epithelial cells in the initial stages of infection, efface the microvilli, and induce characteristic histological and ultrastructural lesions in animal models (Law, 1994). Knutton *et al.* (1989) showed that these attaching and effacing (A/E) lesions are composed of dense concentrations of actin microfilaments in the apical cytoplasm beneath the attached bacteria and that both O55 : H7 and O157 : H7 strains produce similar A/E lesions in human tissue cultures. Furthermore, the production of A/E lesions requires neither plasmid-encoded products nor expression of cytotoxins (Knutton *et al.*, 1989; Tzipori, Gibson & Montanaro, 1989) and the A/E mechanism is determined, in part, by the expression of intimin, and other products of the chromosomal *eae* gene complex (Donnenberg, Yu & Kaper, 1993; Law, 1994). Presumably the *eae* gene complex was present in the most recent ancestor of the O55 : H7 and O157 : H7 strains because of the overall close genomic relatedness of these strains.

Evolution scenerio of divergence

The above observations suggest the hypothesis that the O157 : H7 pathogenic clone emerged when an O55 : H7-like progenitor, already possessing a mechanism for adherence to intestinal cells, acquired secondary virulence factors (Shiga-like cytotoxins and plasmid-encoded adhesins) via horizontal transfer and recombination. The working model is that, first, an ancestral *E. coli* evolved the chromosomally encoded gene products that mediate attaching and effacing adherence. This attribute alone may be sufficient for bacteria to cause diarrhoeal disease in infants, as is the case of the contemporary O55 : H7 clone. Secondly, an O55 : H7-like progenitor cell, already able to cause disease with the A/E mechanism, acquired secondary virulence factors, such as the genes encoding Shiga-like toxins and adhesins, via horizontal genetic transfer from other strains. With these genes expressed in the A/E chromosomal background, a new pathogenic lineage causing a new type of disease emerged – the O157 : H7 clone.

Besides SLT-production, O157 : H7 strains exhibit several phenotypic attributes that are rare in the *E. coli* population as a whole. For example, O157 : H7 strains do not ferment D-sorbitol rapidly, in contrast to about 95% of other *E. coli* strains, thus providing a useful phenotype for screening of O157 : H7 strains based on sorbitol indicator medium (Farmer & Davis, 1985). Recently, SLT-producing sorbitol-fermenting O157 : H– were discovered in an outbreak of HUS in Upper Bavaria, Germany, and Gunzer *et al.* (1992) have shown they are the most prevalent pathogen in children associated with HUS in Germany. Karch *et al.* (1993) showed that isolates of sorbitol (SF) O157 : H– were nearly identical in *Xbal* restriction patterns in pulsed-field gel electrophoresis and differed markedly from SF O157 : H45 and non-SF O157 : H7. They further showed that SF O157 : H– carry sequences related to the *eae* and *sltll* gene, and are lysogenized by toxin-converting phages. From these observations, Karch and coworkers posit that the SF O157 : H– represent a new clone with similar virulence properties to non-SF O157 : H7 (Karch *et al.*, 1993).

Comparison of the multilocus enzyme genotypes of the SF O157 : H– strains from Germany reveals that this distinct clone is part of the O157 : H7 clone complex (Whittam and Karch, unpublished data). In conjunction with the phenotypic attributes of strains, the relatedness of genomes, as indicated by similarity in multilocus genotypes, allows one to begin to reconstruct the steps involved in the emergence and radiation of the O157 : H7 clone complex. The evolutionary scenerio is outlined in Fig. 4 as a series of bifurcations in a clonal phylogeny. The model begins with an ancient common ancestor, labelled A0, of the EPEC O127 : H6 strain (isolate E2348/69) and the lineage that gave rise to EHEC O157 : H7 (Fig. 4). Because both the O127 : H6 and O157 : H7 have the *eae* gene, which is 87% similar in nucleotide sequence (Yu & Kaper, 1992), it is assumed that this gene was present in the common ancestor. The numbers by each branch in the phylogeny indicate the minimum number of allele replacements at different loci inferred from the 35-enzyme multilocus genotypes. During the course of the evolution of the EHEC lineage in which at least ten allele

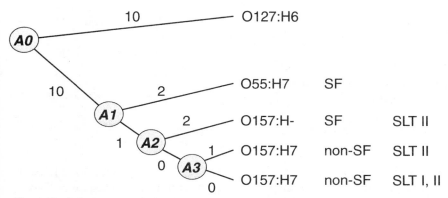

Fig. 4. Evolutionary scenario for the emergence of *E. coli* O157 : H7. The numbers on each branch denote the number of allelic replacements inferred from multilocus enzyme electrophoresis of 35 enzymes. Hypothetical ancestors are designated in the shaded circles. Abbreviations: SF = sorbitol fermenting; non-SF = non-sorbitol fermenting; SLT = Shiga-like toxin gene.

replacements occurred, the cells evolved the H7 flagellar antigen which is hypothesized to be present on the most recent of the O55/O157 : H7 branches (node A1). At this point, cells also fermented sorbitol (SF). In the transition for A1 to A2, the lineage acquired the O157 lipopolysaccharide antigen that occurs in the descendents. Concurrently, this ancestor may have acquired an SLT II gene and the EHEC plasmid (Levine *et al.*, 1987) which is characteristic of the derived O157 : H7 lines. From ancestor A2, the SF-fermenting O157 : H7 clone emerged and accumulated two enzyme allele replacements and lost the H7 antigen (H–). The other branch led to the immediate ancestor (A3) of the non-SF O157 : H7 clones that are common and widespread in North America and Europe. One O157 : H7 clone acquired a second SLT gene (SLT I), and this double toxin producer was in particular high frequency in the north west United States in the early 1980s (Tarr *et al.*, 1989). The other non-SF O157 : H7 clone typically carries only SLT II. But, as noted below, toxin acquisition and loss may have occurred repeatedly so that this interpretation above is too simplistic.

Interestingly, O157 : H7 strains differ from O55 : H7 by an electromorph of 6PGD (encoded by the *gnd* locus) suggesting the possibility that the transition from A1 to A2 involved a recombination event in the *rfb* region with cotransfer of the nearby *gnd* locus. A similar observation was recorded for differences in the somatic antigens and the 6PGD alleles found in three closely related *E. coli* K1 clones associated with invasive disease (Selander *et al.*, 1987).

Evolutionary time scale

To put the scenerio depicted in Fig. 4 on an evolutionary time scale, one can use information about the degree of divergence of *eaeA* alleles and the rate

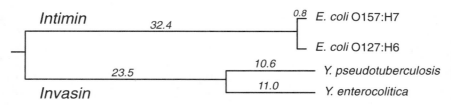

Fig. 5. Gene tree for *E. coli* intimin and *Yersinia* invasin. Neighbour-joining tree based on percentage amino acid divergence in central conserved region.

of evolutionary at the molecular level. The protein product of the *eaeA* locus, is an outer membrane protein called intimin (Donnenberg & Kaper, 1992). Intimin is remarkably similar in amino acid sequence to the invasins of *Yersinia pseudotuberculosis* (Isberg, Vorrhis & Falkow, 1987) and *Y. enterocolitica* (Young *et al.*, 1990) which mediate the binding and penetration of bacteria into eukaryotic cells. The central region of the invasin and intimin genes is relatively well conserved and shows that the *Yersinia* invasins are about ten times more divergent in amino acid sequences than the *E. coli* intimins (Fig. 5).

By estimating the nucleotide changes per nonsynonymous site in the comparison of these genes, one can infer times of divergence for recent branches of the tree. Ochman and Wilson (1987) estimated that *Yersinia* and *Escherichia* split about 200–250 million years ago. For 226 codons in the central region, the number of nonsynonymous differences per site is 0.0116 ± 0.005 between the *E. coli* intimin genes, and 0.503 ± 0.041 and 0.497 ± 0.041 for each intimin gene to the invasin of *Y. pseudotuberculosis*. With the assumption of a uniform rate of amino acid replacement in this region, the time of splitting of the *E. coli eae* alleles is 4.6–5.8 million years ago. Thus, the evolution scenario for the emergence of O157 : H7 strains began about 5 million years ago with the ancestral intimin gene (Fig. 4).

Shiga-like toxin genes

A striking characteristic of O157 : H7 strains is their capacity to produce potent cytotoxins that can kill HeLa and Vero cells and are clearly distinct in effects on cell cultures from the LT and ST enterotoxins of pathogenic *E. coli* (O'Brien & Holmes, 1987; Karmali, 1989). The cytotoxins produced by O157 : H7 strains are functionally related to Shiga toxin produced by *Shigella dysenteriae* and are referred to as Shiga-like cytotoxins (O'Brien & Holmes, 1987). Because of their cytotoxigenic effect on Vero cells, SLT toxins are also referred to as Verocytotoxins. Although the precise role of the Shiga-like toxins (SLT) in the pathogenicity of O157 : H7 strains is unknown, toxin production appears to be an important component of the virulence mechanism of this organism.

Two antigenically distinct types of SLTs are recognized, SLT-I and SLT-II, which share only about 60% sequence similarity. Shiga toxin and SLT-I are nearly identical in sequence (Strockbine *et al.*, 1988), whereas a variant of SLT-II (SLT-IIv) isolated from oedema disease in swine shows 80% similarity with SLT-II (Weinstein *et al.*, 1988).

The hypothesis that the O157 : H7 clone originated from a O55 : H7-like ancestor through horizontal transfer and recombination gains support from the observation that SLTs occur in diverse lineages of the *E. coli* population and can be transferred by phages in laboratory conditions (O'Brien *et al.*, 1984; Newland *et al.*, 1985). The toxin-converting phages are members of a diverse family that appear to be widespread in nature (O'Brien & Holmes, 1987). Related cytotoxins have also been discovered in *Citrobacter freundii* suggesting that SLT genes can transfer laterally among enteric bacteria (Schmidt *et al.*, 1993). Finally, the pattern of codon usage in the SLT genes also differs from that of most *E. coli* protein-encoding genes (Jackson *et al.*, 1987) and the codon adaptation index (Sharp & Li, 1987) for SLT genes falls below values reported for 165 *E. coli* genes. Such a low degree of codon adaptation strongly argues that the SLT genes are foreign to *E. coli* and were relatively recently acquired via horizontal transfer (Whittam & Ake, 1993).

DIVERGENCE OF PATHOGENIC O111 CLONES

E. coli of serogroup O111 were the first strains implicated as the main cause of outbreaks of severe gastro-enteritis in infant nurseries in the United Kingdom in the 1940s (Bray, 1945; Kauffmann & Dupont, 1950 and O111 is now recognized as one of the classic serogroups of enteropathogenic *E. coli* or EPEC (Levine & Edelman, 1984). Since this early work, *E. coli* O111 strains have been implicated in numerous epidemics of serious enteric disease, including 28% of 50 outbreaks of infantile diarrhoea in the United States from 1934–87 (Moyenuddin *et al.*, 1989). An O111 strain was incriminated in an outbreak of prolonged, relapsing diarrhoeal illness in a day-care centre in Seattle, Washington (Paulozzi *et al.*, 1986) and, more recently, an O111 strain caused an extensive community outbreak of diarrhoea involving more than 700 people in Finland (Vicjanen *et al.*, 1990). In addition to clear-out epidemics, *E. coli* O111 strains are prevalent in endemic (or sporadic) cases of diarrhoea. In Canada, Gurwith *et al.* (1978) recovered O111 strains in 3.5% of 418 children with acute diarrhoea, significantly more often than in matching controls, and in Brazil, Gomes *et al.* (1989) found O111 strains in 19% of 500 cases of diarrhoea in infants with only 0.4% in matching controls.

Surveys of O111 strains of diverse origin have shown that there is a substantial amount of genetic and phenotypic variation among bacteria in this O-serogroup. Although motile O111 strains from outbreaks of infantile diarrhoea usually have either H2, H12, or H21 flagellar antigens, in many cases, non-motile (H–) isolates are common and isolates of other specific H types are often cultured (Stenderup & Ørskov, 1983; Scaletsky *et al.*, 1985). Strains with serotypes O111 : H2 and H– typically carry genes specifying the EPEC adherence factor (EAF) that mediates localized adherence (LA) of bacteria to cultured epithelial cells and is a characteristic of strains of the classic EPEC serotypes (Nataro *et al.*, 1985; Gomes *et al.*, 1989). In contrast, O111 : H8 isolates have been classified as enterohaemorrhagic *E. coli* (EHEC) because they lack EAF and possess a plasmid related to one found in O157 : H7 strains that cause haemorrhagic colitis (Levine *et al.*, 1987).

To test the hypothesis that separate outbreaks of infantile diarrhoea were caused by the same bacterial strains, Stenderup and Ørskov (1983) examined O111 isolates of diverse origin and found that many cases of infantile diarrhoea were caused by organisms with identical serotypes, biotypes, and outer-membrane protein patterns, suggesting that, together, these characters marked bacterial clones with widespread distributions. This finding gained support from population genetic analysis by multilocus enzyme electrophoresis (Selander *et al.*, 1986), which showed that O111 : H2 strains of independent origin were clonally related with identical alleles at many enzyme-encoding loci (Ørskov *et al.*, 1990). More recently, Whittam *et al.* (1993) found that O111 isolates are genetically diverse, but that most belong to a limited number of bacterial clones as marked by distinct electrophoretic types.

To identify the major clonal types of the O111 serogroup associated with diarrhoeal disease and assess the variation among clones with regard to specific factors incriminated in virulence, Campos *et al.* (1994) characterized the multilocus genotypes of a collection of O111 strains using enzyme electrophoresis, and examined the distribution of adherence and toxin factors that contribute to the virulence of diarrhoeagenic *E. coli*.

Genetic variation among isolates of *Escherichia coli* O111 obtained from patients with diarrhoea in Brazil was extensive: among 152 isolates, there were 16 distinct ETs which differed on average at 40% of the enzyme loci (Fig. 6). Most isolates belonged to one of four major bacterial O111 clones that differed in virulence properties. ET 12 included the bulk of the enteropathogenic *E. coli* (EPEC) O111 : H2 strains and that typically showed localized adherence and intimate attachment in tissue culture assays. ET 1 was composed of diffusely adhering strains and lacked other virulence markers. ET 9 included strains that showed intimate attachment, but did not have localized adherence or SLT genes. Finally, ET 8 comprised O111 strains that are SLT-producers and have the corresponding traits of enterohaemorrhagic *E. coli* (EHEC). Enteroaggregative strains constituted ET 10 and also occurred in ET 1. Isolates of the major clones were found in South and North America and matched in ET and virulence factors to previously described diarrhoeagenic clones that are widely disseminated in the human population.

Virulence factors

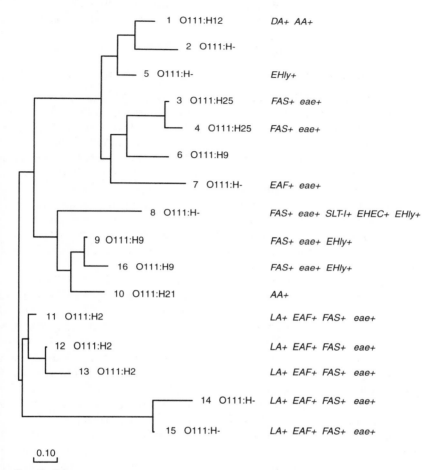

Fig. 6. Clonal phylogeny for 16 electrophoretic types of *E. coli* O111 strains. Genetic distance is measured in terms of the number of detectable codon differences per enzyme locus. The presence of virulence factors in the branches of the tree are given in the right column. All isolates tested were negative with probes for the diffuse adherence factor and Shiga-like toxin II gene. Abbreviations: EAF, enteropathogenic *E. coli* adherence factor; *eae*, *E. coli* attaching-effacing gene; SLT-1, Shiga-like toxin gene I; EHEC, enterohaemorrhagic *E. coli* adherence factor; FAS, fluorescence actin staining; EHly, enterohaemolysin; LA, localized, DA, diffuse, AA, aggregative pattern of adherence of bacteria to HEp-2 cells.

Because the major O111 clones are genetically distantly related and exhibit different combinations of virulence factors, it is likely that they have accumulated many genetic differences and have distinct mechanisms of pathogenesis (Fig. 6). The results show that genetic divergence of bacteria with O111 antigen, as measured by allelic variation in enzyme loci, is accompanied by divergence in virulence properties of clones so that identification and classification of pathogenic E. coli strains cannot be based solely on serotyping or a single virulence factor.

RAPID ANALYSIS AND GENOMIC DIVERGENCE

Many methods have been developed for distinguishing closely related bacterial strains for epidemiological purposes (Maslow, Mulligan & Arbeit, 1993). One method with growing popularity is arbitrary primer PCR, also known as randomly amplified polymorphic DNA or RAPD analysis. This PCR-based method requires only very small quantities of DNA and virtually no prior knowledge about nucleotide sequences. RAPD analysis uses low stringency PCR with a single, short oligonucleotide to generate an array of amplified DNA fragments that can be separated by size on agarose gels. Used in its simplest form, the array of fragment sizes provides a 'fingerprint' for delineating strains differences. Under certain assumptions however, RAPD data can be utilized to infer genetic divergence at the nucleotide level (Clark & Lanigan, 1993).

Wang *et al.* (1993) examined genetic variation detected by RAPD analysis among 75 isolates representing 15 ETs that mark widespread pathogenic clones of E. coli referred to as the DEC (diarrhoeagenic E. coli clones). With five primers, the number of distinct PCR bands scored on agarose gels ranged from 30 to 44 across all DEC clones. The power of single-primer PCR to discriminate among strains was great, as reflected in the observation that individual primers resolved, on average, 58 distinct fragment profiles of types among the 15 original ETs. In total, there were 189 bands and 74 distinct types for all five primers among the 75 strains, and only two O157 : H7 isolates of DEC 3 had identical RAPD fingerprints.

Genetic variation at the nucleotide level can be estimated from RAPD data by the method of Clark and Lanigan (1993). Briefly, for a pair of haploid isolates x and y, we counted the number of bands of identical size (n_{xy}), calculated the fraction of shared fragments as $F = 2n_{xy}/(n_x + n_y)$ where n_x and n_y are the number of bands in x and y, respectively, and iterated the following equation until $P = P_1$ to get an estimate of P (the probability no mutation has occurred at a primer site since the time from a common ancestor of two genomes):

$$P = [F(3 - 2P_1)]^{1/4}$$

where $F^{1/4}$ was used as the initial value of P_1. The number of nucleotide substitution per site can then be estimated as

$$d = -\left(\frac{2}{r}\right) \ln (P)$$

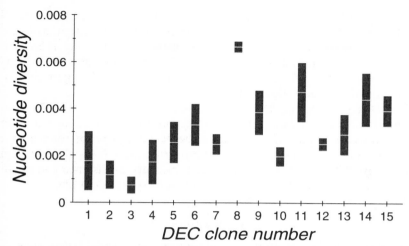

Fig. 7. Nucleotide diversity among strains of diarrhoeagenic *E. coli* (DEC) clones estimated from RAPD analysis. Each bar represents one standard error about the mean diversity.

where *r* is the length of the primer site.

This method can be applied to situations in which bacteria are sampled from separate populations (or groups) and both components of nucleotide diversity (π) within populations and nucleotide divergence (d) between populations can be estimated. To obtain π among isolates of the same ET, one calculates a composite F_c by comparing all pairs of isolates of the same clone (population) and uses the above equations to estimate the number of substitutions per site. The nucleotide divergence (d) between clones was calculated in a similar manner by comparing all pairs of isolates representing two different ETs (see equation 6 of Clark & Lanigan (1993)).

The extent of nucleotide variation among isolates of each DEC clone was estimated from the fraction of shared RAPD bands. For the RAPD data produced by three 10-mer primers, nucleotide diversity (π) among the five isolates of each DEC clone ranged from about 0.001 to 0.007 (Fig. 7). There were five clonal lineages that had relatively low diversity ($\pi < 0.002$), including O55 : H6 strains of DEC 1 and 2, O157 : H7 strains of DEC 3 and 4, and O26 : H11 strains of DEC 10. The greatest within-clone variation was among isolates of three lineages where $\pi > 0.004$, including DEC clones 8 (O111 : H8), 11 (O26 : H11), and 14 (O128 : H21).

Nucleotide divergence between clones showed a wide range of variation from a low level, equivalent to diversity within a clone ($<0.50\%$), to an extensive level representing 3–4% sequence difference. The most closely related DEC clones, based on the RAPD analysis, are DEC 1 and 2 and DEC 3 and 4 which share 95% of the PCR bands, and are 0.18% and 0.19% different, respectively. A slightly greater degree of divergence, in the range of 0.5% to 1.0%, occurs between DEC clones 5, 3, and 4, DEC 8 and 10, and DEC 13 and 14. These groupings are consistent with the relationships predicted in the DEC phylogeny based on multilocus enzyme electrophoresis (Whittam *et al.*, 1993), with the most divergent DEC clones sharing less than a third of the PCR bands and showing more than 4% divergence in nucleotide sequence.

Estimates of nucleotide divergence between ETs based on RAPD analysis with primers 1–3 and the genetic distance inferred from allelic variation were highly correlated. The correlation coefficient of 0.87 for 105 pairwise comparisons of the 15 DEC clones indicates that the DNA polymorphism assessed by PCR with 10-mers provides a measure of distance congruent with allelic differences detected by multilocus enzyme electrophoresis.

Resolving power

To obtain an idea of the limit of resolution of multilocus enzyme electrophoresis relative to amount of genomic sequence divergence as measured by RAPD analysis, one can compare the levels of variation among isolates within and between ET groupings. In principle, the resolution limit should lie between the maximum within-ET diversity and the minimum between-ET divergence. The maximum π per 100 sites was 0.67 for DEC 8 and three smallest d values per 100 sites were 0.77, 0.81, and 0.89. These values suggest a limit of about 0.7% genomic sequence divergence for enzyme electrophoresis with the 20 loci used to identify the DEC clones. For twice as many loci, the resolution limit should be cut in half. The results of analysis of 15 additional loci are consistent with an increase in resolving power. For the nine DEC ETs, the examination of 15 more loci subdivided strains that were identical for 20 enzymes; the average π per 100 sites based on RAPD analysis was 0.36. In contrast, strains of six DEC ETs were not split, and in these cases, the average π per 100 sites was 0.21. This observation suggests a limit of resolution of genomic divergence near 0.3% for 35 loci, which, as expected, is nearly half the value obtained for 20 loci.

In summary, RAPD data reveal that the amount of nucleotide diversity within a DEC clone (based on 20 enzymes) is of the order of 0.5% sequence variation with divergence between clones of up to 3–4% sequence divergence. The RAPD results were highly correlated with genetic distances and confirm a close genetic relationship and relatively recent divergence of O157 : H7 and O55 : H7 strains.

CONCLUSIONS

Although the study of the genetics of bacterial populations is relatively recent and has focused on a limited number of species, a few generalizations have become evident. Bacterial populations harbour extensive amounts of allelic variation. Because reproduction is asexual, alleles are strongly associated in genotypes and populations are organized into distinct cell lineages or clones. Against this clonal structure, mechanisms for the transfer of genetic material recombine alleles into new combinations and create mosaic alleles at low rate each generation. For surface proteins and genes under positive selection, recombination may be more effective as mosaic genes are driven to high frequency.

Evolution of strains implicated in infectious diseases cannot be understood without reference to the clonal structure of populations. Acquisition of powerful virulence factors via horizontal recombinations can cause convergence in disease characteristics for organisms that are distantly related. In the enteropathogenic *E. coli* that cause infantile diarrhoea, there are two distinct clone complexes, both of which exhibit the localized adherence phenotype. Similarly, the Vi antigen occurs in three distantly related phylogenetic lineages of *S. enterica* (typhi, paratyphi C and dublin) indicating convergence through lateral transfer of the *viaB* gene has occurred in the radiation of these enteric clones (Selander *et al.*, 1990). These observations suggest that only certain clonal genotypes possess the appropriate background for virulence factors to be maintained and for the clone to increase in frequency and spread geographically.

The increase in prevalence of a new enteric disease, haemorrhagic colitis, is due to the emergence and spread of a novel pathogenic clone of *E. coli*, called O157 : H7. This pathogenic clone evolved through a series of steps beginning with an ancestral enteropathogenic strain that possessed the ability to intimately adhere to intestinal epithelial cells. Through the acquisition of Shiga-like cytotoxin genes via toxin-converting phages, this lineage became more virulent to humans. Molecular divergence times suggest that the first stages in the radiation of these enteric pathogens was early in the evolution of the hominids, about 5 million years ago.

Finally, it should be noted that some features of pathogenic clones may be irrelevant to adaptation or disease processes because the traits have been driven to high frequency by genetic hitch-hiking. This point was emphasized previously by Achtman and Pluschke (1986) in their analysis of pathogenic *E. coli* implicated in invasive disease. Thus, the high frequency of sorbitol-negative O157 : H7 strains most likely reflects recent spread of an O157 : H7 clone rather than the influence of direct selection on the benefits or costs of sorbitol fermentation. Of course, it is the hitchhiking of such phenotypic traits that underlies the usefulness of many of the markers that identify pathogenic clones.

ACKNOWLEDGEMENTS

This work was supported by US Public Health Service grants AI 24566 and AI 00964 from the National Institutes of Health.

REFERENCES

Achtman, M. & Hakenbeck, R. (1992). Recent developments regarding the evolution of pathogenic bacteria. In *Molecular Biology of Bacterial Infection: Current Status and Future Perspectives*. (C. E. Hormache, C. W. Penn and C. J. Smyth, eds.), Cambridge University Press, New York, NY.

Achtman, M. & Pluschke, G. (1986). Clonal analysis of descent of virulence among selected *Escherichia coli*. *Annual Review of Microbiology*, **40**, 185–210.

Beltran, P., Musser, J. M., Helmuth, R. III, J. J. F. *et al.* (1988). Toward a population genetic analysis of Salmonella: Genetic diversity and relationships among strains of serotypes *S. cholerasuis*, *S. derby*, *S. dublin*, *S. enteritidis*, *S. heidelberg*, *S. infantis*, *S. newport*, and *S. typhimurium*. *Proceedings of the National Academy of Sciences, USA*, **85**, 7753–7.

Biserčić, M., Feutrier, J. Y. & Reeves, P. R. (1991). Nucleotide sequences of the *gnd* genes from nine natural isolates of *Escherichia coli*: evidence of intragenic recombination as a contributing factor in the evolution of the polymorphic gnd locus. *Journal of Bacteriology*, **173**, 3894–900.

Boyd, E. F., Nelson, K., Wang, F.-S., Whittam, T. S. & Selander, R. K. (1994). Molecular genetic basis of allelic polymorphism in malate dehydrogenase (*mdh*) in natural populations of *Escherichia coli* and *Salmonella enterica*. *Proceedings of the National Academy of Sciences, USA*, **91**, 1280–4.

Bray, J. (1945). Isolation of antigenically homogeneous strains of *Bact. coli neapolitanum* from summer diarrhoea of infants. *Journal of Pathology and Bacteriology*, **57**, 239–47.

Brown, A. H. D., Feldman, M. W. & Nevo, E. (1980). Multilocus structure of natural populations of *Hordeum spontaneum*. *Genetics*, **96**, 523–36.

Campos, L. C., Whittam, T. S., Gomes, T. A. T., Andrade, J. R. C. & Trabulsi, L. R. (1994). *Escherichia coli* serogroup O111 includes several clones of diarrheagenic strains with different virulence properties. *Infection and Immunity*, **62**, 3282–8.

Caugant, D. A., Levin, B. R., Lidin-Janson, G. *et al.* (1983). Genetic diversity and relationships among strains of *Escherichia coli* in the intestine and those causing urinary tract infections. *Progress in Allergy*, **33**, 203–27.

Clark, A. G. & Lanigan, C. M. S. (1993). Prospects for estimating nucleotide divergence with RAPDs. *Molecular Biology and Evolution*, **10**, 1096–111.

Cravioto, A., Gross, R. J., Scotland, S. M. & Rowe, B. (1979). An adhesive factor found in strains of *Escherichia coli* belonging to the traditional infantile enteropathogenic serotypes. *Current Microbiology*, **3**, 95–9.

Cryan, B. (1990). Enterohaemorrhagic *Escherichia coli*. *Scandinavian Journal of Infectious Diseases*, **22**, 1–4.

Donnenberg, M. L., Yu, J. & Kaper, J. B. (1993). A second chromosomal gene necessary for intimate attachment of enteropathogenic *Escherichia coli* to epithelial cells. *Infection and Immunity*, **175**, 4670–80.

Donnenberg, M. S. & Kaper, J. B. (1992). Enteropathogenic *Escherichia coli*. *Infection and Immunity*, **60**, 3953–61.

DuBose, R. F., Dykhuizen, D. E. & Hartl, D. L. (1988). Genetic exchange among natural isolates of bacteria: recombination within the *phoA* gene of *Escherichia coli*. *Proceedings of the National Academy of Sciences, USA*, **85**, 7036–40.

Dykhuizen, D. E. & Green, L. (1991). Recombination in *Escherichia coli* and the definition of biological species. *Journal of Bacteriology*, **173**, 7257–68.

Farmer, J. J. & Davis, B. R. (1985). H7 antiserum-sorbitol fermentation medium: a single tube screening medium for detecting *Escherichia coli* O157 : H7 associated with hemorrhagic colitis. *Journal of Clinical Microbiology*, **22**, 620–5.

Gomes, T. A. T., Vieira, M. A. M., Wachsmuth, I. K., Blake, P. A. & Trabulsi, L. R. (1989). Serotype-specific prevalence of *Escherichia coli* strains with EPEC adherence factor genes in infants with and without diarrhea in São Paulo, Brazil. *Journal of Infectious Diseases*, **160**, 131–5.

Gunzer, F., Böhm, H., Rüssmann, H. *et al*. (1992). Molecular detection of sorbitol-fermenting *Escherichia coli* O157 : H7 in patients with hemolytic–uremic syndrome. *Journal of Clinical Microbiology*, **30**, 1807–10.

Gurwith, M., Hinde, D., Gross, R. & Rowe, B. (1978). A prospective study of enteropathogenic *Escherichia coli* in endemic diarrheal disease. *Journal of Infectious Diseases*, **137**, 292–7.

Hall, B. G. & Sharp, P. M. (1992). Molecular population genetics of *Escherichia coli*: DNA sequence diversity at the *celC*, *crr*, and *gutB* loci of natural isolates. *Molecular Biology and Evolution*, **9**, 654–65.

Hartl, D. L. & Dykhuizen, D. E. (1984). The population genetics of *Escherichia coli*. *Annual Review of Genetics*, **18**, 31–68.

Hedrick, P. W. & Thompson, G. (1986). A two-locus neutrality test: applications to humans, *E. coli*, and lodgepole pine. *Genetics*, **112**, 135–56.

Isberg, R. R., Vorrhis, D. L. & Falkow, S. (1987). Identification of invasin: a protein that allows enteric bacteria to penetrate cultured mammalian cells. *Cell*, **50**, 769–78.

Jackson, M. P., Neill, R. J., O'Brien, A. D., Holmes, R. K. & Newland, J. W. (1987). Nucleotide sequence analysis and comparison of the structural genes for Shiga-like toxin I and Shiga-like toxin II enclosed by bacteriophages from *Escherichia coli* 933. *FEMS Lett*. **44**, 109–14.

Jerse, A. E., Yu, J., Tall, B. D. & Kaper, J. B. (1990). A genetic locus of enteropathogenic *Escherichia coli* necessary for the production of attaching and effacing lesions on tissue culture cells. *Proceedings of the National Academy of Sciences, USA*, **87**, 7839–43.

Karaolis, D. K. R., Lan, R. & Reeves, P. R. (1994). Sequence variation in *Shigella sonnei* (Sonnei), a pathogenic clone of *Escherichia coli*, over four continents and 41 years. *Journal of Clinical Microbiology*, **32**, 796–802.

Karch, H., Böhm, H., Schmidt, H. *et al*. (1993). Clonal structure and pathogenicity of Shiga-like toxin producing, sorbitol-fermenting *Escherichia coli* O157 : H–. *Journal of Clinical Microbiology*, **31**, 1200–5.

Karch, H., Heesemann, J., Laufs, R. *et al*. (1987). A plasmid of enterohemorrhagic *Escherichia coli* O157 : H7 is required for expression of a new fimbrial antigen and for adhesion to epithelial cells. *Infection and immunity*, **55**, 455–61.

Karmali, M. (1989). Infection by verocytotoxin-producing *Escherichia coli*. *Clinical Microbiological Reviews*, **2**, 15–38.

Kauffmann, F. & Dupont, A. (1950). Escherichia strains from infantile epidemic gastro-enteritis. *Acta Pathologia Microbiologia Scandinavica*, **27**, 552–63.

Knutton, S., Baldwin, T., Williams, P. H. & McNeish, A. S. (1989). Actin accumulation at sites of bacterial adhesion to tissue culture cells: basis of a new diagnostic test for enteropathogenic and enterohemorrhagic *Escherichia coli*. *Infection and Immunity*, **57**, 1290–8.

Law, D. (1994). Adhesion and its role in the virulence of enteropathogenic *Escherichia coli*. *Clinical Microbiology Reviews*, **7**, 152–73.

Levin, B. L. (1981). Periodic selection, infectious gene exchange, and the genetic structure of *E. coli* populations. *Genetics*, **99**, 1–23.

Levine, M. M. & Edelman, R. (1984). Enteropathogenic *Escherichia coli* of classic serotypes associated with infant diarrhea: epidemiology and pathogenesis. *Epidemiological Reviews*, **6**, 31–51.

Levine, M. M., Xu, J., Kaper, J. B. *et al.* (1987). A DNA probe to identify enterohemorrhagic *Escherichia coli* of O157 : H7 and other serotypes that cause hemorrhagic colitis and hemolytic uremic syndrome. *Journal of Infectious Diseases*, **156**, 175–82.

Lewontin, R. C. (1991). Twenty-five years ago in genetics: electrophoresis in the development of evolutionary genetic: milestone or millstone? *Genetics*, **128**, 657–62.

Li, J., Nelson, K., McWhorter, A. C., Whittam, T. S. & Selander, R. K. (1994). Recombinational basis of serovar diversity in *Salmonella enterica*. *Proceedings of the National Academy of Sciences*, *USA*, **91**, 2552–6.

Li, W.-H. & Graur, D. (1991). *Fundamentals of Molecular Evolution*. Sunderland, Massachusetts, Sinuaer Associates, Inc.

Maslow, J. N., Mulligan, M. E. & Arbeit, R. D. (1993). Molecular epidemiology: application of contemporary techniques to the typing of microorganisms. *Clinical Infectious Diseases*, **17**, 153–64.

Maynard Smith, J., Smith, N. H., O'Rourke, M. & Spratt, B. G. (1993). How clonal are bacteria? *Proceedings of the National Academy of Sciences*, *USA*, **90**, 4384–8.

Milkman, R. & Bridges, M. M. (1990). Molecular evolution of the *Escherichia coli* chromosome. III. Clonal frames. *Genetics*, **126**, 505–17.

Milkman, R. & Bridges, M. M. (1993). Molecular evolution of the *Escherichia coli* chromosome. IV. Sequence comparisons. *Genetics*, **133**, 455–68.

Milkman, R. & Crawford, I. P. (1983). Clustered third-base substitutions among wild strains of *Escherichia coli*. *Science*, **221**, 378–80.

Milkman, R. & Stoltzfus, A. (1988). Molecular evolution of the *Escherichia coli* chromosome. II. Clonal segments. *Genetics*, **120**, 359–66.

Moyenuddin, M., Wachsmuth, I. K., Moseley, S. L., Bopp, C. A. & Blake, P. A. (1989). Serotype, antimicrobial resistance, and adherence properties of *Escherichia coli* strains associated with outbreaks of diarrheal illness in children in the United States. *Journal of Clinical Microbiology*, **27**, 2234–9.

Murphy, P. M. (1993). Molecular mimicry and the generation of host defense protein diversity. *Cell*, **72**, 823–6.

Musser, J. M., Schlievert, P. M., Chow, A. *et al.* (1990). A single clone of *Staphylococcus aureus* causes the majority of cases of toxic shock syndrome. *Proceedings of the National Academy of Sciences*, *USA*, **87**, 225–9.

Nataro, J. P., Scaletsky, I. C. A., Kaper, J. B., Levine, M. M. & Trabulsi, L. R. (1985). Plasmid-mediated factors conferring diffuse and localized adherence of enteropathogenic *Escherichia coli*. *Infection and Immunity*, **48**, 378–83.

Nelson, K. & Selander, R. K. (1992). Evolutionary genetics of the proline permease gene (*putP*) and the control region of the proline utilization operon in populations of *Salmonella* and *Escherichia coli*. *Journal of Bacteriology*, **174**, 6886–95.

Nelson, K., Whittam, T. S. & Selander, R. K. (1991). Nucleotide polymorphism and evolution in the glyceraldehyde-3-phosphate dehydrogenase gene (*gapA*) in natural population of *Salmonella* and *Escherichia coli*. *Proceedings of the National Academy of Sciences*, *USA*, **88**, 6667–71.

Newland, J. W., Strockbine, N. A., Miller, S. F., O'Brien, A. D. & Holmes, R. K. (1985). Cloning of Shiga-like toxin structural genes from a toxin converting phage of *Escherichia coli*. *Science*, **230**, 179–81.

O'Brien, A. D. & Holmes, R. K. (1987). Shiga and Shiga-like toxins. *Microbiological Reviews*, **51**, 206–20.

O'Brien, A. D., Melton, A. R., Schmitt, C. K. *et al.* (1993). Profile of *Escherichia coli* O157 : H7 pathogen responsible for hamburger-borne outbreak of hemorrhagic colitis and hemolytic uremic syndrome in Washington. *Journal of Clinical Microbiology*, **31**, 2799–801.

O'Brien, A. D., Newland, J. W. Miller, S. F. *et al.* (1984). Shiga-like toxin-converting phages from *Escherichia coli* strains that cause hemorrhagic colitis or infantile diarrhea. *Science*, **226**, 694–6.

Ochman, H. & Selander, R. K. (1984). Standard reference strains of *Escherichia coli* from natural populations. *Journal of Bacteriology*, **157**, 690–3.

Ochman, H., Whittam, T. S., Caugant, D. A. & Selander, R. K. (1983). Enzyme polymorphism and genetic population structure in *Escherichia coli* and *Shigella*. *Journal of General Microbiology*, **129**, 2115–726.

Ochman, H. & Wilson, A. C. (1987). Evolution in bacteria: evidence for a universal substitution rate in cellular genomes. *Journal of Molecular Evolution*, **26**, 74–86.

Ørskov, F. & Ørskov, I. (1983). The clone concept in epidemiology, taxonomy, and evolution of the Enterobacteriaceae and other bacteria. *Journal of Infectious Diseases*, **148**, 346–57.

Ørskov, F., Ørskov, I., Evans, D. J. Jr *et al.* (1976). Special *Escherichia coli* serotypes among enterotoxigenic strains from diarrhoea in adults and children. *Medical Microbiology and Immunology*, **162**, 73–80.

Ørskov, F., Whittam, T. S., Cravioto, A. & Ørskov, I. (1990). Clonal relationships among classic enteropathogenic *Escherichia coli* (EPEC) belonging to different O groups. *Journal of Infectious Diseases*, **162**, 76–81.

Paulozzi, L. J., Johnson, K. E., Kamahele, L. M. *et al.* (1986). Diarrhea associated with adherent enteropathogenic *Escherichia coli* in an infant and toddler center, Seattle, Washington. *Pediatrics*, **77**, 296–300.

Reeves, P. (1993). Evolution of *Salmonella* O antigen variation by interspecific gene transfer on a large scale. *Trends in Genetics*, **9**, 17–22.

Riley, L. W., Remis, R. S., Helgerson, S. D. *et al.* (1983). Hemorrhagic colitis associated with a rare *Escherichia coli* serotype. *New England Journal of Medicine*, **308**, 681–5.

Scaletsky, I. C. A., Silva, M. L. M., Toledo, M. R. F. *et al.* (1985). Correlation between adherence to HeLa cells and serogroups, serotypes, and bioserotypes of *Escherichia coli*. *Infection and Immunity*, **49**, 528–32.

Schmidt, H., Montag, M., Bockemuhl, J., Heesemann, J. & Karch, H. (1993). Shiga-like toxin II related cytotoxins in *Citrobacter freundii* strains from human and beef samples. *Infection and Immunity*, **61**, 534–45.

Selander, R. K., Beltran, P., Smith, N. H. *et al.* (1990). Evolutionary genetic relationships of clones of *Salmonella* serovars that cause human typhoid and other enteric fevers. *Infection and Immunity*, **58**, 2262–75.

Selander, R. K., Caugant, D. A., Ochman, H. *et al.* (1986). Methods of multilocus enzyme electrophoresis for bacterial population genetics and systematics. *Applied and Environmental Microbiology*, **51**, 873–84.

Selander, R. K., Caugant, D. A. & Whittam, T. S. (1987). Genetic structure and variation in natural populations of *Escherichia coli*. In *Escherichia coli* and *Salmonella typhimurium: Cellular and Molecular Biology*. (F. C. Neidhardt, J. L. Ingraham, K. B. Low *et al.*, eds.), 1625–1648. American Society for Microbiology, Washington, DC.

Selander, R. K., Li, J., Boyd, E. F., Wang, F.-S. & Nelson, K. (1994). DNA sequence analysis of the genetic structure of populations of *Salmonella enterica* and *Escherichia coli*. In *Bacterial Systematics and Diversity*. (F. G. Priest, A. Ramos-Cormenzana and R. Tindall, eds.), Plenum Press, New York, NY *in press*.

Selander, R. K. & Musser, J. M. (1990). Population genetics of bacterial pathogenesis. In *Molecular Basis of Bacterial Pathogenesis*. (B. H. Iglewski and V. L. Clark, eds.), 11-36. Academic Press, Inc., San Diego, Calif.

Selander, R. K., Musser, J. M., Caugant, D. A., Gilmour, M. N. & Whittam, T. S. (1987). Population genetics of pathogenic bacteria. *Microbial Pathogenesis*, **3**, 1–7.

Sharp, P. M. & Li, W.-H. (1987). The codon adaptation index – a measure of the directional synonymous codon usage bias, and its potential applications. *Nucleic Acids Research*, **15**, 1281–95.

Smith, H. R. & Scotland, S. M. (1988). Vero cytotoxin-producing strains of *Escherichia coli*. *Journal of Medical Microbiology*, **26**, 77–85.

Stenderup, J. & Ørskov, F. (1983). The clonal nature of enteropathogenic *Escherichia coli* strains. *Journal of Infectious Diseases*, **148**, 1019–24.

Strockbine, N. A., Jackson, M. P., Sung, L. M., Holmes, R. K. & O'Brien, A. D. (1988). Cloning and sequencing of the Shiga toxin from *Shigella dysenteriae* type 1. *Journal of Bacteriology*, **170**, 1116–22.

Tarr, P. I., Neill, M. A., Clausen, C. R. et al. (1989). Genotypic variation in pathogenic *Escherichia coli* O157 : H7 isolated from patients in Washington, 1984–1987. *Journal of Infectious Diseases*, **159**, 344–7.

Tzipori, S., Gibson, R. & Montanaro, J. (1989). Nature and distribution of mucosal lesions associated with enteropathogenic and enterohemorrhagic *Escherichia coli* in piglets and the role of plasmid-mediated factors. *Infection and Immunity*, **57**, 1142–50.

Tzipori, S., Karch, H., Wachsmuth, K. I. et al., (1987). Role of a 60-megadalton plasmid and Shiga-like toxins in the pathogenesis of infection caused by enterohemorrhagic *Escherichia coli* O157 : H7 in gnotobiotic piglets. *Infection and Immunity*, **55**, 3117–25.

Tzipori, S., Wachsmuth, I. K., Chapman, C. et al. (1986). The pathogenesis of hemorrhagic colitis caused by *Escherichia coli* O157 : H7 in gnotobiotic piglets. *Journal of Infectious Diseases*, **154**, 712–16.

Viljanen, M. K., Peltola, T., Junnila, S. Y. T. et al. (1990). Outbreak of diarrhoea due to *Escherichia coli* O111 : B4 in schoolchildren and adults: association of Vi antigen-like reactivity. *Lancet*, **336**, 831–4.

Wang, G., Whittam, T. S., Berg, C. M. & Berg, D. E. (1993). RAPD (arbitrary primer) PCR is more sensitive than multilocus enzyme electrophoresis for distinguishing related bacterial strains. *Nucleic Acids Research*, **21**, 5930–3.

Weinstein, D. L., Jackson, M. P., Samuel, J. E., Holmes, R. K. & O'Brien, A. D. (1988). Cloning and sequencing of a Shiga-like toxin type II variant from *Escherichia coli* strain responsible for edema disease of swine. *Journal of Bacteriology*, **170**, 4223–30.

White, D. G., Wilson, R. A., Gabriel, A. S., Saco, M. & Whittam, T. S. (1990). Genetic relationships among strains of avian *Escherichia coli* associated with swollen-head syndrome. *Infection and Immunity*, **58**, 3613–20.

Whittam, T. S. (1992). Sex in the soil. *Current Biology*, **2**, 676–8.

Whittam, T. S. & Ake, S. E. (1993). Genetic polymorphisms and recombination in natural populations of *Escherichia coli*. In *Mechanisms of Molecular Evolution*. (N. Takahata and A. G. Clark, eds.), 223–245. Sinauer Associates, Inc., Sunderland, Mass.

Whittam, T. S., Ochman, H. & Selander, R. K. (1983a). Geographic components of linkage disequilibrium in natural populations of *Escherichia coli*. *Molecular Biology and Evolution*, **1**, 67–83.

Whittam, T. S., Ochman, H. & Selander, R. K. (1983b). Multilocus genetic structure in natural populations of *Escherichia coli*. *Proceedings of the National Academy of Sciences, USA*, **80**, 1751–5.

Whittam, T. S., Wachsmuth, I. K. & Wilson, R. A. (1988). Genetic evidence of clonal descent of *Escherichia coli* O157 : H7 associated with hemorrhagic colitis and hemolytic uremic syndrome. *Journal of Infectious Diseases*, **157**, 124–33.

Whittam, T. S. & Wilson, R. A. (1988*a*). Genetic relationships among pathogenic strains of avian *Escherichia coli*. *Infection and Immunity*, **56**, 2458–66.

Whittam, T. S. & Wilson, R. A. (1988*b*). Genetic relationships among pathogenic *Escherichia coli* of serogroup O157. *Infection and Immunity*, **56**, 2467–73.

Whittam, T. S., Wolfe, M. L., Wachsmuth, I. K. *et al.* (1993). Clonal relationships among *Escherichia coli* strains that cause hemorrhagic colitis and infantile diarrhea. *Infection and Immunity*, **61**, 1619–29.

Whittam, T. S., Wolfe, M. L. & Wilson, R. A. (1989). Genetic relationships among *Escherichia coli* isolates causing urinary tract infections in humans and animals. *Epidemiology and Infection*, **102**, 37–46.

Woodward, J. M., Connaugton, I. D., Fahy, V. A., Lymbery, A. J. & Hampton, D. J. (1993). Clonal analysis of *Escherichia coli* of serogroups O9, O20, and O101 isolated from Australian pigs with neonatal diarrhea. *Journal of Clinical Microbiology*, **31**, 1185–8.

Young, V. B., Miller, V. L., Falkow, S. & Schoolnik, G. K. (1990). Sequence, localization and function of the invasin protein of *Yersinia enterocolitica*. *Molecular Microbiology*, **4**, 119–28.

Yu, J. & Kaper, J. B. (1992). Cloning and characterization of the *eae* gene of enterohaemorrhagic *Escherichia coli* O157 : H7. *Molecular Microbiology*, **6**, 411–17.

POPULATION GENETICS OF PHASE VARIABLE ANTIGENS

J. R. SAUNDERS

Department of Genetics and Microbiology, University of Liverpool
L69 3BX, UK

INTRODUCTION

Many bacterial pathogens are subject to phase (on↔off or volume) and/or antigenic (qualitative) variation in crucial surface antigens. Phase-variable systems are somewhat arbitrarily separated conceptually from conventional mutation phenomena. In reality, the rates of change observed, some of the molecular events involved, and the outcomes in phase variable systems may show little distinction from other forms of 'mutation'. The molecular mechanisms generating variation range from single base changes through to complex recombinational rearrangements (Dybvig, 1993). Although some bacteria are recognized as being particularly variable, it is clear that bacterial species in general can be regarded as multicellular populations that continually generate individual cells with variant phenotypes. This allows rapid adaptation to changes in the environment, but equally may mean that some individuals may be at a disadvantage under prevailing conditions.

Variation in principal surface components is of great potential significance to pathogenic microorganisms. Exposed surface structures are principal targets for the immune response and other defence mechanisms of the host. Much attention has therefore been focused on the advantages of phase and antigenic phase variation in permitting pathogens to avoid host immune surveillance. Although the immune response of the host has an undoubted effect on the population structures of pathogenic bacteria, particularly if they are obligate pathogens, other consequences of variation are often neglected. Variation in the surface layers of bacterial cells has profound effects on critical properties of bacteria, notably on their adherence phenotype and ability to interact with other cells. The interaction of pathogenic bacteria with leukocytes, for example, can be profoundly affected by the nature of outer membrane proteins and other surface structures, such as lipopolysaccharides. It is likely that similar changes in the surfaces of non-pathogenic bacteria could, for example, also affect their ability to form biofilms or their palatability for protozoa or other predators. Hence, phase variation could also influence survival of free-living species inhabiting the open environment.

PROBLEMS ASSOCIATED WITH THE POPULATION BIOLOGY OF PHASE VARIABLE SYSTEMS

Changes in the regulation of a wide variety of genes are necessary in order to optimize growth of pathogens at sites of colonization and/or infection. Although some of the molecular mechanisms driving variation are known, the selective pressures encountered by pathogens during the course of infections, and their consequences for the population biology of bacteria are less well understood.

The details of population genetics of phase variable antigens are likely to be complicated or even obscured by the sampling regimes used to isolate the pathogens concerned. The distribution of genes within organisms recovered from clinical, veterinary or public health laboratories may only reflect the genotype of the predominant bacterial variants present at the time and site of sampling. Generally, this will be during the later stages of infection, for example, when isolation is from the bloodstream or a characteristic lesion. Less commonly, isolates are obtained early during infection or colonization stages, for example, by throat or other forms of swabbing of carriers or contacts. Sampling may also be complicated by the programmed nature of expression of some variable antigens, in for example, bacteria causing relapsing fevers. For phase variations represented by simple on↔off switches or expression changes involving individual genes, any sampling bias may be unimportant because the relevant genotype will be retained in the genomes of all isolates. However, for those variation phenomena involving one way on↔off switches or requiring complex genome rearrangements, particularly if these involve non-reciprocal recombination events, then the genotype sampled will not necessarily be representative. For many of the more complicated variable loci, there may also be no such thing as a wild-type allele as a reference point. In such cases, the variable locus present in the isolate may differ radically from that in the original organism infecting the patient or carrier. Furthermore, many phase variable antigens are characteristically unstable and virulence-related genetic characteristics may be lost permanently at high frequency during laboratory culture. In certain cases, colonial morphology may be associated with particular phases and allow subculture of those phase variants typical of *in vivo* grown bacteria. Moreover, even supposedly pure cultures of clinical isolates may well be heterogeneous with respect to any particular phase variable system. However, the risk remains that culture techniques will obscure the true diversity of variable antigens in a natural population.

It would perhaps be helpful to consider populations of bacterial pathogens on two levels (1) local (to the infected host) and (2) global (that infecting a population of hosts). Populations of medically important bacteria are usually viewed at the global level since sampling is based on single, or occasionally sequential isolates from series of individual patients or contacts. Comparisons are then made between isolates from different patients in different geographical locations. For important pathogens rather more is therefore known about global population genetics (see Chapters by Spratt *et al.* and Maiden & Feavers, this volume), than about genetic variation that might occur within the confines of an infected host.

Biological fractionation based on adhesion and other phenotypes is a frequent consequence of growth of a pathogen when colonizing and subsequently infecting a host species. The tightly adhered subpopulations seen in many bacterial pathogens may differ genotypically as well as phenotypically from free, unattached forms. Free bacteria might not only be responsible for infection of the next host, but also be more likely to be sampled by conventional techniques. Phase variation may also result in phenotypes responsible for compartmentalising bacteria at particular sites or tissues. Thus, variants with particularly avid adherence to host cell receptors might have different fates to those that are non adherent and be subject differentially to sampling regimes. For example, the pathogens *Streptococcus gordanii* and *Staphylococcus aureus* (Baselga *et al.*, 1993), undergo a phase transition from either making polysaccharide slime (SP) or not doing so (NSP). The SP forms grow attached to surfaces in a matrix of insoluble polysaccharide which probably facilitates colonisation, whereas the NSP variants are unattached and may allow dispersal. Attached SP variants may aggregate in microcolonies that are more resistant to phagocytosis and the action of antibiotics, leading to chronic infections (Baselga *et al.*, 1993). In chronic mastitis caused by *S. aureus* the dormant stages of the disease may be due to SP variants in microcolonies, whereas unattached NSP variants would be responsible for the acute episodes of the disease that occur sporadically in infected individuals (Baselga *et al.*, 1993).

MECHANISMS OF PHASE VARIATION

Phase variations are achieved by a variety of molecular switching mechanisms that vary enormously in different bacteria. At the DNA level these range from single base substitutions, deletions or additions up to dramatic genome rearrangements (some examples are shown in Table 1). Generally, the genetic changes involved are restricted spatially to specific regions of the genome. However, some phase changes involve switching of global regulators that have effects on multiple sets of genes. For the most part, phase changes are randomly generated genetically and particular phase variants are selected *in vivo* either as a consequence of advantageous or disadvantageous functional properties conferred, or through immune selection by the host.

Table 1. *Some examples of phase variable genetic systems*

Variation	Property affected	Species	Mechanism
On→Off			
cap locus	Production of capsular polysaccharide	*H. influenzae*	Recombinational deletion
pilE	Pilus-facilitated adhesion	*N. gonorrhoeae/N. meningitidis*	Recombinational deletion (5' end of *pilE*)
On→Off→On			
bvgS	BvgS global virulence regulator	*B. pertussis*	Strand slippage – translation
cap locus	Production of capsular polysaccharide	*H. influenzae*	Deletion, transformation and recombination
dae operon	F1845 fimbriae	*E. coli*	Methylation and Lrp control of transcription
fae operon	K88 fimbriae	*E. coli*	Methylation and Lrp control of transcription
fimA	Type 1 fimbriae	*E. coli*	Methylation and Lrp control of inversion
lic operons	Lipopolysaccharide epitopes	*H. influenzae*	Strand slippage – translation
hif gene cluster	LKP-fimbrial adhesin	*H. influenzae*	Strand slippage – translation
opa	PII outer membrane proteins	*N. gonorrhoeae/N. meningitidis*	Strand slippage – translation
pap operon	Pap fimbrial adhesin	*E. coli*	Methylation and Lrp control of transcription
opc	Adhesion/invasion	*N. meningitidis*	Strand slippage – transcription
pilC	Pilus-facilitated adhesion	*N. gonorrhoeae/N. meningitidis*	Strand slippage – translation
pilE	Pilus-facilitated adhesion	*N. gonorrhoeae/N. meningitidis*	Frame-shift in pilin structural gene
sfa operon	S fimbriae	*E. coli*	Methylation and Lrp control of transcription
yopA	YopA invasin protein	*Yersinia pestis*	Strand slippage – translation
vlp	Variable lipoproteins	*M. hyorhinis*	Strand slippage – transcription
On$_a$/Off$_b$→Off$_a$/On$_b$			
gin	Host specificity of tail fibres	Mu phage	Site-specific inversion
hin	Flagellar variation	*Salmonella* spp	Site-specific inversion
piv	Type 4 pilin variation (Q↔I)	*M. bovis* and *M. lacunata*	Site-specific inversion
On$_a$→On$_x$			
pilV shuffton	Recipient specificity in conjugation	IncI plasmid R64	Multiple site-specific inversion
opa	PII outer membrane proteins	*N. gonorrhoeae/N. meningitidis*	Recombination
pilE	Pilus-facilitated adhesion	*N. gonorrhoeae/N. meningitidis*	Reciprocal recombination or gene conversion
pilE	Pilus-facilitated adhesion	*N. meningitidis* – class II piliated	Recombinational rescue – heterologous pilin gene
pilE	Pilus-facilitated adhesion	*N. gonorrhoeae/N. meningitidis*	Post translational glycosylation
VMP genes	Major exposed proteins	*B. hermsii*	Non-reciprocal recombination/translocation
Volume			
cap locus	Amount of capsular polysaccharide	*H. influenzae*	Amplification of *cap* operon *in situ*
fim	Pilus-facilitated adhesion	*B. pertussis*	Strand slippage - transcription
opc	Adhesion/invasion	*N. meningitidis*	Strand slippage - transcription

Strand-slippage mechanisms resulting in phase variations

Phase variation may be achieved by alterations in reiterated sequences of bases that lie either upstream of a gene and affect transcription, or within the gene and alter the translational reading frame. Regions of genomes that contain reiterated bases may be hot spots for transient mispairing during the local denaturation that accompanies passage of the replication fork (Streisinger & Owen; 1985; Levinson & Gutman, 1987). This process of slipped strand mispairing, occurs independently of homologous recombination functions and adds an additional element of randomness into the expression of a number of genes critical in pathogenesis. DNA containing tracts of polypurines or pyrimidines is able to adopt a number of configurations that differ from the more normal β-form. Under conditions of strain imposed by superhelical coiling, β-form DNA may undergo transitions to H-DNA comprising a triple-stranded and a single-stranded region (Htun & Dahlberg, 1989). H-DNA can stimulate gene expression when positioned on the 5' side of a promoter region (Kohwi & Kohwi-Shigematsu, 1991) and the presence of single-stranded DNA thus generated, albeit transiently, may promote slipped-strand mispairing (Belland, 1991). Phase variations arising through such mechanisms can be compared by analogy with a slipping clutch. Most probably the events that modulate such events are random and not programmed. However, randomness of this type leads to the generation of phase variants that may have advantageous phenotypes at different stages of colonisation or disease.

Translational control of phase variation by slipped strand mispairing

Significant variation is found in the opacity (Opa) proteins of *N. gonorrhoeae* and *N. meningitidis*. These polypeptides are essential for gonococcal invasion of epithelial cells and in the interaction of these bacteria with human neutrophils. Opa proteins are encoded by a family of *opa* genes that are located at scattered sites on the neisserial genome. There are 11–12 such genes in *N. gonorrhoeae* and three or four in *N. meningitidis* (Bhat *et al.*, 1991; Aho *et al.*, 1991). Some of the Opa variants can promote adhesion of gonococci and non capsulate meningococci to human epithelial and endothelial cells and may be important in the early stages of infection and invasion (Makino, Van Putten & Meyer, 1991; Simon & Rest, 1992; Kupsch *et al.*, 1993; Virji *et al.*, 1993a). The neisserial *opa* genes are highly conserved at the nucleotide level, but contain a short semi-variable sequence at the N-terminal coding region and two hypervariable regions termed HV1 and HV2 in the region encoding the central, surface-exposed domain of the polypeptide (Aho *et al.*, 1991; Connell, Shaffer & Cannon, 1990). Antigenic variation is achieved in this case by the interaction of phase variations within the different copies of the *opa* genes. This means that one or more different *opa* loci (or even none) may be expressed at any one time. In addition, further variation is possible by recombination between existing *opa* loci to create novel *opa* gene sequences (Connell *et al.*, 1988). Phase variation in the form on↔off is achieved at each *opa* locus at the translational level by alterations in the reading frame of the signal peptide region. The DNA encoding the signal peptide of each *opa* gene contains a series or pentameric repeats (CRs) of the sequence 5'-CTCTT-3'. Alteration in the numbers of repeats occurs in the absence of *recA* function and shifts the reading frame of the signal region of the Opa polypeptide. In one frame (nine or other multiples of three repeats) the protein is in frame with the coding region for the mature polypeptide, but in the other two (eight and ten repeats, etc.) the protein is abnormal and truncated (Bhat *et al.*, 1991; Stern & Meyer, 1987; Stern *et al.*, 1986). The addition of repeats in this crucial region is brought about by slipped strand mispairing of the repeated tracts (Belland *et al.*, 1989; Belland, 1991). This would occur during replication, possibly as a consequence of transient formation of single-stranded DNA caused by H-DNA formation within the CR region (Belland, 1991).

   A related slipped strand mispairing mechanism may account for variable expression of lipopolysaccharide (LPS) epitopes in *Haemophilus influenzae* which is significant in the modulation of pathogenicity of this organism (Maskell *et al.*, 1992; High, Deadman & Moxon, 1993; Weiser, 1993). The LPS of *H. influenzae*, other *Haemophilus* species such as *H. somnus* (Inzana, Gogolewski & Corbeil, 1992) and the pathogenic *Neisseria* spp (Van Putten, 1993) lack conventional O side chains and are more correctly called oligosaccharides (LOS). The LOS/LPS of these species are heterogeneous and known to undergo variation in both structural and antigenic properties during the course of infections (Inzana *et al.*, 1992; Van Putten, 1993). In *H. influenzae* a tetrameric motif of 5'-CAAT-3' is present in tandem direct repeat in the 5' ends of the coding regions of three different *lic* genes which are required for the expression of phase variable LPS epitopes. Again, the number of repeats modulates the downstream part of the gene into and out of frame with respect to the open reading frame of the functional gene product. The number of repeats necessary for expression of the *lic2A* locus appears to be strain-dependent (e.g. in strain RM7004 the number of repeats necessary for the 'non' phenotype was 16) and is probably not the only controlling factor (High *et al.*, 1993).

   Translational control of phase variation achieved by slippage at repeated sequences has been observed in a number of other bacterial pathogens (Table 1). For example, the *bvgS* gene of *Bordetella pertussis* encoding a global regulator of virulence genes contains a tract of C residues within the open reading frame, such that a change from six (on) to seven (off) Cs shifts the downstream coding region out of frame (Stibitz *et al.*, 1989). The *yopA* gene found in *Yersinia pestis* contains a tract of As and removal of one of these results in a frame shift that truncates the gene product (Rosqvist, Skurnik & Wolfwatz, 1988).

Translational frame-shifting also occurs at a tract of G residues in the gonococcal *pilC* gene, which encodes a polypeptide necessary for the assembly and normal adhesion function of neisserial pili (Jonsson, Nyberg & Normark, 1991; Jonsson, Pfeifer & Normark, 1992). When PilC is not expressed, gonococci and meningococci produce few pili or none at all and exhibit reduced adhesion (Jonsson *et al.*, 1992; Rudel *et al.*, 1992). This may provide a mechanism for desorption from human cells and assist spread to alternative colonisation sites.

Transcriptional Modulation of Phase Variation

N. meningitidis produces an outer membrane protein called Opc which shows only limited sequence homology to the Opa family of polypeptides. Opc is associated with adhesion and invasion phenotypes in non capsulate meningococci (Virji *et al.*, 1992*a,b*). The Opc protein is subject to a phase variation of the on↔off type, but with a volume control. Cells in the off phase (Opc^-) are unable to adhere by this mechanism, but meningococci expressing Opc at high (Opc^{++}) or intermediate level (Opc^+) can do so. Bacteria with an Opc^{++} phenotype alone are also able to invade both human epithelial and endothelial cells (Virji *et al.*, 1993*a*). It is possible that Opc-down regulated variants may be selected during the later stages of invasive disease by *N. meningitidis*, since Opc^{++} bacteria are more frequently isolated from nasopharyngeal sites than either the bloodstream or cerebrospinal fluid (Olyhoek *et al.*, 1991; Virji *et al.*, 1992*b*). Although the Opc protein is highly immunogenic, only Opc^{++} bacteria appear to be susceptible to bactericidal antibodies against Opc: Opc^- and Opc^+ variant meningococci are not killed by such antibodies (Rosenqvist *et al.*, 1993). Hence, once invasion has been accomplished, there may be a strong pressure in favour of variants that are down-modulated at the *opc* locus.

Unlike Opa phase changes, the Opc variation is achieved by transcriptional regulation. The *opc* gene contains an unusual promoter region which includes a conventional −10 region, but lacks a typical −35 consensus sequence. The promoter region does however contain a variable tract of adjacent C residues that modulate the phase change. When the number of Cs is ≤10 or ≥15 the gene produces no detectable *opc* mRNA and expresses an Opc^- phenotype; when the number is 11 or 14 moderate amounts of transcript are produced and the phenotype is Opc^+; and when the number is 12 or 13, 5–10 times more *opc* mRNA is produced and the phenotype is Opc^{++} (Achtman, 1994; Sarkari *et al.*, 1994).

Transcriptional control of phase variation is found in a number of other systems. The LKP fimbriae of *H. influenzae* are required for adhesion of these bacteria to human cells (Kar, To & Brinton, 1990). These fimbriae (pili) assist in colonization of the nasopharynx as a prelude to localized infections, or to systemic infections in the form of meningitis. However, they are not required for invasion of the mucosa and may indeed hamper this process (Farley *et al.*, 1992). Therefore, phase variation in this antigen is a major advantage to the pathogen. The fimbrial gene cluster in this case is coregulated and consists of *hifA* (fimbrial subunit), *hifB* (periplasmic chaperone), *hifC* (outer membrane usher) and *hifD* and *hifE* (both minor subunits necessary for biogenesis) (van Ham *et al.*, 1993). The *hifB*, *hifC*, *hifD* and *hifE* genes are coregulated with the *hifA* gene through combined, divergently oriented promoters (van Ham *et al.*, 1993). Transcription in either direction is regulated by the number of TA residues in a tract of poly TA that spans the -10 and -35 sequences (van Ham *et al.*, 1994).

A C-tract upstream of the pilin (*fim*) gene of *B. pertussis* regulates expression at the transcriptional level, probably by altering the interaction of the promoter region with RNA polymerase and regulatory proteins in a manner that is dependent on the number of C residues (Willems *et al.*, 1990). The promoters of the variable lipoprotein (*vlp*) genes found in *Mycoplasma hominis* contain a tract of A residues that may also be subject to strand slippages affecting transcriptional efficiency (Yogev *et al.*, 1991).

Phase Variations mediated by the effects of differential methylation on transcription

Methylation, notably at the GATC sequence represented by DNA adenine methylation (Dam) sites, is frequently involved in regulation of gene expression, DNA repair and replication. Phase variation of the pyelonephritis-associated (Pap) pili of *E. coli* is typified by an on↔off switching of production that involves the differential methylation of two GATC sequences, $GATC_{1028}$ and $GATC_{1130}$ (Braaten *et al.*, 1994). These Dam sites lie between the divergent promoters for *papI* (p_I) and *papB* (p_B) in the *pap* regulatory region. The pattern of methylation at these sites provides a simple on↔off switching system and requires both Dam methylase and the global regulator leucine-responsive regulatory protein (Lrp) (Nou *et al.*, 1993). When $GATC_{1028}$ is unmethylated and $GATC_{1130}$ is methylated the *pap* phase is on. Conversely, when $GATC_{1028}$ is methylated and $GATC_{1130}$ is unmethylated the *pap* phase is off. PapI is a regulatory protein that modulates the switch between the two *pap* phases by interacting with Lrp (Vanderwoude, Braaten & Low, 1992). In the absence of PapI, Lrp inhibits methylation of $GATC_{1130}$ by binding close to this site. Methylation of the $GATC_{1028}$ site prevents binding of PapI-modified Lrp and maintains the off state until DNA replication generates hemi-methylation at $GATC_{1028}$ (Nou *et al.*, 1993). Lrp–PapI can bind to hemi-methylated $GATC_{1028}$ whereupon this site can become protected from methylation to form the on phase methylation pattern. Similar Dam methylation-regulated phase variation has been shown to occur in operons controlling production of other pili such as *fae* (K88) (Huisman *et al.*, 1994), *dae* (F1845) and *sfa* (S) (Vanderwoude & Low, 1994).

Phase Variations effected by DNA inversions

Regions of bacterial, plasmid and phage genomes ranging from 100 bp to more than 35 kb may exhibit inversions at frequencies typically of about 10^{-3} to 10^{-4} (Dybvig, 1993). The site-specific inversion of DNA segments provides a simple on↔off switch for genes contained within, or adjacent to, the invertible region by altering the spatial relationship of that region to a promoter or other regulatory element (Plasterk & Vandeputte, 1984).

The type 1 fimbriae found in both commensal and pathogenic *E. coli* are phase variable and apparently subject to selection during infections, since clinical isolates exhibit marked heterogeneity in fimbrial phenotype. The ability to change from the fimbriate to the afimbriate phase may also contribute to the development of peritonitis caused by these bacteria. Type 1 fimbriae promote adherence to mannose-containing receptors present on macrophages and certain epithelial cell types. Type 1 fimbriae may protect bacteria from macrophages but are potent immunogens and by non-specific binding to Tamm-Horsfall protein and secretory IgA may block attachment of the bacteria to epithelial surfaces.

Phase variation of Type 1 fimbriae involves the inversion of a 314 bp sequence located upstream of *fimA*, the gene for the major fimbrial structural subunit. In one orientation (on), *fimA* is transcribed from a promoter within the invertible element; but in the opposite orientation (off), the structural gene is divorced from its promoter. *Fim* inversion involves the global regulatory polypeptides histone-like protein (H-NS), integration host factor (IHF) and Lrp. The *fim* inversion *per se* is performed by two proteins of the lambda integrase family, FimB and FimE, which share 48% sequence homology with each other. The *fim* switch is locked in either the on or off position in the absence of these two gene products. FimB promotes both on–off and off–on inversions, whereas FimE is involved in on–off inversions only (Gally *et al.*, 1993; McClain, Blomfield & Eisenstein, 1991). Furthermore, the activities of the *fimB* and *fimE* genes in switching are affected both by growing temperature and composition of media, suggesting environmental regulation of the phase change. In wild-type cells the switching rate can be as high as 0.75 per generation. Inversion mediated by both *fimB* and *fimE* is stimulated by alanine and branched chain amino acids acting together with Lrp (Blomfield *et al.*, 1993). Lrp may be involved in bending DNA and, in concert with IHF, may be responsible for aligning the terminal inverted repeats of the *fim* element prior to recombination (Blomfield *et al.*, 1993). *fimB*-promoted inversion increases with temperature up to an optimum between 37 and 40°C (Gally *et al.*, 1993). In contrast, *fimE*-promoted inversion declines as the temperature is raised from 28 to >40°C. This favours switching to the On position at growth temperatures appropriate to growth in the human host.

Although invertible systems may be used to control expression of bacterial surface components, they also have profound roles in determining the host range of extrachromosomal elements such as plasmids or phages (Plasterk & Vandeputte, 1984). Notable among these is the Gin (G inversion) system of bacteriophage Mu (Plasterk, Brinkman & Vandeputte, 1983; Giphartgassler, Plasterk & Vandeputte, 1982). In this case inversion of the G region of the Mu genome leads to the expression of two different sets of proteins that determine the host range of the phage. In one orientation the phage produces two proteins that allow it to absorb to *E. coli* and in the other to absorb to *Citrobacter freundii*. Bacteriophage P1 of *E. coli* has a related invertible system (Cin) regulating its host range C region (Plasterk & Vandeputte, 1984). The consequence of phase variation in these cases is alteration in the host specificity of a proportion of progeny phage particles. These may subsequently be subjected to different evolutionary influences in the alternative host(s).

The flexible sex pili of ColI1 conjugative plasmids such as R64, are composed of pilin subunits where C-terminal sequence changes can determine species-preference in recipients. These are modulated by a complex series of genetic rearrangements in the coding region for the C-terminus, called a shufflon (Kim & Komano, 1992; Komano *et al.*, 1990). In this case, site-specific recombination events invert and reorder four different segments of coding DNA to produce up to seven different C-terminal sequence variants. The specificity for recipient so conferred on an individual pilus variant may determine the subsequent evolutionary fate of the encoding plasmid by compartmentalizing it (at least temporarily) in a particular bacterial species.

Complex recombination systems responsible for phase variation

More complex phenotypes, notably leading to antigenic or other forms of qualitative variation, are made possible by recombinational exchanges in or between variable loci. Such phase variable systems are typically metastable and often subject to deletional events producing variants that are disabled for the locus concerned. In many instances the accumulation of disabled phase variants within populations appears to be prevented by gene transfer events that allow repair of the deleted region.

Capsule expression in H. influenzae *type b*

Capsular serotype b strains are responsible for >95% of invasive disease caused by *H. influenzae*. The b type capsular polysaccharide is an important virulence determinant that allows these bacteria to survive in the intravascular environment. Loss of the ability to produce capsule occurs as a consequence of the structure of the chromosomal *cap* locus (Kroll & Moxon, 1989). This consists of a potentially unstable duplicated tandem repeat of two identical 17 kbp regions separated by a unique bridge region of 1.2 kbp.

Capsule-deficient variants can arise by a two-stage process that initially involves reduction by homologous recombination of the number of *cap* copies to one. This event substantially reduces viability, with only about 1 in 200 cells surviving to produce very slow growing bacteria (class I variants) that internally accumulate capsular polysaccharide which they are unable to export. Second stage rescue mutations that abolish the ability to produce capsular polysaccharide are necessary to produce class II variants which grow at the normal rate (Kroll *et al.*, 1990; Brophy *et al.*, 1991). Once fully deficient class II variants have been generated, the only mechanism for reversion to Cap$^+$ is by transformation with wild-type *cap* DNA. Class II variants exhibit enhanced adherence and invasion of cultured human epithelial cells when compared with their capsulate parents. These properties would be expected to favour colonization and persistence in the human respiratory tract, whereas the capsulate form would be better equipped to survive in the bloodstream. It is perhaps surprising therefore that class II variants are only found occasionally amongst non-typeable haemophili colonising the respiratory tract. Class I variants have such a low viability that they might provide a bottleneck which checks loss of the Cap$^+$ phenotype from within the population (Brophy *et al.*, 1991).

Phase variation by recombination in the *cap* locus provides an on↔off switch by altering the number of copies of *cap* from 2 to 1 (or even 0) and vice versa. The system also has a volume control, since amplification of the *cap* locus can occur *in situ* to produce three, four or more tandem copies per chromosome (Kroll, 1992). Indeed, the number of copies is proportional to the amount of capsular material produced and clinical isolates tend to have more, rather than less copies on primary isolation.

Translocational rearrangement in phase changes of Borreliae

In the spirochete *Borrelia hermsii*, which causes relapsing fevers in rodents, surface proteins are changed in order presumably to avoid the host immune response (Barbour, 1990). Antigenic variation in the surface-exposed variable major proteins (VMPs) of this bacterium accounts for fluctuations in the numbers of borreliae in the bloodstream of the host. The animal produces antibodies against the VMP serotype of *B. hermsii* causing the initial infection and begins to clear it from the bloodstream. Relapses occur when a new variant is generated that expresses a VMP serotype to which the host has manufactured no protective antibodies. Switching between VMP types, which can occur in the absence of antibody selection, occurs at about 10^{-4} per cell per generation. In mice, relapse serotypes emerge in an approximate order with certain types predominating during the early stages of infection. This indicates a degree of genetic programming that is apparently atypical for phase variable antigens. Phase changes of VMPs in Borreliae are reversible, since serotypes eliminated by neutralising antibodies in one host may reappear in the relapse populations infecting a second host. Antigenic variation of VMPs is achieved by translocation of VMP gene sequences from a silent locus carried on one of a number of linear storage plasmids to a subtelomeric expression site carried on a linear expression plasmid (Kitten & Barbour, 1990; Plasterk, Simon & Barbour, 1985).

Recombinational phase and antigenic variation of pili in pathogenic Neisseria

The pathogenic *Neisseria* exhibit an extreme form of variability that is manifested in multiple phase and antigenic variations in critical surface-located virulence determinants (Meyer, Gibbs & Haas, 1990; Robertson & Meyer, 1992). Some of these such as *opa*, *opc* and *pilC* are controlled by relatively simple genetic switches (see above). However, the most extreme form of variation is seen in pili which are required for virulence in *N. gonorrhoeae* and *N. meningitidis*. Meningococci can produce one of two types of pili, class I which are homologous to those produced by all gonococci, and class II which exhibit antigenic, biochemical and functional differences to those of *N. gonorrhoeae* (Virji *et al.*, 1989). The distribution of class I and II piliated strains in different clinical isolates of *N. meningitidis* is approximately 60/40 (Saunders *et al.*, 1993). Piliated gonococci and meningococci of both classes adhere only to cells of human origin (Virji *et al.*, 1991) indicating that they either act as adhesins *per se* or that they are presenting some other entity (most probably a pilus-associated protein such as PilC) that acts as adhesin. Gonococcal and class I meningococcal pili belong to the N-methyl-Phe (class 4) group of pili produced by other pathogens, including enteropathogenic *Escherichia coli* (Giron, Ho & Schoolnik, 1991). *Dichelobacter (Bacteroides) nodosus*, *Moraxella* spp. and *Pseudomonas aeruginosa* (Patel *et al.*, 1991; Tonjum *et al.*, 1991). *Moraxella bovis* and *Moraxella lacunata* exhibit a simple invertible switching system that leads to the expression of either of two alternative pilins, Q and I (Rozsa & Marrs, 1991). However, the pathogenic *Neisseria* spp. are unique in exhibiting extreme variability in the biochemical and antigenic nature of their pili.

The mature pilin subunit of gonococcal and Class I meningococcal pili is divided into three domains: the first 53 N-terminal amino acids are conserved (C), amino acids 54 to the first of two cysteine residues at position approximately 120 constitute a semivariable (SV) region characterized by amino acid substitutions, and a hypervariable (HV) region from position 120 to the carboxy terminus at about 160 characterized by insertions and deletions of amino acids, as well as substitutions. Such pili exhibit significant variation both within and between strains. Both phase (on↔off↔on) and antigenic pilin variation occur at high rate in gonococci and both class I and II pilin-producing meningococci during laboratory cultures and in the course of natural infections (Tinsley & Heckels, 1986). Pilus phase variation in

Neisseria may permit the non-piliated, non-attaching phase to desorb from initial sites of infection and allow movement to other locations. The non-piliated state may also enable organisms to be transmitted from one host to another provided that reversion to the piliated form is possible before a new infection site is reached (Virji & Heckels, 1984).

The structural gene for the pilin subunit in *Neisseria* (*pilE*) gene is notably unstable, with frequent deletions occurring, particularly at the 5' end of the gene which contains the promoter. Deletions within *pilE* provide one mechanism for phase variability in pilus production. Some such deletion events can be repaired by recombination with intact *pil* sequences (see below). Others lead to variants that are unable to revert to producing pilin. This is because sufficient of the 5' end of the *pilE* gene is removed, preventing recombination with homologous sequences in silent (*pilS*) loci. Pilus phase variation also results from a variety of alternative molecular mechanisms, including frame shifting within the *pilE* coding sequence to produce non-functional (non-assembled) pilins (Bergstrom *et al.*, 1986), the presence or absence of accessory gene products such as PilC (which is itself subject to phase variation, see above), and *recA*-dependent recombination events that place a coding sequence for a non-functional pilin at *pilE* (Koomey *et al.*, 1987; Haas, Schwarz & Meyer, 1987). Pilus antigenic variation is also effected by *recA*-dependent recombination events that involve replacement of pilin sequences in *pilE* with sequences contained in one of up to 19 or so *pilS* loci. These contain partial pilin sequences that are truncated to varying degrees from the 5' end (Koomey *et al.*, 1987; Haas, Veit & Meyer, 1992). Pilin sequences may be transferred intragenomically from *pilS* loci to *pilE* by non-reciprocal recombination (gene conversion). This process may be primarily responsible for generating variation within the pilin sequences of neisserial populations. Alternatively, novel *pilS* or *pilE* sequences can be acquired externally from lysed neisserial cells by transformation (Gibbs *et al.*, 1989; Siefert *et al.*, 1988). This second route involves reciprocal recombination, is essentially conservative, and usually leads to replacement of contiguous *pil* sequences. It may be important in ensuring the spread of particularly useful pilin sequences within populations of *N. gonorrhoeae* and *N. meningitidis*. The exact mechanisms for recombinational switching at *pilE* remain controversial. Mixing and matching might occur by *recA*-dependent recombination involving swapping of 'mini-cassettes' of variable sequence that lie between short conserved nucleotide sequences within *pil* loci (Robertson & Meyer, 1992). Alternatively/additionally, incoming *pil* sequences may act as templates for *recA*-dependent 'deletion-repair' of naturally occurring deletions within *pilE* (Hill, Morrison & Swanson, 1990). It is known that *pil* genes can be exchanged between neisserial strains grown under co-cultivation (Frosch & Meyer, 1992) and presumably transformation may allow exchange of pilin sequence information in natural mixed infections. The propensity for individual strains to exhibit deletions in class I *pilE* loci may explain the existence of class II-piliated meningococci. These still retain partial non-expressed class I pillin sequence information as *pilS* loci (Perry, Nicolson & Saunders, 1988). The class II meningococcal pilus is antigenically related to pili produced by the commensal species *Neisseria lactamica* (Saunders *et al.*, 1993). It seems likely therefore that *N. meningitidis* strains that have suffered the misfortune to lose all or part of their class I *pilE* gene could acquire a functional replacement through transformation with DNA containing class II pilin sequences from *N. lactamica*. This would be possible during their coexistence on the mucosal surface of the human nasopharynx.

Each pathogenic neisserial strain is capable of producing a different pilin sequence at any one time by varying the coding sequence at *pilE*. Indeed, it has been calculated that over 10^6 combinations of pilin sequence are possible in *N. gonorrhoeae* (Rudel *et al.*, 1992). This is the result of mixing and matching of different combinations of the repertoire of pilin sequence information stored at *pilE* and in the *pilS* loci. The antigenic changes that result from alterations in pilin primary amino acid sequence provide an evident advantage for these bacteria in avoiding host immune surveillance. Perhaps more importantly, antigenic variation is accompanied by functional changes with respect to both the efficiency and specificity of adhesion to human epithelial and endothelial cells. Pili are necessary for adherence of gonococci and meningococci to both human epithelial and endothelial cells (Virji *et al.*, 1991). Natural genetic variation in the meningococcus generates variants that are fully piliated and capable of adherence to endothelial cell lines, but which fail to adhere to epithelial cells (Virji *et al.*, 1993*b*). This implies that the mechanisms of adherence to endothelial and epithelial cells are different and that variation in the pilus may determine tissue specificity of individual members of the bacterial population during infections. Antigenic variation in gonococcal and meningococcal pili is accompanied by changes in apparent molecular weight of the pilin subunit (Virji *et al.*, 1992*a*), indicating that adherence is affected, at least in part, by fundamental changes in the primary amino acid sequence of pilin.

The ease with which adhesion variants of markedly differing qualitative and quantitative affinity for various human cell lines can be isolated from supposedly isogenic class I piliated meningococci (Nassif *et al.*, 1993; Virji *et al.*, 1993*b*) and gonococci (Rudel *et al.*, 1992) suggests that the extent of functional variation is substantial. Sequence variation in class I meningococcal pili (and probably also gonococci) has an additional consequence in promoting or preventing post-translational modification of pilin (Virji *et al.*, 1993*b*). This occurs as a consequence of the presence or absence of N-glycosylation sequons (Asn-Xaa-Ser/Thr) within the SV and/or HV regions of mature pilin (Virji *et al.*, 1993*b*). It appears that glycosylation probably acts as a down-modulating influence on adhesion. In general, the more N-glycosylation motifs are present on a pilin sequence, the less avidly the

Epithelial surface

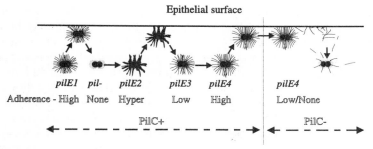

Fig. 1. Possible functional consequences of pilus variation in capsulate *Neisseria meningitidis*. Capsulate meningococci are dependent on pili for adherence to epithelial or endothelial surfaces. A pilus variant expressing a normal high adherence (moderate glycosylation) phenotype (*pilE1*) may give rise to a non-piliated (*pil⁻*) variant that desorbs from the epithelial surface and migrates elsewhere. This, in turn, may spawn a second Pil⁺ variant (*pilE2*) with a hyperadherent (low glycosylation) phenotype (*Note*: Pili on such bacteria tend to be present in pronounced bundles.) This, in turn, may produce a variant (*pilE3*) with low adherence (high glycosylation) phenotype that can desorb from the site of attachment. A further change may generate a variant (*pilE4*) with normal adherence/glycosylation phenotype that re-adheres at a different site. Independent phase changes in *pilC* may further modulate adhesion phenotype. Even strains expressing highly adherent pilins may desorb from cell surfaces following a PilC⁺ to PilC⁻ phase change which results in substantial reduction in the numbers of pili produced per cell.

producing strain adheres to human epithelial cell lines. Post-translational modification of pilin provides an additional, untemplated source of variability. For example, N-glycosylation sites, if present, could be glycosylated or not, or could involve attachment of different carbohydrate groups. This could alter the interaction of pili directly with some sort of receptor on the surface on the eukaryotic cell, or indirectly by modulating the interaction of pilin molecules with other proteins such as PilC that might be the adhesin *per se*. Thus, individual meningococci or gonococci, even if they shared the same primary amino acid sequence in their pilin molecules, could express functionally distinct adhesion phenotypes. Adhesion properties of piliated meningococci are further modulated by phase changes resulting in variable expression of *pilC* (Virji *et al.*, 1993*b*). In the course of colonisation or infection at an epithelial or endothelial surface, a variety of functionally variant pili could be generated by the bacteria and their offspring. This would allow sequential adhesion to and detachment from different sites on cell surfaces and tissues (Fig. 1). Superimposed on meningococcal pilus variation (*pilE* and *pilC*) would be simultaneous phase variation in other virulence determinants, including *opa*, *opc* and LOS genes.

THE SIGNIFICANCE OF PHASE VARIATION IN NATURAL POPULATIONS

The assumption is often made that bacteria are supremely adapted by evolution to survive in their natural environments. However, the phase variable systems present in many bacteria are frequently responsible for apparently deleterious or disabling genetic changes. These may be of an irreversible nature, as exemplified by the loss of Cps production in *H. influenzae* (Kroll, 1992) and non-reverting *pil⁻* variants in the pathogenic *Neisseria* (Swanson *et al.*, 1987). Such phase variants are relatively easily isolated and maintained in the laboratory. It is not clear whether they would necessarily survive for appreciable periods in natural populations, since they would have no apparent future as pathogens or even commensals whilst burdened with such disabilities. It may well be, however, that the formation of such disabled variants as a small proportion of the total population is tolerated and is indeed an inevitable consequence of a phase variable life style. With large populations and complicated tissue composition in a mammalian host, this may not be disadvantageous to the bacterial species concerned. In some instances, for example, capsule variants of *H. influenzae*, it would be possible for the genetic damage caused by a phase variable life style to be repaired by gene transfer events. Interestingly, this is clearly not possible in the case of non-reverting Pil⁻ variants in the pathogenic *Neisseria*, since these are non-transformable (Swanson *et al.*, 1986, 1990) which could be a device to prevent the potentially disastrous spread of *pil⁻* genotypes within neisserial populations.

Genetic transfer events may be responsible in some cases for the reversion of off phase variants to the on phase and for generating diversity. However, conversely gene transfer may have a dampening effect on the variation observed in a population. For example, transformation in *Neisseria*, whilst allowing substantial genetic exchange and the formation of mosaic genes, can lead to the predominance of particular genotypes in clinical populations (see Chapter by Spratt). This is probably caused by the spread of particularly beneficial genotypes, for example, pilin sequences conferring specific adhesion properties. This homogenizing effect may occur within local populations due to the selection of a predominating genotype, or as a consequence of coinfection.

For reversible phase changes, loss by individual cells of crucial virulence determinants may be compensated, at least in part, by advantages in avoiding host defence mechanisms. Surface components in particular are likely to be highly immunogenic and present prime targets for protective antibodies. Capsular polysaccharides may mask adhesins or other components protecting against antibodies or phagocytosis but reducing adhesion and/or invasion. Structures such as pili may also predispose bacteria to phagocytosis. This could lead to selection for genetic systems that switch off, or at least down-regulate expression, for a proportion of the population at any one time. Moreover, it is possible that different phase variants of the same strain could co-operate during the colonization and infection processes.

Some bacteria, such as the pathogenic *Neisseria*, can exhibit a bewildering array of phenotypes due to complex multiple genetic switches. This is reinforced in this particular group by an extremely high rate of genetic exchange which is probably untypical for bacteria in general. At the single cell level within the local bacterial population infecting a single host individual such genetic variation at multiple loci might approximate to individuality as understood in higher organisms.

REFERENCES

Achtman, M. (1994). Clonal spread of serogroup A meningococci – a paradigm for the analysis of microevolution in bacteria. *Molecular Microbiology*, **11**, 15–22.

Aho, E. L., Dempsey, J. A., Hobbs, M. M., Klapper, D. G. & Cannon, J. G. (1991). Characterization of the *opa* (class-5) gene family of *Neisseria meningitidis*. *Molecular Microbiology*, **5**, 1429–37.

Barbour, A. G. (1990). Antigenic variation of a relapsing fever *Borrelia* species. *Annual Review of Microbiology*, **44**, 155–71.

Baselga, R., Albizu, I., Delacruz, M., Delcacho, E., Barberan, M. & Amorena, B. (1993). Phase variation of slime production in *Staphylococcus* – implications in colonization and virulence. *Infection and Immunity*, **61**, 4857–62.

Belland, R. J., Morrison, S. G., Vanderley, P. & Swanson, J. (1989). Expression and phase variation of gonococcal P-II genes in *Escherichia coli* involves ribosomal frameshifting and slipped-strand mispairing. *Molecular Microbiology*, **3**, 777–86.

Belland, R. J. (1991). H-DNA formation by the coding repeat elements of neisserial *opa* genes. *Molecular Microbiology*, **5**, 2351–60.

Bergstrom, S., Robbins, K., Koomey, J. M. & Swanson, J. (1986). Piliation control mechanisms in *Neisseria gonorrhoeae*. *Proceedings of the National Academy of Sciences, USA*, **83**, 3890–4.

Bhat, K. S., Gibbs, C. P., Barrera, O., Morrison, S. G., Jahnig, F., Stern, A., Kupsch, E. M., Meyer, T. F. & Swanson, J. (1991). The opacity proteins of *Neisseria gonorrhoeae* strain MS11 are encoded by a family of 11 complete genes. *Molecular Microbiology*, **5**, 1889–901.

Blomfield, I. C., Calie, P. J., Eberhardt, K. J., Mcclain, M. S. & Eisenstein, B. I. (1993). Lrp stimulates phase variation of type-1 fimbriation in *Escherichia coli* K-12. *Journal of Bacteriology*, **175**, 27–36.

Braaten, B. A., Nou, X. W., Kaltenbach, L. S. & Low, D. A. (1994). Methylation patterns in *pap* regulatory DNA control pyelonephritis-associated pili phase variation in *Escherichia coli*. *Cell*, **76**, 577–88.

Brophy, L. N., Kroll, S. J., Ferguson, D. J. P. & Moxon, E. R. (1991). Capsulation gene loss and rescue mutations during the Cap+ to Cap− transition in *Haemophilus influenzae* type-b. *Journal of General Microbiology*, **137**, 2571–6.

Connell, T. D., Black, W. J., Kawula, T. H., Barritt, D. S., Dempsey, J. A., Kverneland, K., Stephenson, A., Schepart, B. S., Murphy, G. L. & Cannon, J. G. (1988). Recombination among protein-II genes of *Neisseria gonorrhoeae* generates new coding sequences and increases structural variability in the protein-II family. *Molecular Microbiology*, **2**, 227–36.

Connell, T. D., Shaffer, D. & Cannon, J. G. (1990). Characterization of the repertoire of hypervariable regions in the protein-II (*opa*) gene family of *Neisseria gonorrhoeae*. *Molecular Microbiology*, **4**, 439–49.

Dybvig, K. (1993). DNA rearrangements and phenotypic switching in prokaryotes. *Molecular Microbiology*, **10**, 465–71.

Farley, M. M., Whitney, A. M., Spellman, P., Quinn, F. D., Weyant, R. S., Mayer,

L. & Stephens, D. S. (1992). Analysis of the attachment and invasion of human epithelial cells by *Haemophilus influenzae* biogroup *aegyptius*. *Journal of Infectious Diseases*, **165**, 111–14.

Frosch, M. & Mayer, T. F. (1992). Transformation-mediated exchange of virulence determinants by cocultivation of pathogenic *Neisseriae*. *Fems Microbiology Letters*, **100**, 345–9.

Gally, D. L., Bogan, J. A., Eisenstein, B. I. & Blomfield, I. C. (1993). Environmental-regulation of the *fim* switch controlling type-1 fimbrial phase variation in *Escherichia coli* K-12 – effects of temperature and media. *Journal of Bacteriology*, **175**, 6186–93.

Gibbs, C. P., Reimann, B. Y., Schultz, E., Kaufmann, A., Haas, R. & Meyer, T. F. (1989). Reassortment of pilin genes in *Neisseria gonorrhoeae* occurs by two distinct mechanisms. *Nature, London*, **338**, 651–2.

Giphartgassler, M., Plasterk, R. H. A. & Vandeputte, P. (1982). G inversion in bacteriophage-Mu – a novel way of gene splicing. *Nature, London*, **297**, 339–42.

Giron, J. A., Ho, A. S. Y. & Schoolnik, G. K. (1991) An inducible bundle-forming pilus of enteropathogenic *Escherichia coli*. *Science*, **254**, 710–13.

Haas, R., Schwarz, H. & Meyer, T. F. (1987). Release of soluble pilin antigen coupled with gene conversion in *Neisseria gonorrhoeae*. *Proceedings of the National Academy of Sciences, USA*, **84**, 9079–83.

Haas, R., Veit, S. & Meyer, T. F. (1992). Silent pilin genes of *Neisseria gonorrhoeae* MS11 and the occurrence of related hypervariant sequences among other gonococcal isolates. *Molecular Microbiology*, **6**, 197–208.

High, N. J., Deadman, M. E. & Moxon, E. R. (1993). The role of a repetitive DNA motif (5′-CAAT-3′) in the variable expression of the *Haemophilus influenzae* lipopolysaccharide epitope alpha-gal(1-4)beta-gal. *Molecular Microbiology*, **9**, 1275–82.

Hill, S. A., Morrison, S. G. & Swanson, J. (1990). The role of direct oligonucleotide repeats in gonococcal pilin gene variation. *Molecular Microbiology*, **4**, 1341–52.

Htun, H. & Dahlberg, J. E. (1989). Topology and formation of triple-stranded H-DNA. *Science*, **243**, 1571–6.

Huisman, T. T., Bakker, D., Klaasen, P. & Degraaf, F. K. (1994). Leucine-responsive regulatory protein, IS1 insertions, and the negative regulator *faeA* control the expression of the *fae* (K88) operon in *Escherichia coli*. *Molecular Microbiology*, **11**, 525–36.

Inzana, T. J., Gogolewski, R. P. & Corbeil, L. B. (1992). Phenotypic phase variation in *Haemophilus somnus* lipooligosaccharide during bovine pneumonia and after *in vitro* passage. *Infection and Immunity*, **60**, 2943–51.

Jonsson, A. B., Nyberg, G. & Normark, S. (1991). Phase variation of gonococcal pili by frameshift mutation in *pilC*, a novel gene for pilus assembly. *EMBO Journal*, **10**, 477–88.

Jonsson, A. B., Pfeifer, J. & Normark, S. (1992). *Neisseria gonorrhoeae pilC* expression provides a selective mechanism for structural diversity of pili. *Proceedings of the National Academy of Sciences, USA*, **89**, 3204–8.

Kar, S., To, S. C. M. & Brinton, C. C. (1990). Cloning and expression in *Escherichia coli* of LKP pilus genes from a nontypeable *Haemophilus influenzae* strain. *Infection and Immunity*, **58**, 903–8.

Kim, S. R. & Komano, T. (1992). Nucleotide sequence of the R721 shufflon. *Journal of Bacteriology*, **174**, 7053–8.

Kitten, T. & Barbour, A. G. (1990). Juxtaposition of expressed variable antigen genes with a conserved telomere in the bacterium *Borrelia hermsii*. *Proceedings of National Academy of Sciences, USA*, **87**, 6077–81.

Kohwi, Y. & Kohwi-Shigematsu, T. (1991). Altered gene expression correlates with DNA structure. *Genes and Development*, **5**, 2547–54.

Komano, T., Funayama, N., Kim, S. R. & Nisioka, T. (1990). Transfer region of IncI1 plasmid R64 and role of shufflon in R64 transfer. *Journal of Bacteriology*, **172**, 2230–5.

Koomey, M., Gotschlich, E. C., Robbins, K., Bergstrom, S. & Swanson, J. (1987). Effects of *recA* mutations on pilus antigenic variation and phase transitions in *Neisseria gonorrhoeae*. *Genetics*, **117**, 391–8.

Kroll, J. S., Loynds, B., Brophy, L. N. & Moxon, E. R. (1990). The *bex* locus in encapsulated *Haemophilus influenzae* – a chromosomal region involved in capsule polysaccharide export. *Molecular Microbiology*, **4**, 1853–62.

Kroll, J. S. (1992). The genetics of encapsulation in *Haemophilus influenzae*. *Journal of Infectious Diseases*, **165**, 93–6.

Kroll, J. S. & Moxon, E. R. (1989). Genetic and phenotypic variation in the capsulation and virulence of culture collection strains of *Haemophilus influenzae* type-b. *Microbial Pathogenesis*, **7**, 449–57.

Kupsch, E. M., Knepper, B., Kuroki, T., Heuer, I. & Meyer, T. F. (1993). Variable opacity (*opa*) outer-membrane proteins account for the cell tropisms displayed by *Neisseria gonorrhoeae* for human leukocytes and epithelial cells. *EMBO Journal*, **12**, 641–50.

Levinson, G. & Gutman, G. A. (1987). Slipped-strand mispairing – a major mechanism for DNA-sequence evolution. *Molecular Biology and Evolution*, **4**, 203–21.

Makino, S., Van Putten, J. P. M. & Meyer, T. F. (1991). Phase variation of the opacity outer-membrane protein controls invasion by *Neisseria gonorrhoeae* into human epithelial cells. *EMBO Journal*, **10**, 1307–15.

Maskell, D. J., Szabo, M. J., Deadman, M. E. & Moxon, E. R. (1992). The *gal* locus from *Haemophilus influenzae* – cloning, sequencing and the use of *gal* mutants to study lipopolysaccharide. *Molecular Microbiology*, **6**, 3051–63.

McClain, M. S., Blomfield, I. C. & Eisenstein, B. I. (1991). Roles of *fimB* and *fimE* in site-specific DNA inversion associated with phase variation of type-1 fimbriae in *Escherichia coli*. *Journal of Bacteriology*, **173**, 5308–14.

Meyer, T. F., Gibbs, C. P. & Haas, R. (1990). Variation and control of protein expression in *Neisseria*. *Annual Review of Microbiology*, **44**, 451–77.

Nassif, X., Lowy, J., Stenberg, P., Ogaora, P., Ganji, A. & So, M. (1993). Antigenic variation of pilin regulates adhesion of *Neisseria meningitidis* to human epithelial cells. *Molecular Microbiology*, **8**, 719–25.

Nou, X. W., Skinner, B., Braaten, B., Blyn, L., Hirsch, D. & Low, D. (1993). Regulation of phylonephritis-associated pili phase variation in *Escherichia coli*-binding of the *papL* and the *lrp* regulatory proteins is controlled by DNA methylation. *Molecular Microbiology*, **7**, 545–53.

Olyhoek, A. J. M., Sarkari, J., Bopp, M., Morelli, G. & Achtman, M. (1991). Cloning and expression in *Escherichia coli* of *opc*, the gene for an unusual class-5 outer-membrane protein from *Neisseria meningitidis* (meningococci surface-antigen). *Microbial Pathogenesis*, **11**, 249–57.

Patel, P., Marrs, C. F., Mattick, J. S., Ruehl, W. W., Taylor, R. K. & Koomey, M. (1991). Shared antigenicity and immunogenicity of type-4 pilins expressed by *Pseudomonas aeruginosa*, *Moraxella bovis*, *Neisseria gonorrhoeae*, *Dichelobacter nodosus*, and *Vibrio cholerae*. *Infection and Immunity*, **59**, 4674–6.

Perry, A. C. F., Nicolson, I. J. & Saunders, J. R. (1988). *Neisseria meningitidis* C114 contains silent, truncated pilin genes that are homologous to *Neisseria gonorrhoeae pil* sequences. *Journal of Bacteriology*, **170**, 1691–7.

Plasterk, R. H. A., Brinkman, A. & Vandeputte, P. (1983). DNA inversions in the chromosome of *Escherichia coli* and in bacteriophage-Mu – relationship to other site-specific recombination systems. *Proceedings of the National Academy of Sciences, USA, Biological Sciences*, **80**, 5355–8.

Plasterk, R. H. A., Simon, M. I. & Barbour, A. G. (1985). Transposition of structural genes to an expression sequence on a linear plasmid causes antigenic variation in the bacterium. *Borrelia hermsii. Nature*, **318**, 257–63.

Plasterk, R. H. A. & Vandeputte, P. (1984). Genetic switches by DNA inversions in prokaryotes. *Biochimica et Biophysica Acta*, **782**, 111–19.

Robertson, B. D. & Meyer, T. F. (1992). Genetic variation in pathogenic bacteria. *Trends in Genetics*, **8**, 422–7.

Rosenqvist, E., Hoiby, E. A., Wedege, E., Kusecek, B. & Achtman, M. (1993). The 5c protein of *Neisseria meningitidis* is highly immunogenic in humans and induces bactericidal antibodies. *Journal of Infectious Diseases*, **167**, 1065–73.

Rosqvist, R., Skurnik, M. & Wolfwatz, H. (1988). Increased virulence of *Yersinia pseudotuberculosis* by two independent mutations. *Nature, London*, **334**, 522–5.

Rozsa, F. W. & Marrs, C. F. (1991). Interesting sequence differences between the pilin gene inversion regions of *Moraxella lacunata* ATCC-17956 and *Moraxella bovis* EPP63. *Journal of Bacteriology*, **173**, 4000–6.

Rudel, T., Van Putten, J. P. M., Gibbs, C. P., Haas, R. & Meyer, T. F. (1992). Interaction of 2 variable proteins (*pilE* and *pilC*) required for pilus-mediated adherence of *Neisseria gonorrhoeae* to human epithelial cells. *Molecular Microbiology*, **6**, 3439–50.

Sarkari, J., Pandit, N., Moxon, E. R. & Achtman, M. (1994). Variable expression of the Opc outer membrane protein in *Neisseria meningitidis* is caused by size variation of a promoter containing poly-cytidine. *Molecular Microbiology*, (in press).

Saunders, J. R., Wakeman, J., Sims, G., O'Sullivan, H., Hart, C. A. & Virji, M. (1993). Piliation in *Neisseria meningitidis* and its consequences. *Journal of Medical Microbiology*, **39**, 7–9.

Seifert, H. S., Ajioka, R. S., Marchal, C., Sparling, P. F. & So, M. (1988). DNA transformation leads to pilin antigenic variation in *Neisseria gonorrhoeae. Nature, London*, **336**, 392–5.

Simon, D. & Rest, R. F. (1992). *Escherichia coli* expressing a *Neisseria gonorrhoeae* opacity-associated outer-membrane protein invade human cervical and endometrial epithelial cell lines. *Proceedings of the National Academy of Sciences, USA*, **89**, 5512–16.

Stern, A., Brown, M., Nickel, P. & Meyer, T. F. (1986). Opacity genes in *Neisseria gonorrhoeae* – control of phase and antigenic variation. *Cell*, **47**, 61–71.

Stern, A. & Meyer, T. F. (1987). Common mechanism controlling phase and antigenic variation in pathogenic *Neisseriae. Molecular Microbiology*, **1**, 5–12.

Stibitz, S., Aaronson, W., Monack, D. & Falkow, S. (1989). Phase variation in *Bordetella pertussis* by frameshift mutation in a gene for a novel two-component system. *Nature, London*, **338**, 266–9.

Streisinger, G. & Owen, J. E. (1985). Mechanisms of spontaneous and induced frameshift mutation in bacteriophage-T4. *Genetics*, **109**, 633–59.

Swanson, J., Bergstrom, S., Robbins, K., Barrera, O., Corwin, D. & Koomey, J. M. (1986). Gene conversion involving the pilin structural gene correlates with pilus and reversible pilus⁻ changes in *Neisseria gonorrhoeae. Cell*, **47**, 267–76.

Swanson, J., Robbins, K., Barrera, O., Corwin, D., Boslego, J., Ciak, J., Blake, M. & Koomey, J. M. (1987). Gonococcal pilin variants in experimental gonorrhea. *Journal of Experimental Medicine*, **165**, 1344–57.

Swanson, J., Morrison, S., Barrera, O. & Hill, S. (1990). Piliation changes in transformation-defective gonococci. *Journal of Experimental Medicine*, **171**, 2131–9.

Tinsley, C. R. & Heckels, J. E. (1986). Variation in the expression of pili and outer-membrane protein by *Neisseria meningitidis* during the course of meningococcal infection. *Journal of General Microbiology*, **132**, 2483–90.

Tonjum, T., Marrs, C. F., Rozsa, F. & Bovre, K. (1991). The type-4 pilin of *Moraxella nonliquefaciens* exhibit unique similarities with the pilins of *Neisseria gonorrhoeae* and *Dichelobacter* (*Bacteroides*) *nodosus*. *Journal of General Microbiology*, **137**, 2483–90.

Van Ham, S. M., van Alphen, L., Mooi, F. R. & van Putten, J. P. M. (1993). Phase variation of *Haemophilus influenzae* fimbriae – transcriptional control of two divergent genes through a variable combined promoter region. *Cell*, **73**, 1187–96.

Van Ham, S. M., van Alphen, L., Mooi, F. R. & van Putten, J. P. M. (1994). The fimbrial gene cluster of *Haemophilus influenzae* type b. *Molecular Microbiology*, **11**.

Vanderwoude, M. W., Braaten, B. A. & Low, D. A. (1992). Evidence for global regulatory control of pilus expression in *Escherichia coli* by Lrp and DNA methylation – model building based on analysis of *pap*. *Molecular Microbiology*, **6**, 2429–35.

Vanderwoude, M. W. & Low, D. A. (1994). Leucine-responsive regulatory protein and deoxyadenosine methylase control the phase variation and expression of the *sfa* and *daa* pili operons in *Escherichia coli*. *Molecular Microbiology*, **11**, 605–18.

Van Putten, J. P. M. (1993). Phase variation of lipopolysaccharide directs interconversion of invasive and immuno-resistant phenotypes of *Neisseria gonorrhoeae*. *EMBO Journal*, **12**, 4034–51.

Virji, M., Heckels, J. E., Potts, W. J., Hart, C. A. & Saunders, J. R. (1989). Identification of epitopes recognized by monoclonal antibodies SM1 and SM2 which react with all pili of *Neisseria gonorrhoeae* but which differentiate between 2 structural classes of pili expressed by *Neisseria meningitidis* and the distribution of their encoding sequences in the genomes of *Neisseria* spp. *Journal of General Microbiology*, **135**, 3239–51.

Virji, M., Kayhty, H., Ferguson, D. J. P., Alexandrescu, C., Heckels, J. E. & Moxon, E. R. (1991). The role of pili in the interactions of pathogenic *Neisseria* with cultured human endothelial cells. *Molecular Microbiology*, **5**, 1831–41.

Virji, M., Alexandrescu, C., Ferguson, D. J. P., Saunders, J. R. & Moxon, E. R. (1992*a*). Variations in the expression of pili – the effect on adherence of *Neisseria meningitidis* to human epithelial and endothelial cells. *Molecular Microbiology*, **6**, 1271–9.

Virji, M., Makepeace, K., Ferguson, D. J. P., Achtman, M., Sarkari, J. & Moxon, E. R. (1992*b*). Expression of the Opc protein correlates with invasion of epithelial and endothelial cells by *Neisseria meningitidis*. *Molecular Microbiology*, **6**, 2785–95.

Virji, M., Makepeace, K., Ferguson, D. J. P., Achtman, M. & Moxon, E. R. (1993*a*). Meningococcal *opa* and *opc* proteins – their role in colonization and invasion of human epithelial and endothelial cells. *Molecular Microbiology*, **10**, 499–510.

Virji, M., Saunders, J. R., Sims, G., Makepeace, K., Maskell, D. & Ferguson, D. J. P. (1993*b*). Pilus-facilitated adherence of *Neisseria meningitidis* to human epithelial and endothelial cells – modulation of adherence phenotype occurs concurrently with changes in primary amino acid sequence and the glycosylation status of pilin. *Molecular Microbiology*, **10**, 1013–28.

Virji, M. & Heckels, J. E. (1984). The role of common and type-specific pilus antigenic domains in adhesion and virulence of gonococci for human epithelial cells. *Journal of General Microbiology*, **130**, 1089–95.

Weiser, J. N. (1993). Relationship between colony morphology and the life-cycle of *Haemophilus influenzae* – the contribution of lipopolysaccharide phase variation to pathogenesis. *Journal of Infectious Diseases*, **168**, 672–80.

Willems, R., Paul, A., Vanderheide, H. G. J., Teravest, A. R. & Mooi, F. R. (1990). Fimbrial phase variation in *Bordetella pertussis* – a novel mechanism for transcriptional regulation. *EMBO Journal*, **9**, 2803–9.

Yogev, D., Rosengarten, R., Watsonmckown, R. & Wise, K. S. (1991). Molecular basis of mycoplasma surface antigenic variation – a novel set of divergent genes undergo spontaneous mutation of periodic coding regions and 5′ regulatory sequences. *EMBO Journal*, **10**, 4069–79.

POPULATION GENETICS AND GLOBAL EPIDEMIOLOGY OF THE HUMAN PATHOGEN *NEISSERIA MENINGITIDIS*

MARTIN C. J. MAIDEN AND IAN M. FEAVERS

Division of Bacteriology, National Institute for Biological Standards and Control, Blanche Lane, South Mimms, Potters Bar, Herts, EN6 3QG, UK

INTRODUCTION

Variation among bacterial isolates is addressed by both population genetics and epidemiology. However, differences in the perspectives of each discipline can hinder the transfer of concepts between these two complementary fields. Epidemiological studies concentrate on measuring the spread of pathogenic bacteria and assessing risks to human populations. In contrast, population genetical studies determine the rates and mechanisms of genetic changes in bacterial populations, with the ultimate aim of understanding how new strains or species arise. Thus, to an epidemiologist, strains of a given species that do not cause disease may be of limited interest, even if pathogenic strains are atypical of the species as a whole. A population geneticist surveying the same species will include all possible isolates and may regard a rare pathogenic strain as of little significance. The perception of time by each of these disciplines is also different. A strain responsible for disease that persists in a particular locale for a period of several years will be thought of as long-lasting epidemiologically, but may be regarded as transitory in terms of the history of the species, which will be measured in centuries or millennia.

In this chapter, recent data from the human pathogen *Neisseria meningitidis* will be used to illustrate that (i) a complete understanding of the epidemiology of pathogenic bacteria requires the study of their population genetics and (ii) extensive bacterial strain collections, assembled for epidemiological purposes, can provide an excellent resource for the study of bacterial population genetics and evolution. The chapter will also outline how a detailed knowledge of the population genetics of pathogenic bacteria contributes to the rational development and assessment of public health measures.

Population structures

The basic concepts of bacterial population genetics are the subjects of other chapters of this book and will be only briefly summarized here. Genetical material can be passed between bacteria in two ways: vertically, from mother cell to daughter cell on cell division; and horizontally, between two otherwise unrelated cells. Whereas vertical transmission is a process that must occur during asexual reproduction by binary fission, horizontal genetical exchange requires a mechanism and an opportunity for DNA transfer and is essentially a chance event.

Populations in which vertical transmission predominates are clonal. As bacteria divide and spread, new alleles that arise by mutation are limited to the direct descendants of the cell in which they arose and cannot reassort with alleles of other genes into novel genetic combinations. The process of selection acting on a strictly asexual species results in a population consisting of a limited number of clones, each clone comprising closely related bacteria that share a common ancestor cell and a set of characteristic alleles (Selander & Levin, 1980). Such clonal populations exhibit linkage disequilibrium (non-random assortment of alleles; Selander et al., 1986; Maynard Smith et al., 1993).

Horizontal genetical exchange provides a mechanism for sexual transmission of genetic material, at least for small pieces of the chromosome, allowing the reassortment of mutant alleles among strains, increasing the possible combinations of alleles and thereby the genetical diversity of the species (Maynard Smith, Dowson & Spratt, 1991). This process works against the natural tendency of an asexual population to be clonal. There is a spectrum of possible population structures that bacterial populations can possess ranging from highly clonal to panmictic (Maynard Smith et al., 1993), the point on the spectrum that a given species will occupy being dependent on the frequency of horizontal genetical exchange relative to multiplication and spread of the organism (Spratt et al., this volume).

In population genetical studies, the genetical relationships among members of the population are inferred by analysis of the largest possible sample of chromosomal genes from the largest possible number of strains. There is no perfect way of achieving this, as it is not possible to examine the entire chromosome of every possible isolate, and all studies rely on sampling first the bacterial population and second the bacterial chromosome. For a pathogenic bacterium, the sample of the population is normally represented in the strain collections that have been assembled by epidemiologists. These collections are very useful for population genetic analyses as they usually contain isolates spanning relatively long periods of time and the date and location of isolation of the strains are usually known. Such strain collections, however, have the drawback that they principally comprise isolates from diseased patients. Although epidemiological studies will frequently include strains from 'carriers' (healthy individuals colonized with the pathogenic bacterium) these are rarely equal or even comparable in number with isolates from diseased patients. Therefore, epidemiological strain collections will only be accurate representations of the bacterial population when most colonizations result in disease. If the attack rate of the organism (the ratio of disease to harmless colonization) is low, the epidemiological strain collection may not be representative of the population of the 'pathogen' and population genetical studies based on this biased sample may be misleading.

The most widely used technique for sampling bacterial chromosomes and estimating the genetical relatedness of isolates is multi-locus enzyme electrophoresis (MLEE, Selander *et al.*, 1986). In this technique, alleles encoding a number of chromosomal proteins (usually enzymes that are easily visualized with staining procedures) are inferred by analysis of their gene products by starch gel electrophoresis. The MLEE data are then used to group the isolates into 'electrotypes' or ETs (Selander *et al.*, 1986). The members of a single ET share the same alleles at each of the loci examined and are often inferred to be 'clones'. However, as ETs are the result of a sample of data from the chromosome, all members of ETs may not be clones in the sense of being identical, although all clones (identical strains) will belong to the same ET. In addition, ETs can include unrelated bacteria that have chance associations of alleles, as discussed by Spratt *et al.* (this volume).

Many DNA-based methods for determining the relationships between isolates have been developed in the last decade and it is likely that data from these approaches will play an increasingly important role in clinical strain characterization. These include pulsed field gel electrophoresis (PFGE) fingerprinting (Bygraves & Maiden, 1992; Arbeit *et al.*, 1990), restriction fragment length polymorphism (RFLP) (Fox *et al.*, 1991), particularly of conserved genes such as those encoding 16s rRNA (Stull, Lipuma & Edlind, 1988; Woods *et al.*, 1992; Jordens & Pennington, 1991), and direct nucleotide sequence analysis of chromosomal genes employing the polymerase chain reaction (PCR) (Embley, 1991).

The meningococcus and meningococcal disease

There are a number of compelling reasons for studying *N. meningitidis*, the meningococcus. Medically, it is a pathogen of worldwide significance, causing an acute and dangerous disease (Peltola, 1983; Schwartz, Moore & Broome, 1989). All age groups can contract meningococcal meningitis and septicaemia, disease syndromes that occur either separately or together, but they are principally childhood diseases. Systemic meningococcal disease develops rapidly, is difficult to diagnose, and can result in the death of an otherwise healthy individual within a matter of hours of symptoms presenting. The case fatality rate is high, being between 5–25% with antibiotic treatment, rising to about 40% for fulminant meningococcal septicaemia, even when intensive supportive therapy is available. In addition, there are a number of potential complications resulting from infection by *N. meningitidis*: meningococcal septicaemia can cause a severe skin rash known as 'purpuric rash', sometimes resulting in permanent tissue damage, digit, or even limb loss; meningococcal meningitis can cause neurological sequelae, most commonly deafness (Peltola, 1983).

These features ensure that meningococcal disease has a high public profile and make it a priority for the health care systems of many countries. Effective infant immunization would, without doubt, be the most effective means of controlling the disease, but unfortunately there is no comprehensive childhood vaccine available. The commercially produced vaccines, based on capsular polysaccharides and developed in the 1960s (Gotschlich, Godschneider & Artenstein, 1969), are ineffective in protecting infants and provide poor long-term immunity in adults (Reingold et al., 1985; Lepow et al., 1986; Zollinger & Moran, 1991; Peltola et al., 1985; Gotschlich et al., 1972; Lepow et al., 1977).

Irrespective of its medical importance, the bacterium has a number of biological features that make it an interesting subject for study. In common with its close relative N. gonorrhoeae, the meningococcus is antigenically highly variable and possesses a variety of genetical mechanisms for mediating antigenic change in surface antigens (Poolman, Hopman & Zanen, 1982; Vedros, 1987; Aho et al., 1991; Feavers et al., 1992; Hu, 1987). These variable surface antigens are important in the interaction of the bacterium with the host, including immune reactions, adhesion, and invasion (Virji et al., 1991, 1992).

Despite a long history and much research effort, the mechanisms of pathogenicity of N. meningitidis remain poorly understood. There are sixteenth-century accounts of disease resembling meningococcal septicaemia and meningitis but the first probably authentic description of an epidemic of 'malignant purpuric fever' was made in Geneva in 1805 (Vieusseaux, 1805). The meningococcus itself, a Gram negative diplococcus, was first identified in 1887 in Vienna (Weichselbaum, 1887). The severity of this disease and its high mortality, resulted in the infection being a 'notifiable disease' in many countries from the first decades of this century and in some cases much earlier (Peltola, 1983). Epidemiological data of varying detail is therefore available in these countries for the last 60 years. Reliable strain collections required the development of modern microbiological methods for the storage of isolates. Strains dating back to the 1930s or earlier are available, but the unambiguous association of strains with particular disease outbreaks is only possible for isolates collected from the 1950s onwards (Caugant et al., 1987b; Achtman, 1994).

Characterization of meningococcal isolates

As the MLEE techniques described above are not suitable for routine application in clinical laboratories, serological methods remain the principal means for determining the epidemiological relationships between clinical

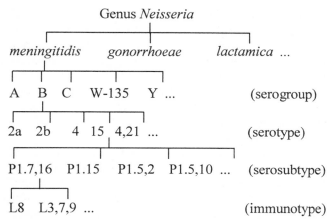

Fig. 1. Meningococcal typing scheme. *Neisseria meningitidis* is divided into serogroups (A, B, C, etc) on the basis of the reaction of its capsular polysaccharide with specific antibodies. These groups are further subdivided by the serological reactions of their class 2/3 OMPs into serotypes and their class 1 OMP into serosubtypes. Meningococcal immunotypes are determined by the reaction of the lipooligosaccharide with specific antibody reagents. In the serological designation each of these characteristics is separated by a semicolon, for example: B:15:P1.7,16:L3,7,9.

isolates of meningococci. These methods rely on reagents, generally either polyclonal or monoclonal antibodies, that react specifically with surface antigens of the meningococcus (Frasch, Zollinger & Poolman, 1985; Poolman *et al.*, 1982; Frasch & Chapman, 1972; Branham, 1953; Vedros, 1987; Frasch, 1987). The antigens most widely used for meningococcal strain characterization are the capsular polysaccharides, which define the serogroup (Branham, 1953; Vedros, 1987), the mutually exclusive class 2 and 3 OMPs, which define the serotype and the class 1 OMPs which define the serosubtype (Tsai, Frasch & Mocca, 1981; Poolman *et al.*, 1982). The typing scheme is summarized in Fig. 1. Unfortunately, these serological tests, particularly serotype and serosubtype, do not necessarily provide reliable indications of the genetical relationships between strains (Achtman, 1994; Maiden, 1993; Caugant *et al.*, 1986*b*, 1987*b*).

The serogroup is the most commonly determined property of a meningococcal strain and, although this is quite crude and can be misleading, it gives some indication of the genetical relationships between isolates. There are 13 identifiable meningococcal serogroups (Vedros, 1987), but more than 90% of isolates from diseased patients are serogroup A, B, or C (Frasch, 1989).

POPULATION STRUCTURE AND EPIDEMIOLOGY OF *NEISSERIA MENINGITIDIS*

The meningococcus, in common with other members of the genus *Neisseria*, is competent for DNA uptake and is readily transformed by foreign DNA

(Biswas, Thompson & Sparling, 1989; Maynard Smith *et al.*, 1991), providing a mechanism for horizontal genetical exchange both within and between *Neisseria* species. This enables *Neisseria* species to behave sexually (Feavers *et al.*, 1992; Spratt *et al.*, 1989; Spratt *et al.*, 1992; Gibbs *et al.*, 1989), with the result that their population structures are not necessarily strictly clonal. In fact, different meningococci appear to have different population structures (Spratt *et al.*, this volume). Interestingly, meningococci associated with different epidemiological forms of disease have different population structures, raising the intriguing question of whether the population genetics of an organism imposes a particular epidemiology or *vice versa*. All systemic meningococcal disease is associated with characteristic syndromes of meningitis and septicaemia, but the disease occurs in a number of epidemiologically distinct forms: (i) epidemic and pandemic disease outbreaks; (ii) localized outbreaks; (iii) hyper-endemic disease; and (iv) endemic disease.

Several strain collections of meningococci have been examined by MLEE (Caugant *et al.*, 1986*a,b*; Moore *et al.*, 1989; Caugant *et al.*, 1990; Wang *et al.*, 1992, 1993, Achtman, 1994) and the original interpretation of these data was that the species was clonal. More recent analyses have shown that strain collections of meningococci comprise isolates that belong to: (i) highly uniform ETs, representing meningococcal clones that have been stable over the period that strains have been reliably collected (about 50 years); (ii) complexes of closely related ETs, representing clones that are changing over the period represented by the strain collections; (iii) diverse ETs, any one of which occurs rarely in strain collections. In other words, over the parts of the world sampled by epidemiologists for the last 50 years or so, most meningococcal infections are caused by organisms that belong to one of a few ETs or complexes of related ETs.

The first group of strains, the highly uniform ETs, are all related to each other and are characterised by the possession of the polymannosamine phosphate (serogroup A) capsule. These 'subgroup A' bacteria have been divided into nine 'serogroups' comprising related ETs. In the strain collections so far examined most of the isolates in one subgroup belong to one or at most two closely related ETs (Wang *et al.*, 1992). Calculation of the index of association (I_A) for these data has indicated that these ETs represent clones (Spratt *et al.*, this volume). These bacteria occasionally cause sporadic endemic disease but the majority of isolates come from epidemics or pandemics of meningococcal disease.

There are three complexes of bacteria with related ET that account for many of the remaining meningococcal isolates: the A4 complex (Caugant *et al.*, 1987*b*); the ET-5 complex (Caugant *et al.*, 1986*b*); and the ET-37 complex (Wang *et al.*, 1993). Most of these bacteria express either serogroup B (poly[(2-8)-α-*N*-acetylneuraminic acid]) or serogroup C (poly[(2-9)-α-*N*-acetylneuraminic acid]) capsules, although other capsular types do occur. Unlike the subgroups of serogroup A, bacteria belonging to these com-

Table 1. *The relationship of meningococcal population structure to epidemiology. Examples are given for each of the population structures that occur in* Neisseria meningitidis, *together with their epidemiology*

Population structure (Maynard Smith *et al.*, 1993)	Disease associated with this population structure (Schwartz *et al.*, 1989)	MLEE designation given to meningococcal example	Reference
Clonal	Epidemic/pandemic	'Subgroups' I–VIII of serogroup A	(Wang *et al.*, 1992)
Epidemic	Localized outbreak	ET-37 complex	(Wang *et al.*, 1993)
	Hyper-endemic	ET-5 complex	(Caugant *et al.*, 1986*b*)
Panmictic (non-clonal)	Endemic disease	—	(Caugant *et al.*, 1990)

plexes are much more diverse; each complex comprising a number of distinct ETs that have been responsible for disease outbreaks. The bacteria from the 'serogroup B' and 'serogroup C' complexes are not associated with pandemics but are associated with other epidemiological forms of meningococcal disease: members of the ET-37 complex are frequently associated with localized outbreaks of meningococcal infection while bacteria belonging to the ET-5 complex are associated with hyper-endemic disease.

The isolates not associated with the main subgroups or complexes are of diverse ET that do not cluster closely with each other. In terms of the population genetical classification of Maynard-Smith *et al.* (1993), these data suggest that subgroups of the 'serogroup A' bacteria are clonal, while ET-5 and ET-37 are 'epidemic clonal' (a potentially ambiguous term that should not be confused with an epidemic *disease* outbreak or the organisms that cause it). The remainder of the meningococcal population can be regarded as panmictic. The relationships and classification of meningococcal subgroups and complexes to their population genetics and epidemiology are summarized in Table 1.

EPIDEMIC AND PANDEMIC DISEASE

The occasional large-scale epidemic disease outbreaks caused by 'serogroup A' meningococci are the most serious public health problem caused by the meningococcus (Moore, 1992). They cover large geographical areas extremely rapidly, sometimes spreading over several continents to constitute a pandemic outbreak in periods of weeks (Achtman, 1990; Wang *et al.*, 1992). Since the Second World War, such outbreaks have largely been limited to developing countries, particularly the 'meningitis belt' of sub-Saharan

Africa and China, where they occur at approximately 10-year intervals (Moore, 1992). Incidence rates, the number of cases of meningococcal disease that occur in a given period annually, are high and can reach more than 1000/100 000 (Lapeyssonnie, 1963; Hu, 1987; Moore, 1992), children aged 8–10 years being the most vulnerable age group (Peltola, 1983).

The MLEE analyses of 500 epidemiologically defined serogroup A strains, carried out in Berlin by Achtman and colleagues, have shown that there are 84 ETs, which have been classified into nine subgroups (I, II, III, IV-1, IV-2, V, VI, VII, VIII) (Olyhoek, Crowe & Achtman, 1987; Wang et al., 1992). Many of the ETs are rare and represented in the collection by one or a few isolates, the great majority of isolates from a given epidemic belonging to one ET of one subgroup. The subgroups are highly uniform both for the chromosomal enzymes used in the MLEE studies and surface antigens (Wang et al., 1992; Suker et al., 1994). A unique feature of this work is that the genetical data can be related to precise epidemiological data on date and location of isolation, which enables the spread of clones of serogroup A meningococci around the world to be mapped and the evolution of the serogroup A bacteria to be investigated (Olyhoek et al., 1987; Achtman, 1994).

Pandemic outbreaks of meningococcal disease, caused by one particular serogroup A subgroup, subgroup III, provide instructive examples of how global spread of a pathogen occurs, how human behaviour can contribute to the rate of bacterial spread, and how the combination of epidemiological analysis with population genetical data analyses enable the progress of epidemics to be measured precisely. In the last 30 years bacteria belonging to this subgroup have spread globally on two occasions, causing pandemics referred to as the 'old wave pandemic' and the 'new wave pandemic' (Achtman, 1990, 1994; Moore, 1992).

The old wave pandemic began with very large epidemics in China in the mid-1960s. The subgroup III bacteria then appeared to spread, with epidemics occurring in Moscow (1969–1971), Norway (1973), and Finland (1975), the last outbreak being in Brazil in 1974–1976 (Achtman, 1994; Wang et al., 1992; Fig. 2). Subgroup III bacteria caused no further epidemics until the mid-1980s when new outbreaks occurred in China and Nepal. Shortly afterwards a subgroup III outbreak became established among the pilgrims to the Hadj in Mecca in 1987. As the Hadj involves the pilgrimage of people from all over the world, the disease was then exported rapidly to many countries by the returning pilgrims (Fig. 2; Jones & Sutcliffe, 1990; Moore et al., 1989; Salih et al., 1990; Riou et al., 1991). This resulted in epidemics in East Africa but, although subgroup III bacteria were isolated from returning Hadjis and their close contacts, no epidemic outbreaks occurred in industrialized countries (Jones & Sutcliffe, 1990; Moore et al., 1988).

The subgroup III bacteria that caused the two pandemics were antigeni-

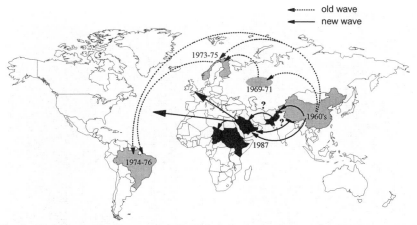

Fig. 2. Global spread of serogroup A meningococci of subgroup III: old and new wave pandemics. The old wave pandemic began in China in the mid-1960s and spread with epidemics occurring in Moscow followed by Norway, Finland, and finally Brazil. The new wave pandemic once again began in China spreading, probably via Nepal and Pakistan, and establishing an outbreak amongst pilgrims to the Hadj in Mecca in 1987. Isolated cases, but no epidemic outbreaks, of meningococcal disease caused by this subgroup were reported amongst pilgrims returning to industrialized countries.

cally uniform and belonged to the same ET (ET-48; Wang *et al.*, 1992) but further detail on the genetical relationships among subgroup III strains was revealed by PFGE fingerprinting, which is sufficiently sensitive to detect small changes between strains (Bygraves & Maiden, 1992). This technique involves the digestion of complete meningococcal chromosomes with restriction endonucleases that cut the chromosome infrequently and resolution of the large chromosomal fragments produced by PFGE to give fingerprint patterns that are characteristic for a particular strain. For the meningococci, the restriction endonuclease *Sfi*I gives a fingerprint characteristic for the subgroups and complexes of ETs, as defined by MLEE, but fingerprints generated with the endonucleases *Nhe*I or *Spe*I, which sample the chromosome more frequently, producing about 15 bands, resolve minor differences within ETs (Fig. 3).

Analysis of the old wave and new wave strains by PFGE fingerprinting showed that, although closely related, the strains causing each of these waves were distinct and further that most of the new wave strains were identical, presumably reflecting the rapid spread of this clone (Fig. 4; Bygraves & Maiden, 1992). In addition, while the ET-48 strains associated with the first stages of the proposed 'new wave pandemic' (Nepal and China) were similar but not identical to the 'old wave pandemic' strains, the strains that spread from Mecca were uniform. This epidemic was caused by a single new variant strain that spread through the pilgrim population in Mecca and

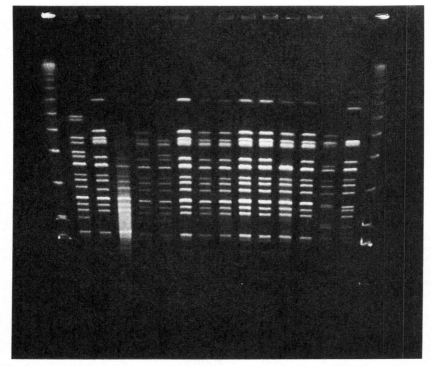

Fig. 3. Pulsed field gel electrophoresis fingerprint analysis of subgroup III isolates. Chromosomal DNA from 13 serogroup A, subgroup III isolates was digested with the rare cutting restriction endonuclease *Spe*I and the fingerprints were resolved by pulsed field gel electrophoresis. Lanes 1 and 16 are molecular size markers, lanes 2 to 12 are the fingerprints of strains isolated from various geographical locations following the 1987 Hadj; lanes 13 and 14 are 'old wave' strains isolated in China (1966) and Brazil (1976) respectively.

which was then disseminated very rapidly throughout the world by the mass movement of the pilgrims.

HYPER-ENDEMIC DISEASE

The term 'hyper-endemic' is used to describe outbreaks of a lower intensity but longer duration than the epidemic and pandemic outbreaks described above. Typically, these outbreaks have annual incidence rates of 10–50/100 000 and may be localized in a relatively small area or restricted to a particular population. A number of such outbreaks have been documented in various countries (Caugant *et al.*, 1986*b*; 1987*a*; 1990; Sacchi *et al.*, 1992*a*). Hyper-endemic disease outbreaks are caused by a novel meningococcal strain with an elevated attack rate becoming established in a

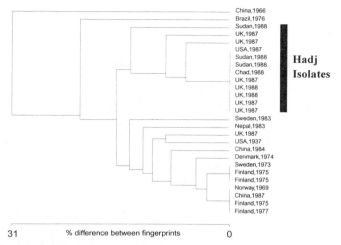

Fig. 4. Tree showing the relationship between old wave and new wave strains determined by pulsed field gel electrophoresis. The differences between the *Nhe*I fingerprints of 27 serogroup A, subgroup III meningococcal strains are shown. The percentage similarity between finger-prints was calculated on the basis of the number of bands shared between each possible pair of fingerprints and shows the uniformity of the stains associated with the Hadj outbreak (new wave pandemic).

population. Since the mid-1970s, meningococci belonging to ET-5 complex have been responsible for hyper-endemic disease in a number of countries: these are the best characterised hyper-endemic meningococci.

The ET-5 complex comprises 22 closely related ETs, all of the which are of recent origin, suggesting that these bacteria have only become a prominent part of meningococcal populations in the last 20 years. During this period they have spread over several continents causing hyper-endemic disease in some countries and larger scale epidemics in others (Caugant *et al.*, 1987*b*). In Norway, they were responsible for a major outbreak amongst teenage school children over the whole country (Caugant *et al.*, 1989), in the UK they caused a prolonged hyper-endemic outbreak in the village of Stone-house, Gloucestershire, also predominantly attacking teenagers (Cart-wright, Stuart & Nash, 1986); larger outbreaks have occurred in Cuba (Caugant *et al.*, 1986*b*), Brazil (Sacchi *et al.*, 1992*a*) and Chile (Caugant *et al.*, 1991).

Unlike the subgroups of the serogroup A bacteria, which have remained antigenically and genetically stable during epidemic spread and where bacteria of only one ET have been associated with disease worldwide, the members of the ET-5 complex are genetically and antigenically diverse. Most ET-5 bacteria are serogroup B, although it is not uncommon to iso-late serogroup C strains, but the subcapsular antigens are more variable. In Stonehouse in the UK, for example, most strains were serologically B:15:

P1.7,16 but at least three other serological types have been associated with the ET-5 complex isolated in the UK.

LOCALIZED EPIDEMIC OUTBREAKS

Localized outbreaks of meningococcal disease occur periodically in many countries. These are usually of shorter duration than a hyper-endemic outbreak and vary in size and incidence, but never reach the proportions of a 'serogroup A' epidemic or pandemic outbreak. Such outbreaks are frequently associated with institutions such as schools or military training establishments, and in these cases there is usually little or no disease in the general population, even if the same meningococcal strain is present (Jones, 1988; Sacchi *et al.*, 1992*b*; Wang *et al.*, 1993).

Many localized outbreaks of meningococcal disease are caused by members of the ET-37 complex. These bacteria are globally distributed and have been isolated from many countries in Europe, North and South America and Africa, the oldest isolates dating back to early in this century. In all cases where population genetical data is available, one or two closely related ETs from the complex have caused the outbreak; however, different outbreaks are caused by different ETs and there are no significant epidemiological links between outbreaks (Wang *et al.*, 1993). Antigenically, the ET-37 bacteria are less stable than the subgroups of serogroup A but more stable than the ET-5 complex. They are normally serogroup C, although serogroup B, W-135, Y and non-groupable isolates have been described. Their subcapsular antigens are either serotypes 2a or 2b and variants of the subtypes P1.5 and P1.2 or its close relative P1.10.

ENDEMIC DISEASE

Endemic disease is the predominant form of meningococcal disease that has occurred in industralized countries in the last 50 years, at annual incidence rates of 1–5/100 000. In the UK, for example, this results in some 1500–2500 cases of disease and about 150 deaths annually, predominantly in children of under 4 years old (the age-specific incidence rate peaks at about 50/100 000 in children aged 6 months). Disease is very rare compared to carriage of meningococcal strains, which in the UK is between 10 and 15% of the population at any one time (Cartwright *et al.*, 1987). Endemic disease peaks in the winter months and may be associated with other respiratory infections.

A MLEE-based study of endemic strains from the Netherlands isolated between 1958 and 1986 by was performed by Caugant and colleagues. In this study, 278 strains from cases of meningococcal disease were examined, chosen to avoid the peak period of disease in the winter. In this collection there were 145 distinct ETs only 21% of which were represented by more than one strain. About half the strains belonged to ETs clustering into two lineages one of which (designated lineage IV) comprised the predominant ETs (40% of isolates) isolated between 1958 and 1966, the other (designated lineage II) comprised the predominant ETs (48% of isolates) isolated between 1967 and 1978 (Caugant *et al.*, 1990).

Similar patterns of diversity were found in a survey of meningococcal strains from endemic cases and carriers of meningococcal disease in England and Wales by PFGE fingerprinting. Out of 95 strains examined there was a wide variety of *Sfi*I fingerprints implying a large number of different ETs in this strain collection. The largest group of similar patterns (18/95) were fingerprint patterns characteristic of bacteria belonging to the ET-5 complex (Bygraves & Maiden, 1991 and J. A. Bygraves and M. C. J. Maiden, unpublished observations).

It can be concluded that the human populations of industrialized countries with high carriage rates for meningococci carry a wide range of different meningococci that can cause sporadic cases of disease. Within this population there will be complexes and clones of bacteria of varying stability that persist for a period of time before becoming unrecognizable. Some of these complexes will comprise strains that are more invasive than other meningococci, therefore having the potential to cause either hyper-endemic disease or localised disease outbreaks. As sampling from diseased patients is not the most reliable way of sampling the carried meningococci, which are the great majority of the meningococcal population, it is probable that endemic meningococci are more diverse than most of the strain collections that have been examined so far.

THE POPULATION GENETICS OF A *N. MENINGITIDIS* OUTER MEMBRANE PROTEIN

More detailed information on the processes involved in the evolution of bacterial strains can be obtained by the comparative sequence analysis of individual genes. For the meningococcus, the most complete set of data of this type available is for the outer membrane protein (OMP) genes. The major OMPs of the meningococcus have attracted study for two main reasons: they are potential vaccine components and are also the targets for serotyping and serosubtyping reagents. All isolates express a number of OMPs, subdivided into five classes (1–5) on the basis of their size on SDS–PAGE and immunological reactivities (Tsai *et al.*, 1981). The class 1, and the mutually exclusive class 2 and 3 proteins are porins that are antigenically stable within a given isolate, not undergoing phase variation. They do, however, vary between isolates, a property that makes them ideal antigens for typing (Poolman *et al.*, 1982; Abdillahi & Poolman, 1988; Maiden *et al.*, 1992). As these proteins are major cell-surface components it is presumed that the variation observed is the result of immune selection.

The class 1 OMP, encoded by the *porA* gene, has many antigenically distinct variants and the nucleotide sequences of the genes encoding many of these have been determined (Barlow, Heckels & Clarke, 1989; McGuiness *et al.*, 1990; Maiden *et al.*, 1991; McGuiness, Lambden & Heckels, 1993; Suker *et al.*, 1994). The predicted secondary structure of the protein has eight cell-surface exposed loops (I–VIII) with the most variation occurring in loops I (or variable region 1; VR1) and IV (or variable region 2; VR2) (McGuiness *et al.*, 1990; Maiden *et al.*, 1991; van der Ley *et al.*, 1991). The meningococcal subtypes, determined by the amino acid sequences of these VRs, are designated 'P1.' followed by two numbers representing epitopes located in the VR1 and VR2 amino acid sequences (Suker *et al.*, 1994). Variation from the prototype sequence within a VR family is indicated by a lower case letter. Molecular genetical analyses of the class 1 OMP genes provide insights into the population genetics of the strains associated with each of the different epidemiologies. The behaviour of one particular class 1 OMP gene, that encoding the P1.5,10 epitopes, will be used to illustrate some of the differences in the population genetics of each of the meningococcal examples described above.

Particular class 1 OMP genes from a globally disseminated gene pool are associated with each of the subgroups of the serogroup A meningococci (Wang *et al.*, 1992; Suker *et al.*, 1994). The P1.5c, 10 gene is the most widely distributed and occurs in four of the nine serogroup A subgroups. This gene is very stable in serogroup A and is identical in strains isolated in the 1930's and 1980's and from different parts of the world, indicating either that mutations occur infrequently, or that there is no selective pressure favouring mutational changes, which are subsequently lost during epidemic spread. Only two variants of this gene have been seen in serogroup A meningococci isolated at diverse global locations: a deletion of the DNA encoding VR2 in a strain isolated in Algeria in 1969; and a change encoding the P1.5c epitope to the P1.5 epitope in the *porA* gene of two strains isolated in Russia in 1989 (Fig. 5).

Serogroup B and C strains exhibit a wider range of serosubtypes than the serogroup A strains. This diversity is generated both by horizontal genetical exchange, mosaic gene structures frequently occuring in the class 1 OMP genes of serogroup B and C meningococci (Feavers *et al.*, 1992), and other mutational processes. A recent survey of serogroup B and C strains isolated in the UK over a two year period, revealed numerous variants of the P1.5, 10 gene, the result of point mutations, insertions and deletions of the *porA* gene (Fig. 6, Feavers, Maiden and Fox, unpublished data). Given the high rates of carriage in the UK, there is ample opportunity for recombination to occur among endemic strains and the evidence suggests that there is selective pressure, presumably imposed by the host immune system, for mutants that result in changes in the surface exposed loops of the class 1 OMP.

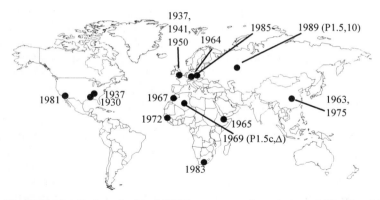

Fig. 5. Worldwide distribution of a class 1 OMP in serogroup A meningococci. The class 1 OMP in serogroup A meningococci is stable over time and distance; the map shows the geographical and temporal distribution of strains expressing the P1.5c,10 class 1 OMP. Only two isolates contained mutants relative to the prototype P1.5c,10 *porA* gene: one isolated in Russia (1989) which held the P1.5 epitope rather than P1.5c and one isolated in Algeria (1969) from which the VR2 sequence encoding the P1.10 epitope was deleted.

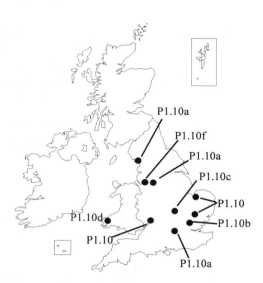

Fig. 6. Distribution of class 1 OMP variants in UK. The class 1 OMP in serogroups B and C is much more variable than in serogroup A; the map shows the locations of meningococci expressing different variants of the P1.10 protein isolated between 1989 and 1991 within the UK alone.

Meningococci of the ET-5 complex express diverse class 1 OMPs (Caugant *et al.*, 1987*b*), suggesting that recombination between these and other strains occurs frequently, but only one particular serosubtype is associated with each hyper-endemic outbreak: P1.5c,10 (Worcester, UK, M.C.J. Maiden, I.M. Feavers and A.J. Fox, unpublished observations); P1.7,16b (Stroud, UK; McGuiness *et al.*, 1991); and P1.15. This indicates that the *porA* gene in hyper-endemic strains is likely to be subject to continual selective pressure from the host's immune response establishing a much wider range of variation than serogroup A strains, although in a particular outbreak only a single clone spreads. Bacteria belonging to the ET-37 complex are restricted to two variants, P1.2 and P1.2b (formerly known as P1.y; Wang *et al.*, 1993), from the possible five P1.2 variants characterised to date. Only one of the two variants is ever seen in a particular outbreak.

A MODEL FOR MENINGOCOCCAL MOLECULAR EPIDEMIOLOGY

Synthesis of the data discussed into a consistent model for the population genetics and associated epidemiology of *N. meningitidis* is problematic, as different strains within the species behave differently and the available data is not complete, particularly for bacteria not commonly associated with disease. However, some general principles can be adduced. An important feature of the meningococci, often obscured by a strictly medical or epidemiological perspective, is that they are mostly asymptomatic colonisers of the upper respiratory tract of humans that only occasionally cause infection. There is much interest in identifying strains that are particularly invasive or 'virulent', but this is complicated by the relatively low attack rates even of 'virulent' meningococci and by the fact that a number of host-related factors including immune status, socioeconomic conditions, and other respiratory infections may promote the development of meningococcal carriage into meningococcal disease.

Attack rates for the meningococcus are difficult to measure precisely but in the endemic situation are around 1 case per 10 000 colonizations. Even for 'virulent' strains in epidemics or pandemics this rate probably rarely exceeds 1 case per 100 colonizations. All strain collections are heavily biased towards isolates from the blood or cerebrospinal fluid of patients with invasive meningococcal disease. One MLEE study, for example, included 601 isolates from patients compared with only 87 isolates from healthy carriers (Caugant *et al.*, 1987*b*). To obtain a representative sample of the population of a bacterium with an attack rate of 1 per 100 colonization, 100 carried strains should be analysed for each disease-causing strain examined. Thus, in the study mentioned above, strains from cases were over-represented by at least 1000 fold and, given that normally the attack rate of the meningococcus is much lower than 1 per 100, the over-representation of strains from diseased patients is probably much more.

It can be argued that it is only the strains that cause disease that are important and require study but, in an organism such as the meningococcus in which DNA is readily exchanged between strains, it is important to understand the population biology of the whole species as the more invasive bacteria arise from the population of less invasive organisms. Similarly, it is important to study the related *Neisseria* species that colonize the human nasopharynx but cause disease even less frequently as they also can contribute genes to the meningococcal gene pool and could in this way influence the invasive potential of novel meningococcal strains.

Neisseria meningitidis is a diverse species both genetically and antigenically and, because of horizontal genetical exchange, the species is largely panmictic sharing gene pools with other pathogenic *Neisseria* (Maynard Smith *et al.*, 1991; Campos *et al.*, 1992; Zhang *et al.*, 1990; Feavers *et al.*, 1992; Suker *et al.*, 1994). Within this panmictic population there will be a number of transitory clones, the stability of which will be variable if the strains so far examined are a reliable guide for the whole species. From time-to-time strains with an increased propensity to cause disease will arise by the random assortment of alleles encoding such processes as adhesion to cell surfaces or evasion of host mucosal immunity. This increased propensity for invasion may increase the attack rate from, say 1 per 10 000 colonizations to 1 per 1 000 colonizations. This may not be important in terms of the survival of the new strain and have little impact on the biology of the species but, as the descendants of this strain spread through a human population, there will be a tenfold increase in the normal level of disease. As most of the meningococcal strains collected are from diseased patients, this strain will outnumber other strains in the collection by at least ten to one. Such strains are often referred to as 'successful'; however, as success for a meningococcus is an asymptomatic colonization of the host these strains are not necessarily successful in the biological sense. They are successful in being isolated and getting into strain collections; such collections therefore mostly comprise a limited number of ETs, the sporadic cases caused by less invasive meningococci giving a glimpse of the overall diversity of the species.

Outbreaks of disease occur as a result of the combination of the appearance or spread of strains with an increased capacity for invasion together with other environmental factors that increase the susceptibility of members of a population to invasion by meningococci; infants are more susceptible, probably because of a lower immunity to meningococci; teenagers in a socially active group probably come into contact with a greater variety of meningococcal strains; military recruits often live in crowded dormitories and engage in unaccustomed physical activity; the populations as a whole may suffer an influenza outbreak.

Epidemics and pandemics of 'serogroup A' meningococci are a special case, caused by a combination of at least two factors (i) the introduction of an immunologically different subgroup to an immunologically naïve population and (ii) environmental conditions. In Africa serogroup A epidemic outbreaks occur in the dry season, when it is hot and dusty, conditions that presumably promote damage of the mucosa and bacterial invasion, but in China serogroup A epidemics are associated with the wet season. Large-scale epidemic disease is generally associated with populations of low socio-economic status and risk factors probably include overcrowding and poor nutrition. Thus, the worst outbreaks occur seasonally in the developing world. If the same clone is introduced into a developed country with a temperate climate either no outbreak occurs (as happened in the USA, UK, and other industrialized countries when the Hadj strain was introduced) or a less intense outbreak with lower attack rates (<15 cases/100 000) is established, generally confined to lower socio-economic groups such as that seen in Finland in the 1970s (Moore, 1992). These epidemics probably end because of a change in weather and/or the population building up a herd immunity to the serogroup A capsule and can be successfully interrupted by vaccination with anti-serogroup A polysaccharide vaccines.

IMPLICATIONS OF POPULATION GENETICAL DATA FOR MENINGOCOCCAL VACCINE DESIGN AND TYPING

A primary aim of studies on the meningococcus is the elimination of meningococcal disease which is best achieved by the development of effective childhood vaccines. Meningococcal epidemiology and population genetics contribute to this goal by improving our understanding of the strains responsible for disease and identifying potential vaccine components. Detailed epidemiological analyses are also an essential component of phase three (placebo-controlled double-blinded) trials designed to determine the effectiveness of novel vaccines. Day-to-day epidemiology, aimed at providing the public health physician with advice on outbreaks, requires reliable identification of clinical isolates. Studies of population genetics permit the reliability of typing tools to be assessed and the rational design of new typing reagents.

In the absence of a good animal model, it is not a simple matter to design new vaccines against the meningococcus and much of the information required has to be inferred from epidemiological and population genetic analyses. Of these, the simplest observation is that the great majority of meningococcal disease is caused by three serogroups (A, B and C) with a relatively minor component from two other capsular types (Y and W-135). The capsule is encoded by an operon (Frosch, Weisgerber & Meyer, 1989) which is much less easily transferred between strains than the, usually single, genes that encode protein antigens. Newly introduced conjugate capsular polysaccharide vaccines have recently been highly effective in eliminating meningitis caused by *Haemophilus influenzae* type b (Murphy *et al.*, 1993), a similar problem to meningococcal meningitis in some respects. There are a number of problems with the development of similar vaccines against the meningococcus, including poor immunogenicity (Gotschlich *et al.*, 1972) and possible cross-reactivity of antibodies against the serogroup B capsular polysaccharide, which resembles antigens that occur in human tissue (Hayrinen, Bitter Suermann & Finne, 1989). This has led to an intensive search for other potential vaccine components including the major OMPs.

The nucleotide sequence analysis of class 1 OMP genes from many strains shows that although only a limited number of antigenic types of the class 1 OMP are associated with serogroup A meningococci, it is much more variable in serogroups B and C (Suker *et al.*, 1994). As there are many variants of the class 1 OMP associated with serogroup B and they appear to mutate rapidly, it is likely that even a multivalent vaccine would be only partially effective in countering endemic disease caused by serogroup B meningococci. It is likely that other protein antigens will behave similarly. This perhaps indicates that more priority should be accorded to overcoming the problems associated with vaccines based on the serogroup B capsule.

During a disease outbreak, epidemiologists are expected to provide rapid and accurate information to public health physicians. Such advice is important in terms of containing outbreaks and can have a very high profile in the community, particularly with an alarming childhood disease such as meningitis. In these cases it is rarely possible to conduct a thorough analysis of isolates such as would be carried out in a population genetical study. There is a great premium in producing rapid and easily interpreted data. Generally, the typing methods used for epidemiological analysis have been developed in advance of studies of the population genetics of the species in question.

In the case of *N. meningitidis*, the effectiveness of the serological typing scheme in determining genetical relationships among isolates varies with the epidemiological form of the disease. The serogroup A strains responsible for epidemic and pandemic disease are antigenically stable and in some cases subgroups possess characteristic class 1 OMP genes. For example, a serogroup A isolate with serotype 4,21 serosubtype P1.20,9, will almost certainly be a subgroup III bacterium. The ET-37 bacteria are antigenically more variable but characteristically are serogroup C, subgroup 2a or 2b and one of a limited number of variants of P1.5,2. In both these cases an epidemiologist could make a reasonably reliable judgement on the serology, particularly in the context of additional information such as geographical location, time of year, age group of patient. Other meningococci are much more difficult to identify by these means. A number of serotypes and serosubtypes have been associated with the ET-5 complex bacteria and these bacteria appear to be changing so rapidly that serological typing is at best an unreliable method for identifying these strains except in the short term. New antigenic forms of this and other hyper-endemic complexes may be in the population but not detected by the current serological reagents.

CONCLUSIONS

The study of infectious disease was revolutionized in the last century by the development of the concept that particular infective agents are associated with each disease. Since then the characterization of clinical isolates of microorganisms has been of central importance in the study of infectious disease and in the development of public health measures, prophylaxis, and treatments. As bacterial populations have proved to be more adaptable to medical intervention than was anticipated in the years immediately following the widespread introduction of antibiotics, the epidemiology and population genetics of bacterial pathogens is likely to be important for many years to come.

Modern molecular technologies have substantially enhanced the volume, rapidity, and level of detail attainable in the characterization of disease-causing bacteria and enhanced studies of their population genetics. Such data permit us to design antibiotics and vaccines more rationally and to anticipate how a bacterial population may change in response to these interventions. They also provide detailed background information for the interpretation of epidemiological information by improving our understanding of (i) the genetical mechanisms involved in the evolution of pathogens and (ii) how pathogens spread through human populations.

Studies of the population genetics of the meningococcus illustrate the following general principles. First, there are no infallible rules in the synthesis of population genetics and epidemiology. In the case of the meningococcus neither its epidemiology nor its population genetics are uniform throughout the species: all meningococci have the potential to cause the same disease syndromes but different meningococci are more or less clonal and cause disease of varying epidemic intensity, ranging from rare endemic disease to large-scale pandemics. Secondly, while it is important that population genetic analyses are performed on rationally chosen and epidemiologically significant strains, due regard should be paid to the fact that less pathogenic strains may contribute to gene pools shared by pathogens. Thirdly, in a recombining population, the genes encoding surfaces antigens are potentially mobile and probably under strong immune selection, calling into question the reliability of serological tests in the determination of the genetical relationships between isolates.

ACKNOWLEDGEMENTS

We would like to thank all our colleagues who have worked with us over the past five years, particularly Jane Bygraves and Janet Suker. Especial thanks goes to Mark Achtman of the Max-Planck-Institut-für-molekulare-Genetik, Berlin for provision of strains and for his invaluable encouragement, advice, and enthusiastic collaboration. Thanks are due also to Professor Brian

Spratt of the University of Sussex and Dennis Jones and Andrew Fox of the Public Health Laboratory, Withington Hospital, Manchester for their assistance with various of our studies into the meningococcus. We are grateful to Per Olcén, Örebro Medical Center Hospital, Sweden for his comments on the manuscript.

REFERENCES

Abdillahi, H. & Poolman, J. T. (1988). Definition of meningococcal class 1 OMP subtyping antigens by monoclonal antibodies. *FEMS Microbiology and Immunology*, **1**, 139–44.

Achtman, M. (1990). Molecular epidemiology of epidemic bacterial meningitis. *Reviews in Medical Microbiology*, **1**, 29–38.

Achtman, M. (1994). Clonal spread of serogroup A meningococci. A paradigm for the analysis of microevolution in bacteria. *Molecular Microbiology*, **11**, 15–22.

Aho, E. L., Dempsey, J. A., Hobbs, M. M., Klapper, D. G. & Cannon, J. G. (1991). Characterization of the *opa* (class 5) gene family of *Neisseria meningitidis*. *Molecular Microbiology*, **5**, 1429–37.

Arbeit, R. D., Arthur, M., Dunn, R., Kim, C., Selander, R. K. & Goldstein, R. (1990). Resolution of recent evolutionary divergence among *Escherichia coli* from related lineages: the application of pulsed field electrophoresis to molecular epidemiology. *Journal of Infectious Diseases*, **161**, 230–5.

Barlow, A. K., Heckels, J. E. & Clarke, I. N. (1989). The class 1 outer membrane protein of *Neisseria meningitidis*: gene sequence and structural and immunological similarities to gonococcal porins. *Molecular Microbiology*, **3**, 131–9.

Biswas, G. D., Thompson, S. A. & Sparling, P. F. (1989). Gene transfer in *Neisseria gonorrhoeae*. *Clinical Microbiological Reviews*, **2**, S24-8.

Branham, S.E. (1953). Serological relationships among meningococci. *Bacteriological Reviews*, **17**, 175–88.

Bygraves, J. A. & Maiden, M. C. J. (1991). The resolution of clonal types of *Neisseria meningitidis* by pulsed field gel electrophoresis. In *Neisseriae 1990* (Achtman, M. *et al.* (ed.), pp. 25–30, Walter de Gruyter, Berlin.

Bygraves, J. A. & Maiden, M. C. J. (1992). Analysis of the clonal relationship between strains of *Neisseria meningitidis* by pulsed field gel electrophoresis. *Journal of General Microbiology*, **138**, 523–31.

Campos, J., Fuste, M. C., Trujillo, G., Saez Nieto, J., Vazquez, J., Loren, J. G., Vinas, M. & Spratt, B. G. (1992). Genetic diversity of penicillin-resistant *Neisseria meningitidis*. *Journal of Infectious Diseases*, **166**, 173–7.

Cartwright, K. A., Stuart, J. M. & Noah, N. D. (1986). An outbreak of meningococcal disease in Gloucestershire. *Lancet*, **ii**, 558–61.

Cartwright, K. A. V., Stuart, J. M., Jones, D. M. & Noah, N. D. (1987). The Stonehouse survey: nasopharyngeal carriage of meningococci and *Neisseria lactamica*. *Epidemiology and Infection*, **99**, 591–601.

Caugant, D. A., Bovre, K., Gaustad, P., Bryn, K., Holten, E., Hoiby, E. A. & Froholm, L. O. (1986a). Multilocus genotypes determined by enzyme electrophoresis of *Neisseria meningitidis* isolated from patients with systemic disease and from healthy carriers. *Journal of General Microbiology*, **132**, 641–52.

Caugant, D. A., Froholm, L. O., Bovre, K., Holten, E., Frasch, C. E., Mocca, L. F., Zollinger, W. D. & Selander, R. K. (1986b). Intercontinental spread of a genetically distinctive complex of clones of *Neisseria meningitidis* causing epidemic disease. *Proceedings of the National Academy of Sciences, USA*, **83**, 4927–31.

Caugant, D. A., Froholm, L. O., Bovre, K., Holten, E., Frasch, C. E., Mocca, L. F., Zollinger, W. D. & Selander, R. K. (1987a). Intercontinental spread of *Neisseria meningitidis* clones of the ET-5 complex. *Antonie van Leeuwenhoek Journal of Microbiology*, **53**, 389–94.

Caugant, D. A., Mocca, L. F., Frasch, C. E., Froholm, L. O., Zollinger, W. D. & Selander, R. K. (1987b). Genetic structrure of *Neisseria meningitidis* populations in relation to serogroup, serotype, and outer membrane protein pattern. *Journal of Bacteriology*, **169**, 2781–92.

Caugant, D. A., Froholm, L. O., Selander, R. K. & Bovre, K. (1989). Sulfonamide resistance in *Neisseria meningitidis* isolates of clones of the ET-5 complex. *APMIS*, **97**, 425–8.

Caugant, D. A., Bol, P., Holby, E. A., Zanen, H. C. & Froholm, L. O. (1990). Clones of serogroup B *Neisseria meningitidis* causing systemic disease in the Netherlands, 1958–1986. *Journal of Infectious Diseases*, **162**, 867–74.

Caugant, D. A., Froholm, L. O., Sacchi, C. T. & Selander, R. K. (1991). Genetic structure and epidemiology of serogroup B *Neisseria meningitidis*. In *Neisseriae 1990* (Achtman, M., Kohl, P., Marchal, C., Morelli, G., Seiler, A. and Thiesen, B. eds.), pp. 37–42, Walter de Gruyter, Berlin.

Embley, T. M. (1991). The linear PCR reaction: a simple and robust method for sequencing amplified rRNA genes. *Letters in Applied Microbiology*, **13**, 171–4.

Feavers, I. M., Heath, A. B., Bygraves, J. A. & Maiden, M. C. J. (1992). Role of horizontal genetical exchange in the antigenic variation of the Class 1 outer membrane protein of *Neisseria meningitidis*. *Molecular Microbiology*, **6**, 489–95.

Fox, A. J., Jones, D. M., Gray, S. J., Caugant, D. A. & Saunders, N. A. (1991). An epidemiologically valuable typing method for *Neisseria meningitidis* by analysis of restriction fragment length polymorphisms. *Journal of Medical Microbiology*, **34**, 265–70.

Frasch, C. E., Zollinger, W. D. & Poolman, J. T. (1985). Serotype antigens of *Neisseria meningitidis* and a proposed scheme for designation of serotypes. *Reviews of Infectious Diseases*, **7**, 504–10.

Frasch, C. E. (1987). Development of meningococcal serotyping. In *Evolution of meningococcal disease*. (Vedros, N. A. ed.), pp. 39–55, CRC Press Inc., Boca Raton, FL.

Frasch, C. E. (1989). Vaccines for prevention of meningococcal disease. *Clinical Microbiology Reviews*, **2**, S134–8.

Frasch, C. E. & Chapman, S. (1972). Classification of *Neisseria meningitidis* Group B into distinct serotypes. *Infection and Immunity*, **5**, 98–102.

Frosch, M., Weisgerber, C. & Meyer, T. F. (1989). Molecular characterization and expression in *Escherichia coli* of the gene complex encoding the polysaccharide capsule of *Neisseria meningitidis* group B. *Proceedings of the National Academy of Sciences, USA*, **86**, 1669–73.

Gibbs, C. P., Reimann, B.-Y., Schultz, E., Kaufmann, A., Haas, R. & Meyer, T. F. (1989). Reassortment of pilin genes in *Neisseria gonorrhoeae* occurs by two distinct mechanisms. *Nature*, **338**, 651–2.

Gotschlich, E. C., Goldschneider, I. & Artenstein, M. S. (1969). Human immunity to the meningococcus IV. Immunogenicity of group A and group C meningococcal polysaccharides. *Journal of Experimental Medicine*, **129**, 1367–84.

Gotschlich, E. C., Rey, M., Triau, R. & Sparks, K. J. (1972). Quantitative determination of the human immune response to immunization with meningococcal vaccines. *Journal of Clinical Investigation*, **51**, 89–96.

Hayrinen, J., Bitter Suermann, D. & Finne, J. (1989). Interaction of meningococcal group B monoclonal antibody and its Fab fragment with alpha 2-8-linked sialic acid polymers: requirement of a long oligosaccharide segment for binding. *Molecular Immunology*, **26**, 523–9.

Hu, Z. (1987). Epidemiology of meningococcal disease in China. In *Evolution of Meningococcal Disease* (Vedros, N. A. ed.), pp. 19–32, CRC Press Inc, Boca Raton, Fla.

Jones, D. M. (1988). Epidemiology of meningococcal infection in England and Wales. *Journal of Medical Microbiology*, **26**, 165–8.

Jones, D. M. & Sutcliffe, E. M. (1990) Group A meningococcal disease in England associated with the Haj. *Journal of Infection*, **21**, 21–5.

Jordens, J. Z. & Pennington, T. H. (1991). Characterization of *Neisseria meningitidis* isolates by ribosomal RNA gene restriction patterns and restriction endonuclease digestion of chromosomal DNA. *Epidemiology and Infection*, **107**, 253–62.

Lapeyssonnie, L. (1963). La meningite cerebrospinale en Afrique. *Bull. World Health Organ.*, **28** (suppl), 53–114.

Lepow, M. L., Goldschneider, I., Gold, R., Randolph, M. & Gotschlich, E. C. (1977). Persistence of antibody following immunization of children with groups A and C meningococcal polysaccharide vaccines. *Pediatrics*, **60**, 673–80.

Lepow, M. L., Beeler, J., Randolph, M., Samuelson, J. S. & Hankins, W. A. (1986). Reactogenicity and immunogenicity of a quadravalent combined meningococcal polysaccharide vaccine in children. *Journal of Infectious Diseases*, **154**, 1033–6.

McGuinness, B. T., Barlow, A. K., Clarke, I. N., Farley, J. E., Anilionis, A., Poolman, J. T. & Heckels, J. E. (1990). Deduced amino acid sequences of class 1 protein PorA from three strains of *Neisseria meningitidis*. *Journal of Experimental Medicine*, **171**, 1871–82.

McGuinness, B. T., Clarke, I. N., Lambden, P. R., Barlow, A. K., Poolman, J. T., Jones, D. M. & Heckels, J. E. (1991). Point mutation in meningococcal *porA* gene associated with increased endemic disease. *Lancet*, **337**, 514–17.

McGuiness, B. T., Lambden, P. R. & Heckels, J. E. (1993). Class 1 outer membrane protein of *Neisseria meningitidis*: epitope analysis of the antigenic diversity between strains, implications for subtype definition and molecular epidemiology. *Molecular Microbiology*, **7**, 505–14.

Maiden, M. C. J., Suker, J., McKenna, A. J., Bygraves, J. A. & Feavers, I. M. (1991). Comparison of the class 1 outer membrane proteins of eight serological reference strains of *Neisseria meningitidis*. *Molecular Microbiology*, **5**, 727–36.

Maiden, M. C. J., Bygraves, J. A., McCarvil, J. & Feavers, I. M. (1992). Identification of meningococcal serosubtypes by polymerase chain reaction. *Journal of Clinical Microbiology*, **30**, 2835–41.

Maiden, M. C. J. (1993). Population genetics of a transformable bacterium: the influence of horizontal genetical exchange on the biology of *Neisseria meningitidis*. *FEMS Microbiology Letters*, **112**, 243–50.

Maynard Smith, J., Dowson, C. G. & Spratt, B. G. (1991). Localized sex in bacteria. *Nature, London*, **349**, 29–31.

Maynard Smith, J., Smith, N. H., O'Rourke, M. & Spratt, B. G. (1993). How clonal are bacteria? *Proceedings of the National Academy of Sciences, USA*, **90**, 4384–8.

Moore, P. S., Harrison, L. H., Telzak, E. E., Ajello, G. W. & Broome, C. V. (1988). Group A meningococci carriage in travelers returning from Saudi Arabia. *Journal of the American Medical Association*, **260**, 2686–9.

Moore, P. S., Reeves, M. W., Schwartz, B., Gellin, B. G. & Broome, C. V. (1989). Intercontinental spread of an epidemic group A *Neisseria meningitidis* strain. *Lancet*, **ii**, 260–2.

Moore, P. S. (1992). Meningococcal meningitis in sub-Saharan Africa: a model for the epidemic process. *Clinical Infectious Diseases*, **14**, 515–25.

Murphy, T. V., White, K. E., Pastor, P., Gabriel, L., Medley, F., Granoff, D. M. & Osterholm, M. T. (1993). Declining incidence of *Haemophilus influenzae* type b disease since introduction of vaccination. *Journal of the American Medical Association*, **269**, 246–8.

Olyhoek, T., Crowe, B. A. & Achtman, M. (1987). Clonal population structure of *Neisseria meningitidis* serogroup A isolated from epidemics and pandemics between 1915 and 1983. *Reviews of Infectious Diseases*, **9**, 665–82.

Peltola, H. (1983). Meningococcal disease: still with us. *Reviews of Infectious Diseases*, **5**, 71–91.

Peltola, H., Safaray, A., Kayhty, H., Karanko, R. N. & Andre, F. E. (1985). Evaluation of two tetravalent (ACYW-135) meningococcal vaccines in infants and small children—a clinical study comparing immunogenicity of O-acetyl negative and O-acetyl positive group C polysaccharides. *Pediatrics*, **76**, 91–6.

Poolman, J. T., Hopman, C. T. P. & Zanen, H. C. (1982). Problems in the definition of meningococcal serotypes. *FEMS Microbiology Letters*, **13**, 339–48.

Reingold, A. L., Broome, C. V., Hightower, A. W., Bolan, G. A., Adamsbaum, C., Jones, E. E., Phillips, C., Tiendrebeogo, H. & Yada, A. (1985). Age-specific differences in duration of clinical protection after vaccination with meningococcal polysaccharide A vaccine. *Lancet*, **ii**, 114–18.

Riou, J. Y., Caugant, D. A., Selander, R. K., Poolman, J. T., Guibourdenche, M. & Collatz, E. (1991). Characterization of *Neisseria meningitidis* serogroup A strains from an outbreak in France by serotype, serosubtype, multilocus enzyme genotype and outer membrane protein pattern. *European Journal of Clinical Microbiology and Infectious Diseases*, **10**, 405–9.

Sacchi, C. T., Pessoa, L. L., Ramos, S. R., Milagres, L. G., Camargo, M. C., Hidalgo, N. T., Melles, C. E., Caugant, D. A. & Frasche, C. E. (1992*a*). Ongoing group B *Neisseria meningitidis* epidemic in Saõ Paulo, Brazil, due to increased prevalence of a single clone of the ET-5 complex. *Journal of Clinical Microbiology*, **30**, 1734–8.

Sacchi, C. T., Zanella, R. C., Caugant, D. A., Frasch, C. E., Hidalgo, N. T., Milagres, L. G., Pessoa, L. L., Ramos, S. R., Camargo, M. C. & Melles, C. E. (1992*b*). Emergence of a new clone of serogroup C *Neisseria meningitidis* in Saõ Paulo, Brazil. *Journal of Clinical Microbiology*, **30**, 1282–6.

Salih, M. A., Danielsson, D., Backman, A., Caugant, D. A., Achtman, M. & Olcen, P. (1990). Characterization of epidemic and nonepidemic *Neisseria meningitidis* serogroup A strains from Sudan and Sweden. *Journal of Clinical Microbiology*, **28**, 1711–19.

Schwartz, B., Moore, P. S. & Broome, C. V. (1989). Global epidemiology of meningococcal disease. *Clinical Microbiological Reviews*, **2**, S118–24.

Selander, R. K., Caugant, D. A., Ochman, H., Musser, J. M., Gilmour, M. N. & Whittam, T. S. (1986). Methods of multilocus enzyme electrophoresis for bacterial population genetics and systematics. *Applied and Environmental Microbiology*, **51**, 837–84.

Selander, R. K. & Levin, B. R. (1980). Genetic diversity and structure in *Escherichia coli* populations. *Science*, **210**, 545–7.

Spratt, B. G., Zhang, Q.-Y., Jones, D. M., Hutchison, A., Brannigan, J. A. & Dowson, C. G. (1989). Recruitment of a penicillin-binding protein gene from

Neisseria flavescens during the emergence of penicillin resistance in *Neisseria meningitidis*. *Proceedings of the National Academy of Sciences, USA*, **86**, 8988–92.

Spratt, B. G., Bowler, L. D., Zhang, Q. Y., Zhou, J. & Smith, J. M. (1992). Role of interspecies transfer of chromosomal genes in the evolution of penicillin resistance in pathogenic and commensal *Neisseria* species. *Journal of Molecular Evolution*, **34**, 115–25.

Stull, T. L., Lipuma, J. J. & Edlind, T. D. (1988). A broad spectrum probe for molecular epidemiology of bacteria: ribosomal RNA. *Journal of Infectious Diseases*, **157**, 280–6.

Suker, J., Feavers, I. M., Achtman, M., Morelli, G., Wang, J.-F. & Maiden, M. C. J. (1994). The *porA* gene in serogroup A meningococci: evolutionary stability and mechanism of genetic variation. *Molecular Microbiology*, **12**, 253–65.

Tsai, C.-M., Frasch, C. E. & Mocca, L. F. (1981). Five structural classes of major outer membrane proteins in *Neisseria meningitidis*. *Journal of Bacteriology*, **146**, 69–78.

van der Ley, P., Heckels, J. E., Virji, M., Hoogerhout, P. & Poolman, J. T. (1991). Topology of outer membrane proteins in pathogenic *Neisseria* species. *Infection and Immunity*, **59**, 2963–71.

Vedros, N. A. (1987). Development of meningococcal serogroups. In *Evolution of Meningococcal Disease*. (Vedros, N. A. ed.), pp. 33–37, CRC Press Inc., Boca Raton, FL.

Vieusseaux, M. (1805). Memoire sur le maladie qui a régné a Genêve au printemps de 1805. *Journal de Médecine, Chirurgie et Pharmacie*, **II**, 163–5.

Virji, M., Kayhty, H., Ferguson, D.J.P., Alexandrescu, C., Heckels, J. E. & Moxon, E. R. (1991). The role of pili in the interactions of pathogenic *Neisseria* with cultured human endothelial cells. *Molecular Microbiology*, **5**, 1831–41.

Virji, M., Makepeace, K., Ferguson, D. J. P., Achtman, M., Sarkari, J. & Moxon, E. R. (1992). Expression of the Opc protein correlates with invasion of epithelial and endothelial cells by *Neisseria meningitidis*. *Molecular Microbiology*, **6**, 2785–95.

Wang, J.-F., Caugant, D. A., Li, X., Hu, X., Poolman, J. T., Crowe, B. A. & Achtman, M. (1992). Clonal and antigenic analysis of serogroup A *Neisseria meningitidis* with particular reference to epidemiological features of epidemic meningitis in China. *Infectious Immunity*, **60**, 5267–82.

Wang, J.-F., Caugant, D. A., Morelli, G., Koumare, B. & Achtman, M. (1993). Antigenic and epidemiologic properties of the ET-37 complex of *Neisseria meningitidis*. *Journal of Infectious Diseases*, **167**, 1320–9.

Weichselbaum, A. (1887). Uber die aeriologie der akuten meningitis cerebrospinalis. *Fortschritte der Medecin*, **5**, 573–5.

Woods, T. C., Helsel, L. O., Swaminathan, B., Bibb, W. F., Pinner, R. W., Gellin, B. G., Collin, S. F., Waterman, S. H., Reeves, M. W., Brenner, D. J. *et al*. (1992). Characterization of *Neisseria meningitidis* serogroup C by multilocus enzyme electrophoresis and ribosomal DNA restriction profiles (ribotyping). *Journal of Clinical Microbiology*, **30**, 132–7.

Zhang, Q., Jones, D. M., Saez Nieto, J. A., Perez Trallero, E. & Spratt, B. G. (1990). Genetic diversity of penicillin-binding protein 2 genes of penicillin-resistant strains of *Neisseria meningitidis* revealed by fingerprinting of amplified DNA. *Antimicrobial Agents and Chemotherapy*, **34**, 1523–8.

Zollinger, W. D. & Moran, E. (1991). Meningococcal vaccines – present and future. *Transactions of the Royal Society of Tropical Medicine and Hygiene*, **85** Suppl 1, 37–43.

SAMPLING AND DETECTING BACTERIAL POPULATIONS IN NATURAL ENVIRONMENTS

ROGER W. PICKUP

Institute of Freshwater Ecology, Windermere Laboratory, The Ferry House, Far Sawrey, Ambleside, Cumbria, LA22 0LP, UK

SAMPLING

In order to determine the roles played by microorganisms in a particular habitat, some form of sampling procedure has to be undertaken (see Herbert, 1990). There are usually three options available:

(a) to remove a sample from an environment and return it to the laboratory for analysis. This approach is often synonymous with 'destructive sampling' which renders the sample non-representative of the environment from which it was removed (e.g. grab sample from benthic environment; Herbert, 1990);

(b) to remove a sample from the environment whilst attempting to maintain *in situ* conditions during transportation and subsequent laboratory analysis. During the laboratory analysis, the sample can be maintained as close as possible to *in situ* conditions, with one or two parameters being varied for experimental purposes. Further development of this principle leads to systems that are often termed 'microcosms'. These can vary from the simple two phase systems (lakewater/air or soil/air) to the complex (three-phase flow-through sediment/lakewater systems; Morgan, Rhodes & Pickup, 1993). Microcosms have provided model systems with which to study survival, movement, transport, gene transfer and microbial interactions (see Trevors, 1988);

(c) to perform the experiments in the field, with the minimum of disturbance to the habitat. This is the least flexible of the options, and only a limited range of parameters can be measured, for instance, methane flux from upland soils (Baker, Norman & Bland, 1992). These limitations arise owing to logistical constraints particularly where the transportation of delicate equipment to remote locations is required.

INTRODUCTION

The aim of the microbial ecologist is to understand the activities of, and interactions between, microorganisms and their environment. The detection and isolation of a wide range of bacteria are essential components in the study of microbial ecology (for reviews see Austin, 1988; Grigorova & Norris, 1990; Pickup, 1991). In recent times, great emphasis has been placed on the detection and enumeration of groups of bacteria that are indicative of pollution of terrestrial and aquatic habitats. The most common microbial populations monitored, for example, are those associated with sewage and water quality. Their presence can be determined routinely using set procedures and specific bacteriological media. However, this is not so for the vast proportion of bacteria identified through direct counting procedures (Roszak & Colwell, 1987). The adoption of molecular techniques to microbial ecology has increased enormously both the sensitivity and precision of detection and identification of specific organisms and populations in the environment. Studies at the community level constitute one of the most difficult areas of microbial ecology (Brock, 1987). The combination of classical microbial ecology and molecular biology has also afforded the opportunity to study the activity and interaction at either community or population level. Naturally, there is a wide quantitative gap between molecular and the more classical ecological approaches, but they are able to complement each other in different situations. For the microbial ecologists interested in this approach (it is by no means the only approach open to microbial ecologists) such techniques can be directed towards the total community or separate populations (synecology) or to individual species within a population (autecology) (Hall et al., 1990). This chapter will consider how sampling and sample processing are crucial to the studies in microbial ecology and will consider their limitations.

SAMPLE PROCESSING: STUDYING MICROBIAL POPULATIONS

Culturable bacteria

Often population studies focus on enumeration, and no attempt is made to correlate their presence with activities or spatial distribution within a habitat. However, the importance of traditional methods for enumeration should never be underestimated. There are numerous strategies for the detection and isolation of bacteria from environmental samples (see Herbert, 1990). Conventional media-based isolation methods are designed for enumeration of the culturable population. A count of culturable bacteria is obtained after growth on a suitable medium containing carbon and/or other energy sources (Roszak & Colwell, 1987). Since all media are selective to a lesser or greater extent, and only a small proportion of bacteria are recoverable, viable counts are rarely quantitative. Modification of recovery procedures can increase the number of viable bacteria isolated from any sample. In addition to media selectivity, seasonal variations also affect sampling and recovery efficiency (Voloudakis et al., 1991) and incubation temperatures play a significant role in the efficiency of recovery. Temperatures optimized for medically important bacteria (i.e. those above 30°C) select against the majority of environmental isolates. However, for aquatic bacteria, maximum recovery can be obtained following incubation at between 10 and 20°C for long periods (Jones, 1977). Further modifications of isolation procedures are made where appropriate to target specialized groups of bacteria, for example, fluorescent pseudomonads, methanogens and yellow-pigmented bacteria (including *Flavobacterium* and Cytophaga) (see Balows *et al.*, 1992). More accurate counts than those obtained using spread or pour plate methods can be achieved by most-probable-number (MPN) procedures. The MPN technique employs a serial dilution of the sample in appropriate media, and a high degree of replication provides statistically significant enumeration of culturable organisms (Roszak & Colwell, 1987). In general, enumeration methods that rely on cell culture will always underestimate the presence of bacteria (Roszak & Colwell, 1987).

Total bacterial counts: direct non-culture methods

As it is not normally possible to enumerate all the culturable bacteria in an environmental sample, methods for direct counting that do not rely on the culturability of bacteria are more quantitative than viable counting procedures (Fry, 1988). Direct counting of fluorescently stained bacteria on black membrane filters by epifluorescence microscopy has become the most frequently used method for total bacterial population estimates. This technique permits a rapid quantitative estimation of total bacteria numbers (Hall *et al.*, 1990) and it is especially effective in aquatic systems (Jones & Simon, 1975). It is not always necessary to use membranes since this technique can be applied directly in counting chambers or on natural or artificial surfaces (Hall *et al.*, 1990). Epifluorescence is achieved by using a range of nucleic acid or protein stains, the most commonly used include acridine orange and 4′-6′-diamidino-2-phenylidole (DAPI; Fry, 1988). The quality of the microscopic image depends on many factors, such as light source, optical characteristics of the instrument used, the type of fluorochrome employed and the nature of the sample under analysis (Fry, 1988). Although problematic, techniques have been developed to count bacteria directly *in situ* (Jones, 1977; Fry, 1988). However, if the only information required is total numbers, as opposed to information on community structure and spatial distribution, then standard procedures can be employed to remove bacteria from solid surfaces or particles and the direct count can then be performed on a cell suspension (Goulder, 1987; Fry, 1988).

Unfortunately, direct counting procedures that employ simple fluorescent stains are non-discriminatory by nature. Counts of individual genera or species can be achieved only if the target is morphologically distinct or distinguishable from the rest of the indigenous microflora by some other means. Although bacteria have little morphological diversity, it is possible to recognize a limited number of organisms based on cell morphology alone using microscopic techniques. Filamentous organisms are an obvious example, especially since they can comprise up to 50% of the biomass in freshwater sediments (Jones & Jones, 1986; Jones, 1987). Similarly, distinct organisms such as *Ochrobium* spp., *Achromatium* spp. and the putative genes *Metallogenium* can be identified easily and enumerated (Jones, 1987). However, the majority of bacteria do not possess any distinct morphological characteristic and therefore identification requires additional biochemical or immunological techniques.

Immunofluorescence microscopy has been widely applied to the detection and enumeration of particular microorganisms when conventional techniques have proved difficult. Fluorochromes, such as fluorescein isothiocyanate, can be coupled to an antibody that binds directly with a target antigen on the cell or to a second antibody that recognizes an antibody produced against the microorganism (for example, see Chantler & McIllmurray, 1988). Immunofluorescence detection has been used for various bacteria that are difficult to culture, for example methanogenic bacteria (Conway de Macario, Wolin & Macario, 1982), methane oxidizers (Reed & Duggan, 1978), N-fixing organisms (Renwick & Gareth, 1985) nitrifiers in soil and marine habitats (Belser & Schmidt, 1978; Ward & Carlucci, 1985) and acid tolerant *Thiobacillus ferrooxidans* (Apel *et al.*, 1976).

There are several problems associated with the use of fluorescent antibodies to detect cells in the environment. First, it is important that the identifying antigen is a unique cell surface protein which is not only host-specific but is readily available as a target for the antibody. Secondly, it is common to raise antibodies against cells grown under laboratory conditions which do not reflect the physiological status of the organism in the environment. Therefore, it is important to determine whether that antigen is expressed under environmental conditions. Thirdly, reduction in specificity of the antibody may occur owing to interference from particular matter, natural autofluorescence of co-existing material, cross-reactivity with other antigens and failure to reach the target organism due to the presence of extracellular material; and finally, this approach cannot distinguish between viable and non-viable cells and therefore cannot be classed as quantitative with respect to active cells (Ford & Olsen, 1988; Trevors, 1992). However, despite these perceived problems, fluorescent monoclonal antibodies specific for the O1 antigen of *Vibrio cholerae*, used in conjunction with fluorescence microscopy, were shown to be a more sensitive method of assessing water quality than standard culture methods (Brayton *et al.*, 1987).

Fluorescent oligonucleotides also offer a potentially highly discriminating detection system when linked to microscopical and flow cytometric techniques. Fluorescent oligonucleotide probes against 16S ribosomal ribonucleic acid (rRNA) sequences provide a detection system of potentially wide application (Giovannoni *et al.*, 1988, Ward *et al.*, 1993). For example, universal eubacterial or archaebacterial sequences are available as well as those specific for a single species (for example, *Desulphovibrio gigas*, *Escherichia coli* and *Pseudomonas fluorescens*; Amman, Krumholz & Stahl, 1990*a*,*b*; Zarda *et al.*, 1991). Early work has shown such probes to work well with laboratory cultures, and fluorescent cells are detectable after hybridization. Ward and co-workers (1993) described some of the limitations of this approach on ecological samples. These include bias in the original sequence data on which the probes are based and the reactivity of the probe as affected by specificity, sensitivity and accessibility of the target.

Wallner, Amman & Beisker (1993) optimized the staining procedure for bacterial cells (in this case *E. coli*) using fluorescent oligonucleotide probes specific for 16S rRNA molecule. As there is one target site per ribosome, probe-conferred fluorescence should be directly proportional to the ribosome concentration. This was confirmed by labelling cells at different stages during growth (Wallner *et al.*, 1993). Therefore, the growth stage of the cells would influence the signal generated. Theoretically, the rigours of real life in natural environments could reduce rRNA levels to such an extent so as to reduce binding sites for the probes which in turn would mean that cells cannot be made sufficiently fluorescent for detection. Wallner and co-workers (1993) also noted that multiple labelling of probes did not improve the signal but increased the non-specific staining. A variety of micro-organisms have been targeted with fluorescently labelled oligonucleotide probes and detected *in situ* in a number of habitats and it is apparent that their activity is sufficient to provide a target for the fluorescent probes. These include *Pseudomonas cepacia* and *Streptomyces scabies* (sandy loam soil, Hahn *et al.*, 1992), *Shewanella putrefaciens* (water column and sediments, DiChristina & DeLong, 1993), *Desulfovibrio vulgaris* (bioreactor, Kane, Poulsen & Stahl, 1993), *Frankia* spp. (root nodules, Hahn, Strarrenburg & Arkermans, 1990), eubacteria in drinking water and associated biofilms (Hicks, Amann & Stahl, 1992; Manz *et al.*, 1993) and archaeobacterial endosymbionts (Embley *et al.*, 1992).

SAMPLING STRATEGY: LIMITATIONS

The first limitation of population studies, particularly those based on enumeration is experimental design which requires replication of samples and the statistical treatment of the subsequent data (Hall *et al.*, 1990). Fry (1993) emphasized the need for error-free statistical analysis, otherwise results and data obtained with considerable effort would be misinterpreted. It is important to stress that an impractical degree of replication may be required to work to a given level of statistical significance. Appropriate example calculations relating the mean and variance of a given variable to an appropriate level of significance are provided by Elliot (1977). Such a test is worthwhile performing if only to demonstrate how far the practical limits of replication fall short of what is possible. The degree of variability of microbiological data, on both temporal and spatial scales is reported by Jones and Simon (1975) further emphasizing the caution which must be exercised in interpreting data (Hall *et al.*, 1990).

A further limitation to research into microbial populations is an inability to isolate and grow in culture the majority of bacteria. There has always been a discrepancy between cell numbers obtained by direct and viable counting methods. Studies concluded that culturable bacteria represented only 0.01–12.5% of the viable bacterial population in terrestrial and aquatic environments (Jones, 1977; Hoppe, 1978; Ferguson, Buckley & Palumbo, 1984). Furthermore, some bacteria have been shown to become unculturable but retain their viability after exposure to the environment and have been called 'non-culturable but viable' (NCBV; Colwell *et al.*, 1985). This complicates both the detection and enumeration of microorganisms. Microorganisms known to achieve the NCBV state include *E. coli*, *Salmonella typhimurium* and *Vibrio* spp. (Colwell *et al.*, 1985; Roszak & Colwell, 1987). However, the NCBV state may be a specialized strategy exercised by a few specific types of organism. In addition, it is often wrongly attributed to some microorganisms, although confirmatory methods have been developed (Morgan, Cranwell & Pickup, 1991; Morgan *et al.*, 1993). There are two other factors which contribute to this discrepancy. The direct count cannot distinguish between cells that are viable, NCBV, or dead. Conversely, media used for the isolation of viable bacteria may actively select against growth because they are too rich in nutrients or do not supply essential co-factors. Methods have been developed that go some way towards distinguishing viable cells under epifluorescence microscopy (see Roszak & Colwell, 1987). Kogure, Simidu & Taga (1979) showed that elongation of cells provided an indication of viability when a nutrient source supplemented with an inhibitor of cell division was added to environmental samples. However, it is clear that not all viable cells, for example, species that exist as cocci or those naturally resistant to the inhibitor (Morgan *et al.*, 1991), can react by elongation in this way. Also, as with bacterial isolation procedures, it is clear that all experimental conditions are not suitable for all samples. Despite some limitations, this method represents a bridge between counting culturable bacteria and direct counts and has been termed a direct viable count (DVC; Kogure *et al.*, 1979).

Any estimation of the numbers of bacteria in the environment must therefore allow for the fact that a high proportion of the target organisms are non-culturable. The traditional and molecular methods that are available show a dichotomy in the organisms they can detect. Culture techniques impose strict limitations on the amount of sample that can be removed from natural ecosystems and adequately processed in the laboratory. Coupled with the non-representative nature of the population obtained this emphasizes the need for sampling procedures that preclude a requirement for culturing. Detection methods will target either the culturable population or the total population (including viable, NCBV, and dead or dying cells). The sampling strategy therefore determines the route and ultimately the sensitivity of the method being used or developed.

DNA ISOLATION: CIRCUMVENTING THE NEED FOR CULTURING

Microbial ecologists have recently begun to apply molecular techniques to the detection of bacteria from the environment, obviating the need for cell culture. Like traditional methods, these techniques also rely heavily on the sampling strategy for their efficiency. Detection methods that do not rely on culturability but use immunological and/or nucleic acid hybridization techniques (Sayler & Layton, 1990) can employ strategies that involve either isolating total cells followed by DNA extraction (indirect), or extracting total DNA directly from a sample (direct; see Trevors, 1992). Furthermore, comparative sequencing of rRNA has revolutionized systematics. Methods for RNA analysis have been directed towards studies in the environment, particularly probe development and microbial classification (eg. Pace *et al.*, 1987; Ward *et al.*, 1993). Moreover, with the introduction of polymerase chain reaction (PCR), it has become possible to investigate the evolutionary relationships of unculturable organisms (Amman *et al.*, 1991). There is one sampling strategy shared by all these approaches, namely DNA extraction. Methods for DNA extraction can be either indirect (whereby bacterial cells are removed from the sample prior to the extractive process) or direct (where cells are lysed within the sample and the DNA is subsequently purified from that sample).

Indirect DNA extraction

Lakewater is probably the most amenable medium to sample and process. In samples where bacterial numbers are low, large volumes of water can be filtered using traditional filtration techniques with 0.4 or 0.22 μm pore size membranes. A further method is the use of tangential flow filtration with 0.22 μm pore size membranes which permit the rapid concentration of particles from large volumes of water. The DNA from these cell suspensions can be recovered using a variety of extraction procedures (Trevors, 1992). It is more difficult to obtain a representative sample of a bacterial population from solid substrates such as soil or freshwater sediment. Problems arise owing to the spatial distribution of the microorganisms in soil and the strong associations they form with particulate matter (Stotzky, 1986). To obtain total cells, the soil sample can be converted into a suspension by mechanical agitation using vortexing or sonication (Trevors, 1992). Once dislodged, the cells can be recovered by centrifugal sedimentation incorporating washing steps as required to remove contaminating material (Holben *et al.*, 1988). Alternatively, the cells can be removed from the soil using chemical dispersants to dissociate the microorganisms from soil particles (for example, MacDonald, 1986; Hopkins, MacNaughton & O'Donnell, 1991) in single- or multi-stage procedures followed by isolation of cells prior to the extraction of DNA. As is usually the case, no single method is suitable for all conditions (Trevors, 1992). Once extracted, the final use of the DNA will determine the degree to which it has to be purified.

Direct extraction methods

Direct DNA extraction methods do not require isolation of bacterial cells, avoiding fractionation of cells from the sample and involve direct lysis of the cells followed by DNA recovery. A rapid method for the direct isolation of DNA from the aquatic environment comprises filtration of a large volume of water through a cylindrical filter membrane with DNA extraction occurring within the filter housing (Sommerville *et al.*, 1989). In a similar approach, Ogram, Sayler & Barkay (1988) used direct lysis followed by ultracentrifugation and hydroxyapatite chromatography to obtain DNA from freshwater sediments. These methods avoid the need for culturing the organisms yet have limitations in the range of DNA fragment sizes that can be isolated (Pickup, 1991; Trevors, 1992). Steffan *et al.* (1988) directly lysed bacterial cells and recovered the DNA using the procedure of Ogram *et al.* (1988). Modifications to these methods and combining PCR have been applied to soil (Selenska & Klingmuller, 1992; Tsai & Olsen, 1991, 1992; Pillai *et al.*, 1991), sediments (Steffan *et al.*, 1988; Tsai & Olsen, 1992*a,b*) and to the aquatic environment in general (Steffan & Atlas, 1988; Steffan *et al.*, 1989; Bej *et al.*, 1990; Bej *et al.*, 1991*a,b,c*) as highly sensitive methods for detecting very small numbers of target organisms (Steffan & Atlas, 1991).

MODIFICATION OF SAMPLING STRATEGY TO OPTIMIZE MOLECULAR APPLICATIONS

It is apparent that the advantages of applying molecular techniques outweigh the limitations imposed by a range of inherent factors. The following section will describe three examples of sampling strategies that have been optimized to recover specific groups of microorganisms from particular habitats for the application of molecular techniques and to correlate their presence with specific activities or functions. It will concentrate on the types of processes/organisms/habitats under study and the method employed. However, it will not discuss the results in detail as most await acceptance in scientific journals.

Large conspicuous sulphide oxidizing bacteria: Achromatium oxaliferums

The oxidation of sulphide to sulphate is a globally significant biogeochemical process which ranges from the purely chemical to completely biologically mediated. These systems are probably atypical and little is known about the relative contribution of chemical and biological oxidation in more typical sediment systems.

The oxidation of organic carbon linked to the reduction of sulphate is a significant process in both freshwater and marine sediment environments (Bak & Pfennig, 1991). Sulphide produced by sulphate reduction can have several fates such as the formation of poorly soluble metal sulphides in the presence of ferrous iron. The remaining sulphide diffuses from the reduced sediment layers to the oxidized uppermost sediment layer where a combination of chemical and biological oxidation results in its re-oxidation to sulphate.

The sediments in shallow water bodies are such that where sulphide gradients from reduced sediment layers overlap with gradients of oxygen from overlying water a diverse group of microorganisms is found that compete effiently with the chemical oxidation of sulphide. Many of these organisms are chemolithoautotrophic or exhibit facultative chemolithoautotrophic or exhibit facultative chemolithoautotrophic growth (Kuenen & Beudeker, 1982) with sulphide as an energy source and oxygen as a terminal electron acceptor. Some species are known to use nitrate, elemental sulphur and possibly manganese as alternative terminal electron acceptors (e.g. Sweets *et al.*, 1990). In mats where sulphide oxidizing bacteria exist, the sulphide oxidation proceeds up to 10^5 times faster than the rates of chemical oxidation (Jorgensen & Revsbech, 1983). It has been further demonstrated that carbon dioxide fixation coupled with the microbial oxidation of sulphide is potentially a significant process in some sediments (Nelson, Jorgensen & Rwsbech, 1986). Aerobic sulphide-oxidizing bacteria therefore constitute key players in several major biogeochemical cycles and have significance for nutrient and energy fluxes of global significance.

A diverse range of morphologically conspicuous bacteria are also believed to be important oxidizers of reduced sulphur compounds in the biosphere (La Riviere & Schmidt, 1991). Of these it is only mat-forming *Beggiatoa* spp. that have received much attention related to their physiology and ecological niche (e.g. Nelson *et al.*, 1986). Organisms with some morphological similarities to *Beggiatoa* spp. have also been described from sulphide-rich oxic environments (e.g. *Thioploca* spp. and *Thiothrix* spp.) as have other morphologically distinct organisms including *Achromatium* *spp.* and *Thiovulum* spp. These have only rarely been obtained in axenic culture or have eluded laboratory cultivation. Like *Beggiatoa* spp., these organisms are found in large numbers at the sediment water interface of marine and freshwater environments where they are believed to perform a similar ecological function to *Beggiatoa* mats.

Achromatium spp. fit into the category of non-culturable but viable organisms though probably not through some specialized survival strategy exhibited by *Vibrio* spp. (Roszak & Colwell, 1987). However, we are able to circumvent this difficulty by directly isolating the organism from sediments by virtue of its unusual morphology. In selected habitats in the English Lake District populations of *Achromatium oxaliferum* can be detected with relative ease. It is morphologically conspicuous due to its large spherical or ovoid size (up to $150 \times 50 \mu m$) and the intracellular accumulation of massive $CaCO_3$ inclusions which occur together with smaller sulphur granules. In all cases the cells were encountered in or on the bottom mud of shallow freshwater ponds as would be expected from their high specific weight. As it is considered to be an obligate aerobe, its location in the sediment reflects the interface between the oxygen and H_2S gradients. *Achromatium oxaliferum* possesses some means of locomotion although this motility is very slow and restricted to jerky rolling movements on solid surfaces, most probably effected by means of peritrichous filaments moving about in the slime layer that surrounds the cells (de Boer, la Riviere & Houwink, 1971). In this way, the cells are capable of migrating chemotactically through the mud, where, depending on the distribution of localized microsources of H_2S, concentration gradients exist over short distances that are manageable by the organisms. However, we have yet to observe this phenomenon.

Most observations on the organism have been done on material taken directly from nature. No effective laboratory enrichment culture methods are available; at best the natural populations can be kept alive in the laboratory for 6–10 months. Collecting and concentrating *Achromatium* cells was described by de Boer *et al.* (1971) and still remains the most efficient means of isolation despite attempts to use novel sedimentation techniques. In short, we have made few modifications to the original procedure. Material from the upper part of the sediment was collected using a plastic tubing feeding a Buchner flask under vacuum. The advantage of this modification is that only the top surface of the sediment is removed. The sediment is allowed to settle and excess water decanted. The remaining sediment/water was then filtered through two coarse membranes of $100 \mu m$ and $50 \mu m$ pore size in a modified syringe unit. The final sediment/water filtrate was then subjected to concurrent gentle swirling in a tilted beaker rotated slowly along its vertical axis. This treatment led to further fractionation of the particles according to their specific gravity and the heavy *Achromatium* cells soon become visible as a narrow white band. The cells were pipetted off and transferred to a smaller beaker and further purified by the same method until a dense mass of *Achromatium* cells was obtained. This contained few cells of algae, protozoa, or sand particles, and the number of smaller bacteria present had been reduced to a very small percentage, as shown by microscopic examination. The method was obviously only successful for *Achromatium* cells that actually contained appreciable quantities of $CaCO_3$ inclusions but represents a refreshing reinforcement of the value of traditional ecological techniques.

The cells collected are stored in membrane filtered sterile lakewater from the original habitat and the cells remain viable for several months with observable dividing cells. With time cell death, recognized by an increase in cells that have lost their morphological integrity, can be observed. The biomass from each sample, however, is sufficient to study the cellular structure of this organism by such techniques as confocal microscopy, electron microscopy and X-ray diffraction (XRD) and can generate enough DNA for PCR sequencing of 16S rRNA. Preliminary results here suggest that *Achromatium oxaliferum* belongs to the eubacteria and it appears that a common ancestry is shared with some phototrophic sulphide oxidizers such as *Chromatium* in the gamma sub-division of the proteobacteria (Head *et al.*, unpublished data). Microscopic examination has revealed an extensive internal structure probably associated with the granular inclusions (unpublished data). Investigations into the physiology of *Achromatium* and its role in sulphur cycling are continuing.

Ammonia oxidizing bacteria in the water column

Autotrophic ammonia-oxidizing bacteria are an ecologically important and physiologically specialized group and have been the focus for our studies on Esthwaite Water (Lake District, UK). They are responsible for the oxidation of ammonia to nitrite, the reaction which drives the process of nitrification in a wide range of environments (Hall, 1986). The organisms are obligate chemoautotrophs. Autotrophic ammonia-oxidizing bacteria are notoriously difficult to isolate in pure culture and, compared to chemoorganotrophs, grow to low cell densities and biomass yields *in vitro*. This, and their autotrophic physiology, means that they are largely intractable to traditional methods of phenotypic characterization (Koops & Moller, 1991). Consequently, ammonia-oxidizing bacteria are classified largely on the basis of morphological criteria such as cell shape and the arrangement of internal membranes (Watson *et al.*, 1989; Bock, Koops & Harms, 1986). Little is known about active nitrifiers in the environment. Data from 16S rRNA catalogues (Woese *et al.*, 1984, 1985) first demonstrated that there are two phylo-genetically distinct groups of autotrophic ammonia-oxidizing bacteria. One of these contained *Nitrosococcus oceanus*, and was within the gamma-subdivision of the Proteobacteria. The other contained *Nitrosococcus mobilis* and representatives of all of the other described genera of ammonia-oxidizers, and was located within the beta-subdivision of the Proteobacteria. The ammonia oxidizing bacteria in the beta-subdivision formed three deep branches; *Nitrosococcus mobilis*; *Nitrosomonas europaea*; and a third branch containing *Nitrosolobus*, *Nitrosomonas europaea* and *Nitrosospira* strains. Each genus was, however, represented by only a single strain and any taxonomic conclusions were further constrained by the fact that cataloguing methods only recover approximately 40% of the RNA sequence (Head *et al.*, 1993). Furthermore, the development of the PCR (Saiki *et al.*, 1988) has made it possible to recover 16S rRNA gene sequences from uncultured and difficult to culture microorganisms (e.g. Giovannoni *et al.*, 1990). The autotrophic ammonia-oxidizing bacteria are therefore excellent candidates for PCR-based phylogenetic analysis.

Using Esthwaite Water as a model system where nutrification activity can be detected in the stratified water column, the model probes developed by Head *et al.* (1993) were targeted against the total microbial population. Parameters such as O_2 concentration, temperature and ammonia concentration were measured in order to identify the active area within the water column (Hall, 1986). Total organisms were required to be sampled from a depth of 15 m and as it was assumed that the nitrifiers were present in low numbers (enrichments are usually required for their isolation; Hall, 1986) a sampling strategy that incorporated concentration was essential. Traditionally, concentration of particles (biological and non-biological) was carried out by standard filtration techniques (Hall *et al.*, 1990) under vacuum through either 0.2 or 0.4 μm pore size filters. Obvious disadvantages of this technique are the limits placed on total volume of sample that can be processed, rapidity and subsequent damage to cells on the membrane. To bypass these problems tangential flow filtration (TFF) was adopted as it permitted large volumes of water to be processed in the field. Particles greater than 0.2 μm in diameter are concentrated not by retention directly against the filter but within the void volume of the TFF unit which increases in particulate concentration as more water is processed. The final concentrated solution (termed retentate) is retained within the unit by setting up a back-pressure. Once released, the concentrate is flushed through into a suitable container. The particulate matter from 100–200 l of lakewater can be reduced to a volume of approximately 500 ml. If required this concentrate can be resuspended in as little as 10 ml after centrifugation. This represents a concentration factor of approximately 10 000-fold. The cells are now amenable to direct DNA extraction followed by PCR amplification. In addition, enrichment cultures can be set up using the concentrate as an initial inoculum. Using this technique, examination of the nitrifying community has shown that the culture collection type-strains are present and easily enriched from the samples. However, they are not dominant organisms as determined by the application of the probes (Head *et al.*, 1993; Hiorns *et al.*, submitted). This sampling strategy has enabled us to gain an insight into the population structure and seasonal dynamics of nitrifiers previously denied by conventional methods.

Methanogenesis: in situ sampling of upland peat bogs

Peat deposits are characterized by a surface zone of fibrous, well-drained organic material overlying a water-saturated zone of more compacted material. The fibrous nature of peat makes it difficult to obtain intact core samples without distortion of the vertical profile. To accurately determine the vertical distribution of microbial activities within the peat any compaction of the profile during the sampling operations should be avoided.

The transition between well drained aerobic conditions and the water-logged anaerobic horizons is characterized by a steep redox gradient. Such gradients are known to support a considerable diversity, and activity, of bacterial populations (e.g. Jones, 1985) and play an important role in the vertical distribution of bacterial activities. It is therefore important that the sampling technique should not allow ingress of oxygen into the water-logged anaerobic zones of the peat and that the redox potentials are maintained. Moreover, the exposure of some obligate anaerobic bacteria, for example the methanogens, to air, even for short periods, would affect the rate of methane production (Boone, 1991, Fetzer, Friedhelm & Conrad, 1993) and therefore reduce the activity of the population from that before the sample was taken.

Two sampling strategies have been utilized to avoid compaction and oxygen contamination of peat cores to be used for the determination of the vertical distribution of methanogenic activity. One approach is to cut the peat, using a long knife (Weider *et al.*, 1990) or other cutting devices (Clymo, 1988), to the shape and depth of the sample tube which is then inserted into the pre-formed space. The second approach involved a cutting device which is attached to the base of the sample tube that cuts the peat when the tube is rotated in the peat deposit (Williams & Wheatley, 1992). The peat core is subsequently enclosed in the sample tube, reducing the potential for oxygen contamination. Both methods fail to prevent oxygen contamination of the peat during removal. Similarly, a variety of methods have been used to section the peat core but all, at some stage, introduce oxygen into the sample despite preventative actions (Weider *et al.*, 1990; Moore & Knowles, 1990; Fetzer *et al.*, 1993). Even relatively short exposures decrease the initial rate of methane production. Methods which sample and section cores of peat avoiding exposure of the anaerobic layers to oxidizing conditions benefit the interpretation of activity measurements.

A simple technique to obtain intact cores of blanket bog peat whilst maintaining anaerobic conditions has been developed at the Institute of Freshwater Ecology for sampling blanket bog peat from the Moorhouse National Nature Reserve (Hall *et al.*, submitted). The sampler comprises an 7.0 cm internal diameter acrylic core tube with a sinuous toothed cylindrical cutter attached to the bottom. A tightly fitting insert was placed into the top of the core tube into which a horizontal 'tommy' bar was attached. The surface vegetation of the peat bog was carefully cut away and the assembled core tube placed upright on the peat surface. Holding the bar, a smooth back and forth cutting motion, with slight downward pressure, cuts through the peat isolating a core in the sample tube. Any compaction of the peat profile is readily observed through the transparent acrylic tubing and the core is rejected. To remove the corer from the peat, a metal cutter was pushed into the peat, adjacent to the core tube, to a depth just below the cutter and rotated through 360° to cut the peat immediately below the cutting teeth at the base of the sample tube. The lower horizontal section was then located on the cutter and the whole core tube gently eased from the peat. The bottom of the peat core extending from the cutter was trimmed with a sharp knife and a gas tight piston was inserted through the cutter and up into the core tube. The intact peat cores were stored upright for transfer to the laboratory.

An apparatus was also developed which allowed these cores to be sectioned (minimum slice thickness 0.5 cm) whilst excluding oxygen from the sections and the remaining core material (Hall *et al.*, submitted). Using these techniques, the vertical distribution of methanogenesis in peat deposits was determined on a number of occasions. The success of the techniques was determined by (i) comparison of rates of activity in cores sectioned anaerobically and some sectioned in air and (ii) the rate of activity of individual sections before and after a brief exposure to oxygen. There was considerable between-core variability in both the rate and vertical distribution of methane production such that a direct comparison between cores was not possible. However, rates of methanogenesis were always lower (average 41%, range 6.4–73%) from the cores sectioned in air. Similarly, the rate of methane production in peat sections exposed to air was always lower than that determined before being oxidized (average 48%, range 32–63%). These results justify the increased sample handling time required to prevent access of air to the sample material.

Preliminary trials have indicated that there was considerable variability in both the magnitude and vertical distribution of rates of methane production between peat cores. However, consistent trends in the data were observed. The mean rate of methane production from the peat cores kept under anaerobic conditions was always greater than that from cores sectioned using the aerobic technique. From the data presented, it was calculated that the average decrease in the rate of methane production by exposure to air was 42% (range 6–73%). These data also show the between-core variability in that the standard deviation was, on average, 36% (range 1.4–80%) of the mean value. The changes in vertical distribution of methanogenesis between cores was not surprising as the depth of the water table over the sampling site varied by as much as 3.0 cm.

Additional evidence for the inhibitory effect of oxidizing conditions was obtained from the determination of methane production from samples of peat before and after exposure to oxygen. Because this analysis was performed on the same sample material, the variability between cores was eliminated. Exposure of the peat to oxidizing conditions reduced the rate of methane production by an average of 48% (range 33–63%). The similarity of the average reduction in the rate of methane production for the two experimental approaches confirmed the importance of sampling handling techniques for the determination of methanogenesis in peat.

The results show that a short exposure of anaerobic peat to oxidizing conditions inhibits the activity of methanogenic bacteria present in the sample. This technique has shown that a more realistic profile of methanogenic activity can be obtained if oxygen contamination is eliminated from the sampling and sectioning of peat cores. This system is now being used to relate the activities of methanogens to their distribution as determined by standard enrichment procedures and by PCR using archaeobacterial primers.

CONCLUSIONS

This chapter focuses on ecological methods for studying microbial populations in the environment and highlights their limitations not in a pessimistic sense but so as to draw attention to the dangers of misinterpretation. The introduction of molecular techniques complements the classical studies, and the examples described, indicate (optimistically) that the study of population dynamics is all the more relevant when attributed to activities of processes in the environment.

ACKNOWLEDGEMENTS

The systems described here have been developed and applied in a number of collaborative projects between the Institute of Freshwater Ecology and The University of Liverpool and more recently the University of Newcastle. I wish to thank Jon Saunders, Clive Edwards, Alan McCarthy, Barbara Hales, Richard Hastings. Will Hiorns (all from the University of Liverpool) and Grahame Hall, Gwyn Jones (both of IFE) and Ian Head (University of Newcastle) for making data available prior to publication.

REFERENCES

Amman, R. I., Krumholz, L. & Stahl, D. A. (1990a). Fluorescent oligonucleotide probing of whole cells for determinative, phylogenetic, and environmental studies in microbiology. *Journal of Bacteriology*, **177**, 762–70.

Amman, R. I., Binder, B. J., Olsen, R. J., Chisholm, S. W., Devereux, R. & Stahl D. A. (1990b). Combination of 16S rRNA-targeted oligonucleotide probes with flow cytometry for analysing mixed microbial populations. *Applied and Environmental Microbiology*, **56**, 1919–25.

Amman, R., Springer, N., Ludwig, W., Gortz, H. & Schleifer, K. (1991). Identification *in situ* and phylogeny of uncultured endosymbiotic bacteria. *Nature*, **351**, 161–4.

Apel, W. A., Dugan, P. R., Filippi, J. A. & Rheins, M. A. (1976). Detection of *Thiobacillus ferrooxidans* in acid mine environments by direct fluorescent antibody technique. *Applied and Environmental Microbiology*, **32**, 159–65.

Austin, B. (1988). *Methods in Aquatic Bacteriology*. John Wiley & Sons. pp. 425.

Bak, F. & Pfennig, N. (1991). Microbial sulfate reduction in littoral sediments of lake Constance. FEMS *Microbiology Ecology*, **8S**, 31–42.

Baker, J. M., Norman, J. M. & Bland, W. L. (1992). Field scale application of flux measurement by conditional sampling. *Agricultural Forestry and Meteorology*, **62**, 31–52.

Balows, A., Truper, H. G., Dworkin, M., Harder, W. & Schleifer, K. H. (1992). *The Prokaryotes*, 2nd edn, pp. 2625–2637. New York, Fischer-Verlag.

Bej, A. K., Steffan, R. J., Dicesare, J., Haff, L. & Atlas, R. M. (1990). Detection of coliform bacteria in water by polymerase chain reaction and gene probes. *Applied and Environmental Microbiology*, **56**, 307–14.

Bej, A. K., Mahbubani, M. H. & Atlas, R. M. (1991*a*). Detection of viable *Legionella pneumophila* in water by polymerase chain reaction (PCR) and gene probes method. *Applied and Environmental Microbiology*, **57**, 597–600.

Bej, A. K., DiCesare, J. L., Haff, L. & Atlas, R. M. (1991*b*). Detection of *Escherichia coli* and *Shigella* spp. in water by using polymerase chain reaction (PCR) and gene probes for *uid*. *Applied and Environmental Microbiology*, **57**, 1013–17.

Bej, A. K., McCarthy, S. C. & Atlas, R. M. (1991*c*). Detection of coliform bacteria and *Escherichia coli* by multiplex polymerase chain reaction: comparison with defined substrate and plating methods for water quality monitoring. *Applied and Environmental Microbiology*, **57**, 2429–32.

Belser, L. W. & Schmidt, E. L. (1978). Serological diversity within a terrestrial ammonia oxidising population. *Applied and Environmental Microbiology*, **49**, 584–8.

Bock, E., Koops, H.-P. & Harms, H. (1986). Cell biology of nitrifying bacteria. In: *Nitrification, Special Publication of the Society for General Microbiology*, vol. 20, pp. 17–38. (Prosser, J. I., ed.) Oxford, IRL Press.

Boone, D. R. (1991). Ecology of methanogenesis. In: *Microbial Production and Consumption of Greenhouse Gases: Methane, Nitrogen oxides, and Halomethanes*. (Rogers, J. E. and Whitman, W. B., eds) American Society for Microbiology, pp. 57–70.

Brayton, P. R., Tamplin, M. L., Huq, A. & Colwell, R. R. (1987). Enumeration of *Vibrio cholerae* O1 in Bangladesh waters by fluorescent-antibody direct count. *Applied and Environmental Microbiology*, **53**, 2862–5.

Brock, T. D. (1987). The study of microorganisms *in situ*: progress and problems. In: *Ecology of Microbial Communities* (Fletcher, M., Gray, T. R. G. and Jones, J. G., eds.). pp. 1–17. Cambridge University Press.

Chantler, S. & MacIllmurray, M. B. (1988). Labelled antibody methods for detection and identification of microorganisms. *Methods in Microbiology*, **19**, 273–332.

Clymo, R. S. (1988). A high resolution sampler of surface peat. *Functional Ecology*, **2**, 425–31.

Colwell, R. R., Brayton, P. R., Grimes, D. J., Roszak, D. R., Huq, S. A. & Palmer, L. M. (1985). Viable but non-culturable *Vibrio cholerae* and related pathogens in the environment: implications for the release of genetically engineered microorganisms. *Biotechnology*, **3**, 817–20.

Conway de Macario, E., Wolin, M. J. & Macario, A. J. L. (1982). Antibody analysis of relationships among methanogenic bacteria. *Journal of Bacteriology*, **149**, 316–19.

de Boer, W. E., La Riviere, J. W. M. & Houwink (1971). Some properties of *Achromatium oxaliferum*. *Antonie Van Leeuwenhoek Journal of Microbiology and Serology*, **37**, 553–63.

DiChristina, T. J. and Delong, E. F. 1993. Design and application of rRNA-targeted oligonucleotide probes for the dissimilatory iron- and manganese-reducing bacterium *Shewanella putrefaciens*. *Applied and Environmental Microbiology*, **59**, 4152–60.

Elliot, J. M. (1977). Some methods in the statistical analysis of samples of benthic invertebrates. *Freshwater Biological Association Scientific Publication no. 25.* pp. 160.

Embley, T. M., Finlay, B. J. Thomas, R. H. & Dyal, P. L. (1992). The use of rRNA sequences and fluorescent probes to investigate the phylogenetic positions of the anaerobic ciliate *Metopus palaeformis* and its archaeobacterial endosymbiont. *Journal of General Microbiology*, **138**, 1479–87.

Ferguson, R. L., Buckley, E. N. & Palumbo, A. V. (1984). Response of marine bacterioplankton to differential centrifugation and confinement. *Applied and Environmental Microbiology*, **47**, 49–55.

Fetzer, S. B., Friedhelm, B. & Conrad, R. (1993). Sensitivity of methanogenic bacteria from paddy soil to oxygen and desiccation. *FEMS Microbiology Ecology*, **12**, 107–15.

Ford, S. F. & Olson, B. (1988). Methods for detecting genetically engineered microorganisms in the environment. *Advances in Microbial Ecology*, **10**, 45–79.

Fry, J. C. (1988). Determination of biomass. In *Methods in Aquatic Bacteriology* (Austin, B., ed.) pp. 27–72. *Modern Microbiological Methods*, John Wiley & Sons.

Fry, J. C. (1993). *Biological Data Analysis: A Practical Approach*. IRL Press, p. 418.

Giovannoni, S. J., Delong, E. F., Olsen, G. J. & Pace, N. R. (1988). Phylogenetic group-specific oligonucleotide probes for the identification of single microbial cells. *Journal of Bacteriology*, **170**, 720–6.

Giovannoni, S. J., Britschgi, T. B., Moyer, C. L. & Field, K. G. (1990). Genetic diversity in Sargasso Sea bacterioplankton. *Nature, London*, **345**, 60–3.

Goulder, R. (1987). Evaluation of the saturation approach to measurement of V_{max} for glucose mineralization by epilithic freshwater bacteria. *Letters in Applied Microbiology*, **4**, 29–32.

Grigorvova, R. & Norris, J. R. (1990). Techniques in microbial ecology. *Methods in Microbiology*, **22**, Academic Press, p. 627.

Hahn, D., Strarrenburg, M. J. C. & Akkermans, A. L. (1990). Oligonucleotide probes that hybridise with rRNA as tool for the study of *Frankia* in root nodules. *Applied and Environmental Microbiology*, **56**, 1342–6.

Hahn, D., Amann, R. I., Ludwig, W., Akkermans, A. L. & Schleifer, K.-H. (1992). Detection of microorganisms in soil after *in situ* hybridization with rRNA-targeted, fluorescently labelled oligonucleotides. *Journal of General of Microbiology*, **138**, 879–87.

Hall, G. H. (1986). Nitrification in lakes. In *Nitrification, Special Publication of the Society for General Microbiology*, vol. 20, pp. 127–156. (Prosser, J. I., ed.) IRL, Oxford.

Hall, G. H., Jones, J. G., Pickup, R. W. & Simon, B. M. (1990). Methods to study the bacterial ecology of freshwater environments. *Methods in Microbiology*, **23**, 181–210.

Head, I. M., Hiorns, W. D., Embley, T. M., McCarthy, A. J. & Saunders, J. R. (1993). The phylogeny of autotrophic ammonia-oxidising bacteria as determined by analysis of 16S ribosomal RNA gene sequences. *Journal of General Microbiology*, **139**, 1147–53.

Herbert, R. A. (1990). Methods for enumerating microorganisms and determining biomass in natural environments. *Methods in Microbiology*, **19**, 1–40.

Hicks, R. E., Amann, R. & Stahl, D. A. (1992). Dual staining of natural bacterioplankton with 4',6-diamidino-2-phenylindole and fluorescent oligonucleotide probes targeting kingdom-level 16S rRNA sequences. *Applied and Environmental Microbiology*, **58**, 2158–63.

Holben, W. E., Jansson, J. K., Chelm, B. K. & Tiedje, J. M. (1988). DNA probe method for the detection of specific microorganisms in the soil bacterial community. *Applied and Environmental Microbiology*, **54**, 703–11.

Hopkins, D. W., MacNaughton, S. J. & O'Donnell, A. G. (1991). A dispersion and differential centrifugation technique for representative sampling of microorganisms from soil. *Soil Biology and Biochemistry*, **23**(3), 217–25.

Hoppe, H.-G. (1978). Relationships between active bacteria and heterotrophil potential in the sea. *Netherlands Journal of Sea Research*, **12**, 78–98.

Jones, J. G. (1977). The effect of environmental factors on estimated viable and total populations of planktonic bacteria in lakes and experimental enclosures. *Freshwater Biology*, **7**, 61–97.

Jones, J. G. & Jones, H. E. (1986). Benthic filamentous bacteria. In *Perspectives in Microbial Ecology, Proceedings of the 4th International Symposium on Microbial Ecology* (Megusar, F. and Ganter, M., eds) pp. 375–382.

Jones, J. G. & Simon, B. M. (1975). An investigation of errors in direct counts of aquatic bacteria by epifluorescence microscopy, with respect to a new method for dyeing membrane filters. *Journal of Applied Bacteriology*, **39**, 317–29.

Jones, J. G. (1985). Microbes and microbial processes in sediments. *Philosophical Transactions of the Royal Society, Series A*, 3–17.

Jones, J. G. (1987). *Diversity in Freshwater Microbiology – Ecology of Microbial Communities* (Fletcher, M., Gray, T. R. G. & Jones, J. G., eds) Cambridge University Press.

Jorgensen, B. B. (1982). Mineralisation of organic matter in the sea – the role of sulphate reduction. *Nature, London*, **296**, 643–5.

Jorgensen, B. B. & Revsbech, N. P. (1983). Colorless sulfur bacteria, *Beggiatoa* spp. and *Thiovulum* sp. in O_2 and H_2S microgradients. *Applied and Environmental Microbiology*, **45**, 1261–70.

Kane, M. D., Poulsen, L. K. & Stahl, D. A. (1993). Monitoring the enrichment and isolation of sulfate-reducing bacteria by using oligonucleotide hybridization probes designed from environmentally derived 16S rRNA sequence. *Applied and Environmental Microbiology*, **59**, 682–6.

Kogure, K., Simidu, U. & Taga, N. (1979). A tentative direct microscopic method for counting living marine bacteria. *Canadian Journal of Microbiology*, **25**, 415–20.

Koops, H.-P. & Moller, U. C. (1991). The lithotrophic ammonia-oxidizing bacteria. In *The Prokaryotes*, 2nd edn, pp. 2625–2637. (Balows, A., Truper, H. G., Dworkin, M., Harder, W. & Schleifer, K. H., eds). Fischer-Verlag, New York.

Kueunen, J. G. & Beudeker, R. F. (1982). Microbiology of *Thiobacilli* and other sulphur-oxidising, mixotrophs and heterotrophs. *Philosophical Transactions of the Royal Society London B*, Series, **298**, 473–98.

La Riviere, J. W. M. & Schmidt, K. (1992). Morphologically conspicuous sulphur oxidising bacteria. In *The Prokaryotes* 2nd edn pp. 3934–3947, (Ballows, A.,

Truper, H. G., Dworkin, M., Harder, W. & Shleifer, K-H. Fischer-Verlag, New York.

MacDonald, R. M. (1986). Sampling soil microfloras: dispersion of soil by ion exchange and extraction of specific microorganisms from suspension by elutriation. *Soil Biology and Biochemistry*, **18**, 399–406.

Manz, W., Szewzyk, U., Ericsson, P., Amman, R., Schleifer, K.-H. & Stenstrom, T.-A. (1993). *In situ* identification of bacteria in drinking water and adjoining biofilms by hybridization with 16S and 23S-rRNA-directed fluorescent oligonucleotide probes. *Applied and Environmental Microbiology*, **59**, 2293–8.

Moore, T. R. & R. Knowles (1990). Methane emissions from fen, bog and swamp peatlands in Quebec. *Biochemistry*, **11**, 45–62.

Morgan, J. A. W., Cranwell, P. A. & Pickup, R. W. (1991). Survival of *Aeromonas salmonicida* in lake water. *Applied and Environmental Microbiology*, **57**, 1777–82.

Morgan, J. A. W., Rhodes, G. & Pickup, R. W. (1993). The survival of non-culturable *Aeromonas salmonicida* in lake water. *Applied and Environmental Microbiology*, **59**, 874–80.

Nelson, D. C., Jorgensen, B. B. & Revsbech, N. P. (1986). Growth pattern and yield of a chemoautotrophic *Beggiatoa* sp. in oxygen-sulphide microgradients. *Applied and Environmental Microbiology*, **52**, 225–33.

Ogram, A., Sayler, G. S. & Barkay, T. (1988). DNA extraction and purification from sediments. *Journal of Microbiological Methods*, **7**, 57–66.

Pace, N. R., Stahl, D. A., Lane, D. J. & Olsen, G. J. (1987). The analysis of natural microbial populations by ribosomal RNA sequences. *Advances in Microbial Ecology*, **9**, 1–56.

Pickup, R. W. (1991). Molecular methods for the detection of specific bacteria in the environment. *Journal of General Microbiology*, **137**, 1009–19.

Pillai, S. D., Josephson, K. L., Bailey, R. L., Gerba, C. P. & Pepper, I. L. (1991). Rapid method for processing soil samples for polymerase chain reaction amplification of specific gene sequences. *Applied and Environmental Microbiology*, **57**, 2283–6.

Reed, W. M. & Duggan, P. R. (1978). Distribution of *Methylomonas methanica* and *Methylsinus trichosporium* in Cleveland harbor as determined by indirect fluorescent anitbody-membrane filter technique. *Applied and Environmental Microbiology*, **35**, 422–30.

Renwick, A. & Gareth, D. (1985). A comparison of fluorescent ELISA and antibiotic resistance identification techniques for use in ecological experiments with *Rhizobium trifolii*, *Journal of Applied Bacteriology*, **58**, 199–206.

Roszak, D. B. & Colwell, R. R. (1987). Survival strategies of bacteria in the natural environment. *Microbiological Reviews*, **51**, 365–79.

Saiki, R. K., Gelfand, D. H., Stoffel, S., Scharf, S. J., Higuchi, G. T., Horn, G. T., Mullis, K. B. & Erlich, H. A. (1988). Primer directed enzymatic amplification of DNA with a thermostable DNA polymerase. *Science*, **239**, 487–91.

Sayler, G. S. & Layton, A. C. (1990). Environmental application of nucleic acid hybridization. *Annual Reviews of Microbiology*, **44**, 625–48.

Selenska, S. & Klingmuller, W. (1992). Direct recovery and molecular analysis of DNA and RNA from soil. *Microbial Releases*, **1**, 41–7.

Sommerville, C., Knight, I. T. Straube, W. L. & Colwell, R. R. (1989). Simple raid method for the direct isolation of nucleic acids from aquatic environments. *Applied and Environmental Microbiology*, **55**, 548–54.

Steffan, R. J. & Atlas, R. M. (1991). Polymerase chain reaction: amplifications in environmental microbiology. *Annual Review of Microbiology*, **45**, 137–61.

Steffan, R. J., Goksoyr, J., Bei, A. K. & Atlas, R. M. (1988). Recovery of DNA from soils and sediments. *Applied and Environmental Microbiology*, **54**, 2908–15.

Steffan, R. J., Goksoyr, J., Asim, K. B. & Atlas, R. M. (1988). Recovery of DNA from soil and sediments. *Applied and Environmental Microbiology*, **54**, 2908–15.

Steffan, R. J. & Atlas, R. M. (1988). DNA amplification to enhance the detection of genetically engineered bacteria in environmental samples. *Applied and Environmental Microbiology*, **54**, 2185–91.

Steffan, R. J., Breen, A., Atlas, R. M. & Sayler, G. S. (1989). Application of gene probes methods for monitoring microbial populations in freshwater ecosystems. *Canadian Journal of Microbiology*, **35**, 681–5.

Stotzky, G. (1986). Influence of soil mineral colloids on metabolic processes, growth, adhesion and ecology of microbes and viruses. In *Interactions of Soil Minerals with Natural Organics and Microbes*. pp. 305–428. ((Haung, P. M. & Schnitzer, M., eds.) Soil Science Society of America. Madison.

Sweets, P. R. A., DeBeer, D., Nielsen, L. P., Verdouw, J. C. Van den Heuvel, Cohen, Y. & Cappenberg, T. E. (1990). Denitrification of sulfur-oxidizing *Beggiatoa* spp. mats on freshwater sediments. *Nature, London*, **344**, 762–3.

Trevors, J. T. (1988). Use of microcosms to study genetic interactions between microorganisms. *Microbiological Sciences*, **5**, 132–6.

Trevors, J. T. (1992). DNA extraction from soil. *Microbial Releases*, **1**, 3–11.

Tsai, Y.-L. & Olson, B. H. (1992*a*). Detection of low numbers of bacterial cells in soils and sediments by polymerase chain reaction. *Applied and Environmental Microbiology*, **57**, 754–7.

Tsai, Y.-L. & Olson, B. H. (1992*b*). Rapid method for separation of bacterial DNA from humic substances in sediments for polymerase chain reaction. *Applied and Environmental Microbiology*, **58**, 2292–5.

Tsai, Y.-L. & Olson, B. H. (1991). Rapid method for direct extraction of DNA from soil and sediment. *Applied and Environmental Microbiology*, **57**, 1070–4.

Voloudakis, A. E., Gitaitis, R. D., Westbrook, J. K., Phatak, S. C. & McCarter, S. M. (1991). Epiphytic survival of *Pseudomonas syringae* pv. syringae and *Ps. tomato* on tomato transplants in southern Georgia. *Plant Diseases*, **75**, 672–5.

Wallner, G., Amman, R. & Beisker, W. (1993). Optimising fluorescent *in situ* hybridisation with rRNA-targeted oligonucleotide probes for flow cytometric identification of organisms. *Cytometry*, **14**, 136–43.

Ward, B. B. & Carlucci, A. F. (1985). Marine ammonia and nitrite oxidising bacteria: Serological diversity determined by immunofluorescence in culture and in the environment. *Applied and Environmental Microbiology*, **50**, 194–201.

Ward, D. M., Bateson, M. M., Weller, R. & Ruff-Roberts, A. L. (1993). Ribosomal RNA analysis of microorganisms in nature. *Advances in Microbial Ecology*, **12**, 219–86.

Watson, S. W., Bock, E., Harms, H., Koops, H.-P. & Hooper, A. B. (1989). Nitrifying bacteria. In *Bergey's Manual of Systematic Bacteriology*, vol. 3, pp. 1808–1834. (Staley, J. T., Bryant, M. P., Pfennig, N. & Hold, J. G., eds.) Williams & Wilkins, Baltimore.

Weider, R. K., Yavitt, J. B. & Lang, G. E. (1990). Methane production and sulphate reduction in two Appalachian peatlands. *Biogeochemistry*, **10**, 81–104.

Williams, B. L. & Wheatley, R. E. (1992). Mineral nitrogen dynamics in poorly drained blanket peat. *Biology and Fertility of Soils*, **13**, 96–101.

Woese, C. R., Weisberg, W. G., Hahn, C. M., Pasier, B. J., Zaplen, L. B., Lewis, B. J., Macke, T. J., Lugwig, W. & Stackebrandt, E. (1985). The phylogeny of the purple bacteria, the gamma sub-division. *Systematic and Applied Microbiology*, **6**, 25–33.

Woese, C. R., Stackebrandt, E., Weisburg, W. G., Paster, B. J., Madigan, M. T., Fowler, V. J., Hahn, C. M., Blanz, P., Gupta, R., Nealson, K. H. & Fox, G. E. (1984). The phylogeny of purple bacteria: the alpha sub-divison. *Systematic and Applied Microbiology*, **5**, 321–6.

Zarda, B., Amman , R., Wallner, G. & Schleifer, K.-H. (1991). Identification of single bacterial cells using digoxigenin-labelled, rRNA-targeted oligonucleotides. *Journal of General Microbiology*, **137**, 2823–30.

THE SPREAD OF DRUG RESISTANCE

PETER M. BENNETT

Department of Pathology and Microbiology, University of Bristol, Bristol BS8 1TD, UK

INTRODUCTION

Effective treatment of infectious diseases with antibiotics is largely a development of the second half of the twentieth century. The introduction of sulphanilamide and penicillin into clinical use 50–60 years ago marked a decisive break with many of the clinical practices of the past and heralded the dawn of modern medicine, in the sense that, for the first time, infectious diseases that had lethally ravaged human communities and against which doctors had been essentially powerless, could be cured. No longer were many of the great scourges of the past to be feared. Ehrlich's dream of 'magic bullets' had been realised. The future looked bright and attention could be turned to other priorities, or so it seemed.

Bacteria, unfortunately, like any other biological system, are not static systems. They constantly evolve and bacterial pathogens are no exception. The new situation simply presented a change of environment, a new evolutionary challenge that has been met with stunning effectiveness, as witness the multiplicity of resistance genes that have appeared in clinically important bacteria in the last 30 years. Fortunately, the challenge presented by the first emergence of drug resistance was taken up by the then infant pharmaceutical industry, which proceeded to discover and develop hundreds of new antibiotics, many, but by no means all, of which have seen clinical service. Unfortunately, many of these alternatives have fared little better than sulphanilamide and benzylpenicillin, in that in due course resistance to each new antibiotic has appeared in one or other clinically important species of bacteria, often after a disconcertingly short time (Table 1). More disconcerting is the realization that, in principle, resistance may develop anywhere in the bacterial kingdom, indeed may already have evolved at the time a drug is first used in the clinic, but not necessarily in the bacteria against which the antibiotic is primarily targeted and that the genetic information specifying resistance may become associated with an efficient transfer mechanism whereby the resistance gene(s) can be spread widely among different bacteria. Indeed, most clinically significant drug resistance involves the acquisition of an entirely new gene, i.e. the genetic information needed for the cell to be resistant is obtained by transfer from another bacterial cell.

Table 1. *Drug resistance among bacteria*

Bacterium	Condition	Antibiotic treatments no longer appropriate or use threatened
Enterococcus spp.	blood poisoning, surgical infection	aminoglycosides, β-lactams, erythromycin, tetracycline, vancomycin
Haemophilus influenzae	meningitis, ear infection, pneumonia, sinusitis	chloramphenicol, penicillins tetracycline, trimethoprim
Mycobacterium tuberculosis	tuberculosis	aminoglycosides, ethambutol, isoniazid, pyrazinamide, rifampicin
Neisseria gonorrhoeae	gonorrhea	penicillins, spectinomycin, tetracycline
Shigella dysenteriae	severe diarrhoea	ampicillin, chloramphenicol, tetracycline, trimethoprim
Staphylococcus aureus	blood poisoning surgical infection, pneumonia	aminoglycosides, β-lactams, erythromycin, tetracycline, quinolones
Streptococcus pneumoniae	meningitis, pneumonia	aminoglycosides, β-lactams, chloramphenicol, erythromycin tetracycline, trimethoprim

Bacteria have shown, and continue to show, an astonishing capacity to develop resistance to a wide variety of what are essentially bacterial poisons, including a range of heavy metals, antiseptics and disinfectants (Amyes & Gemmell, 1992; Chopra, 1988; Foster, 1983; Leelaporn *et al.*, 1994). This is in part due to the incredible evolutionary flexibility of protein structures, which are usually the targets for antibiotics, coupled to the transfer of DNA between bacteria of the same or of different species and genera (Bennett & Hawkey, 1991). This relatively free flow of genes in bacterial populations greatly accelerates the development of resistant populations. So far, fortunately, there have been relatively few cases where doctors have run out of chemotherapeutic possibilities and have had to let an infection run its course. However, as has been stated, elsewhere,

> *There is a steady progressive erosion of the usefulness of antibiotics by a rising tide of acquired resistance in pathogenic bacteria. Total inability to control resistant bacteria is, so far, a rarity. However, doctors are being driven to use second or third line drugs or antibiotic combinations, rather than the first choice of previous years. A heavy cost in side effects, ecological damage, preventable hospitalisation, morbidity and mortality, and in money, is being paid and will increasingly be paid. (Gruneberg, 1992).*

Similar warnings have been made by others (e.g. see Levy & Novick, eds. 1986).

How has this state of affairs arisen? It is pertinent to look briefly at the various mechanisms which a cell may use to avoid the inhibiting effect(s) of an antibiotic, as this provides insight into how resistance might develop to new antibacterial agents. There are several ways, biochemically, by which, in principle, a bacterial cell can become resistant to an antibiotic. Hence an enzyme may be acquired that will destroy the drug, by hydrolysis or modification, as is the case with beta-lactamases and aminoglycoside modifying enzymes, respectively. Alternatively, the drug may be expelled from the cell, as is often seen in cases of tetracycline resistance in both gram-positive and gram-negative bacteria and, recently, in resistance to the latest generation of quinolone antibiotics such as ciprofloxacin in *Staphylococcus aureus*, or it may simply be excluded from the cell. Another possible resistance mechanism is acquisition of an alternative to the enzyme that is inhibited, specifically one that is not susceptible to inhibition at clinically achievable concentrations of the drug, as is seen in cases of trimethoprim resistance and resistance to sulphonamide antibiotics, when the cell acquires, usually in the form of a plasmid encoded gene, a drug-resistant form of dihydrofolate reductase or dihydopteroate synthase, respectively. Of course, resistance may also arise in particular cases as the consequence of the drug target being altered, either by mutation, as is seen for a number of drugs, including β-lactams, aminoglycosides, quinolones and rifampicin, or by covalent alteration, as exemplified by methylation of 16S rRNA to confer resistance to erythromycin. Finally, and perhaps the most surprising change of all is a subtle modification of a biosynthetic pathway, as is seen in the case of vancomycin resistance in *Enterococcus* spp. where the terminal D-Ala residue of the pentapeptide side chain of the peptidoglycan precursor is substituted by lactate, a modification that requires new biosynthetic activities. This substitution prevents vancomycin binding to the nascent peptidoglycan, an interaction which requires the terminal D-Ala-D dipeptide of the pentapeptide side-chain, but allows the transpeptidation reaction that forges the cross-links in the peptidoglycan structure to proceed courtesy also of a new enzyme. Because the lactate residue is cleaved from the peptide side chain when each cross-link is made, the resulting peptidoglycan network is the same as that created from a precursor that has D-Ala as the ultimate aminoacid of the pentapeptide side chain.

These changes, and other refinements, are known to be responsible for much of the antibiotic resistance in bacteria, and most, if not all, of this genetic information is transmissable between many different bacteria, and in particular, among common pathogens. Three basic mechanisms have evolved to transfer DNA from one bacterial cell to another, namely, conjugation, transformation and transduction.

CONJUGATION

Conjugation is the mechanism whereby DNA is transferred from one bacterial cell to another by a process that requires cell-to-cell contact

(Willetts, 1988; Ippen-Ihler, 1989). Conjugative mechanisms have been reported for both gram-positive and gram-negative bacteria, although details differ between the two types. In the case of gram-negative bacteria, the donor organism invariably produces external proteinaceous filamentous appendages called sex-pili. These structures make the initial contact between a donor bacterium and a prospective recipient cell. The sex-pilus is then retracted into the donor cell by pilus disassembly, so drawing the cells together until they make contact. In contrast, in gram-positive conjugation systems contact between donor and recipient cells is promoted by a protein excreted by the potential recipient cell. This protein causes donor and recipient cells to clump; hence its name, clumping inducing agent (CIA). These chemicals may be thought of as bacterial pheromones (Clewell, 1993b). In both systems, once cell-to-cell contact has been made, the membranes of the participating cells are fused. In both systems the biochemical mechanisms by which fusion is achieved are obscure. A membrane channel is formed through which DNA can pass from donor to recipient. DNA transfer involves only one strand of the duplex from a fixed point on the DNA molecule called the origin of transfer, oriT. The transferred strand is almost certainly displaced from the DNA molecule in the donor cell by rolling circle replication which drives transfer of the displaced strand into the recipient cell, where it serves as a template for synthesis of the complementary strand. Hence, conjugation is a semi-conservative replication process.

Many different conjugation systems have been described, particularly among those found in gram-negative bacteria, although all appear to function in essentially the same way. In most cases, these conjugation systems are encoded by bacterial plasmids (see Clewell, 1993a; Thomas, 1989). The exceptions are the conjugative transposable elements found in some Gram-positive bacteria (Murphy, 1989). Conjugative transfer of bacterial drug resistance genes, often several at a time, as the consequence of plasmid transfer, was first observed in Japan in the mid-1950s (Watanabe, 1963).

TRANSDUCTION

Transduction involves transfer of DNA, mediated by a bacteriophage (Kokjohn, 1989). Two forms of transduction have been described: specialized and generalized. The former is best exemplified by the temperate bacteriophage, lambda (λ); by temperate we mean a bacteriophage that can infect a bacterial cell but then become dormant, a state known as lysogeny. To lysogenize *Escherichia coli*, λ undergoes a site-specific integration into the bacterial chromosome where it is replicated passively, as a passenger. In this form it is known as a prophage. When phage development is activated, the prophage is excised from the chromosome. Most excision events are precise, but the occasional one (about 1 in a 100 000) is aberrant and a hybrid piece of DNA, containing phage and chromosomal DNA is released instead. In certain cases, this can be packaged to form an infective particle. Upon injection of the hybrid DNA, the recipient cell acquires not only all or a substantial part of the phage genome, but also chromosomal DNA from the previous host, i.e. a section of the DNA adjacent to the phage integration site. Some hybrid DNAs are relatively stable and the original hybrid can replicate to give rise to many copies. Each can retransfer the chromosomal DNA sequences acquired from the original host. Alternatively, the non-phage DNA sequences on the hybrid DNA molecule may recombine with an homologous section of the chromosome of the new host.

In contrast, generalized transduction is where intact DNA molecules, e.g. plasmids, or pieces of DNA molecules, e.g. chromosome fragments, are packaged, more or less at random, in place of the phage genome, to form pseudophage particles. These are functional particles able to inject their DNA contents into a recipient. If the DNA injected into a cell is homologous to that in the recipient, then some of the sequences may be preserved in the new host by homologous recombination. Alternatively, if the DNA packaged is itself a replicon, then it may establish in the new host independently of any recombination process. Bacteriophage P1 mediates generalised transduction in *E. coli*; P22 does the same in *Salmonella typhimurium*, but the importance of this for gene transfer in these cells is unknown, although transduction in nature between Gram-negative bacteria, particularly in aqueous environments, has been reported (Morrison, Miller & Sayler, 1978; Saye *et al.*, 1987; Saye & Miller, 1989; Young, 1993).

In most systems, the role of bacteriophages in genetic transfer *in vivo* has been little studied. The main exception is in *Staphylococcus aureus* where phage mediated transfer of drug resistance plasmids has been well established for many years (Lacey, 1975). Given the existence of transducing phages in many other systems, it is likely that transduction plays a role in disseminating both plasmid and chromosomal genes, but how common or how important this mechanism is in nature has not so far been determined.

TRANSFORMATION

Transformation describes the conversion of the phenotype of a cell by a change in the genotype brought about by acquisition of DNA from its environment (Saunders & Saunders, 1988; Stewart, 1989). The mechanism involves uptake of 'naked' DNA. Cells that can take up DNA from the environment are said to be competent. Competence can be a natural state, as in the case of *Streptococcus pneumoniae*, with which the first transformation experiments were performed by Griffiths more than half a century ago; alternatively, competence can be induced by chemical treatment, as in the case of *E. coli* (Cohen, Chang & Hsu, 1972). Whether this ability to have competence induced, i.e. by rapidly chilling actively growing cells and washing them in ice-cold $CaCl_2$, has any natural significance seems doubtful.

Perhaps the most efficient and apparently least specific method of inducing DNA uptake is to subject a cell-DNA mix to a short, sharp electric shock, a procedure known as electroporation (Sambrook, Fritsch & Maniatis, 1989). Again, this is an artificial system, but it is not beyond the bounds of possibility that, on occasion in nature, electrical discharges may promote uptake of free DNA by bacteria that have no natural competence.

In the cases of cells that have a natural state of competence, either throughout their life cycles or during a specific stage of the life cycle, DNA uptake principly involves linear fragments (Stewart, 1989). The genes, or parts of them, on these fragments are incorporated into the genome of the recipient cell by recombination. In general, intact, circular DNA is poorly incorporated into the cell by the natural process. However, plasmid genes may be recovered efficiently if the cell already contains sequences homologous to those on the plasmid, e.g. a similar, but non-identical plasmid. This is in contrast to competence induced by chemical treatment of the cells, procedures which allow uptake of intact, double-stranded DNA replicons and which are one of the mainstays of molecular genetic analysis.

DNA TRANSFER BETWEEN DNA MOLECULES

Conjugation, transduction and transformation can account for much of the rapid spread of resistance genes among pathogenic or potentially pathogenic bacteria. There is a plethora of bacterial plasmids carrying one or more resistance genes. However, the variety of different resistance genes is considerably less than the variety of plasmids that carry them, even given the multiplicity of genes conferring resistance to a number of common antibiotics such as β-lactams (Bush, 1989), aminoglycosides (Davies, 1991; Shaw et al., 1993), tetracycline (Salyers, Speer & Shoemaker, 1990), chloramphenicol (Shaw & Leslie, 1989), erythromycin (Leclercq & Courvalin, 1991), trimethoprim (Amyes, Towner & Young, 1992) and now even vancomycin (Arthur & Courvalin, 1993). It was realized more than 20 years ago that what appeared to be essentially the same gene was often carried on several plasmids that appeared to have little in common, other than the resistance gene itself (e.g. Heffron et al., 1975). This implied a mechanism that allowed genes to migrate from one DNA molecule to another, but which required little, if any, DNA–DNA homology.

DNA rearrangement requires recombination, i.e. phosphodiester bonds must be broken and reformed and in the process DNA sequences are replaced or added. Recombination mechanisms may require the participating sequences to be largely homologous, i.e. basically the same, in which case the mechanism is said to be one of homologous recombination, which requires a number of host-encoded functions and in particular the product of the recA gene. This mechanism operates primarily to replace like with like. In contrast, recombination mechanisms that require little or no homology between the participating DNA sequences are primarily additive. Two such mechanisms are known, transposition and site-specific recombination, and both have important roles to play in the spread of drug resistance genes.

Transposition reflects the activities of transposable elements, of which there are essentially four types: (1) insertion sequences or IS elements, that are small, essentially cryptic DNA sequences that can migrate from one genetic location to another, completely unrelated location on the same or on a different DNA molecule and that encode only those functions necessary for transposition; (2) transposons, which are DNA elements that also can move between unrelated sites on the same or on different DNA molecules and which encode a function(s) other than those required for transposition, e.g. a drug resistance determinant; (3) conjugative transposons that not only mediate genetic transfer from one DNA molecule to another, but which also mediate their own transfer from one cell to another; (4) transposing bacteriophages that replicate via one of the mechanisms of transposition, e.g. bacteriophage Mu. No more will be said about this last type of element; for further information consult Pato (1989).

Most transposable elements have features in common. All are discrete DNA sequences and so each has defined ends, usually but not always, terminal inverted repeats (IRs). These short sequences are recognition sites for the element-encoded enzyme, called a transposase, that executes the transposition. A gene encoding a transposase, together with the appropriate terminal sequences are the minimum requirements for a transposable element, and a number of IS elements are basically this combination (Galas & Chandler, 1989). Other elements encode several transposition proteins, e.g. Tn7 has five and also encodes resistance to trimethoprim and strepto-mycin (Craig, 1989).

The first bacterial transposable elements to be characterized were the elements IS*1*, IS*2* and IS*3*, discovered because of their strong polar effects when inserted in the *gal* and *lac* operons of *E. coli* (Starlinger, 1980). These are small elements, 768bp, 1327bp and 1258bp, respectively, and encode only functions necessary for their own transposition. Since their discovery in the mid 1960s, many different IS elements have been discovered in a wide range of bacterial genera (Galas & Chandler, 1989), including some archae-bacteria (Charlebois & Doolittle, 1989). The relevance of IS elements to the spread of antibiotic resistance genes is that, although these small elements do not themselves specify drug resistance, they can mobilise resistance genes by generating what are termed composite (or compound) transposons (Bennett, 1991). When two copies of the same element occupy sites that are close together on the same DNA molecule, they may co-operate to trans-pose both copies of the IS element and the DNA trapped between them as a single length of DNA. Some common transposons, such as Tn*5*, which confers resistance to kanamycin and neomycin, and Tn*10*, which confers resistance to tetracycline, are two such elements (Berg, 1989; Kleckner,

Table 2. *Some composite bacterial drug resistance transposons*

Transposon	Size (kb)	Terminal elements – orientation	Markers
Gram-negative			
Tn5	5.7	IS50 – IR	KmBlSm
Tn9	2.5	IS1 – DR	Cm
Tn10	9.3	IS10 – IR	Tc
Tn903	3.1	IS903 – IR	Km
Tn1525	4.4	IS15 – DR	Km
Tn2350	10.4	IS1 – DR	Km
Tn2680	5	IS26 – DR	Km
Gram-positive			
Tn3851	5.2	N.D.	CmTbKm
Tn4001	4.7	IS256 – IR	GmTbKm
Tn4003	3.6	IS257 – DR	Tm

Data from Galas & Chandler (1989) and Murphy (1989).
IR, inverted repeat; DR, direct repeat; N.D., not determined; Resistance genes: Bl, bleomycin; Cm, chloramphenicol; Gm, gentamicin; Km, kanamycin; Sm, streptomycin; Tb, tobramycin; Tc, tetracycline; Tm, trimethoprim.

1989) (Table 2). Some of these, such as Tn10, which comprises terminal inverted repeat copies of IS10 flanking a tetracycline resistance determinant, have clearly existed for a considerable time, long enough to have accumulated a number of base pair substitutions in one copy of IS10 relative to the second copy. As a result, only one of the copies of IS10 in Tn10, that referred to as IS10R, produces an active transposase. The inactive element, IS10L, simply provides one end, i.e. an IR sequence, of the resistance transposon (Kleckner, 1989). In contrast, other combinations, e.g. Tn903 which comprises a kanamycin/neomycin resistance determinant flanked by inverted copies of IS903, appear to be relatively new arrangements, because both copies of the terminal IS element are identical, i.e. there has been insufficient time since Tn903 formation for the sequences of the IS elements to diverge (Grindley & Joyce, 1980; Oka, Sugisaki & Takanami, 1981).

In principle, a pair of IS elements can work in tandem to mobilize, by transposition, any piece of DNA. The main limitation seems to be a functional one, in that the larger the DNA sequence bracketed by the IS elements the lower the transposition frequency of the composite structure (Chandler, Clerget and Galas, 1982). What appears to be important is the distance between the IR sequences that define the element. Hence, a large structure on a small plasmid may transpose at a respectable frequency, because although the structure to be transposed is large, the distance between the ends is not (equal to the length of the carrier plasmid), while the same structure on a large plasmid or a bacterial chromosome may transpose only at a very low frequency, if at all. This probably reflects the fact that mechanisms of transposition in general require the ends of the transposable sequence to be brought together in a transposase–DNA complex as an initial step in transposition, and because this is not an actively directed step then the further the ends are apart the less frequently this happens. However, given the ability of IS elements to generate deletions of adjacent DNA (Starlinger, 1980), a potentially large composite structure, such as that bounded by copies of IS*1* on the resistance plasmid NR1 may evolve to a somewhat smaller structure which can transpose more readily, as occurred in the case of the 2.5 kb chloramphenicol resistance transposon, Tn*9* (Iida & Arber, 1980). Hence, in an antibiotic-containing environment the selective pressures may act to minimize the size of a composite element, so that the necessary resistance genes are conserved while unnecessary genetic baggage is jetisoned. Such rearrangements would be expected to optimize transposition frequency.

The tendency of composite transposons, such as Tn*5* and Tn*10*, to retain their structure is termed coherence, i.e. the various components stay together in a set relationship (Bennett, 1991). Composite transposons may have direct or inverted terminal IS repeats, as in the cases of Tn*9* and Tn*5*, respectively. With direct repeats there is always the danger of loss of the drug resistance gene(s) as the result of a recombination event occurring between the copies of the IS element, while with an inverted arrangement recombination between the flanking IS elements simply inverts the DNA encoding the drug resistance gene(s) relative to the rest of the DNA molecule.

Composite drug resistance transposons (Table 2) would appear to be essentially combinations of a drug resistance gene, and occasionally more than one, and a particular IS element that arise by chance as a natural consequence of the transposition activity of the IS element. Not so the second class of bacterial transposons, typified by Tn*3*, which encodes the TEM β-lactamase that confers resistance to ampicillin and a number of other β-lactams in many gram-negative species. The β-lactamase gene on this element is fully integrated into the basic transposable sequence (Heffron *et al.*, 1979). Examples of this type of transposon have been found in both gram-negative and gram-positive bacteria, and related elements have been found on both sides of the positive-negative divide (Table 3), suggesting that the basic transposon may have evolved early in the development of the bacterial kingdom. It is an open question as to whether the association of antibiotic resistance genes with this class of transposon is a relatively recent development or represents recombination events of some antiquity, although in the case of integrons (see later) accretion of drug-resistance genes is clearly a continuing process.

The Tn*3*-like family of transposons has diverged into two major subdivisions (Table 3) (Sherratt, 1989). One of these, that including the type element Tn*3*, comprises a set of elements each of which is basically the same transposon, size approximately 5 kb, with minor variations of sequence, e.g. Tn*1*, originating from resistance plasmid RP4, and Tn*3*, originally found on

Table 3. *Some Tn3-like bacterial drug resistance transposons*

Transposon	Size (kb)	Terminal IRs(bp)	Drug markers	Species of origin
Tn*1*	5	38	Ap	*Pseudomonas aeruginosa*
Tn*3*	5	38	Ap	*Salmonella paratyphi*
Tn*21*	19.6	38	HgSmSu	*Shigella flexneri*
Tn*551*	5.3	35	Em	*Staphylococcus aureus*
Tn*917*	5.3	35	Em	*Enterococcus faecalis*
Tn*1721*	11.4	38	Tc	*Escherichia coli*
Tn*2424*	25	38	HgCmAmSmSu	*Shigella sonnei*
Tn*2425*	22	38	HgSmSu	*Shigella sonnei*
Tn*2603*	22	38	HgOxSmSu	*Escherichia coli*
Tn*4451*	6.2	12	Cm	*Clostridium perfringens*

Data from Murphy (1989), Sherratt (1989) and Grinsted *et al.* (1990).
Resistance genes: Ap, ampicillin; Cm, chloramphenicol; Em, erythromycin; Hg, mercuric chloride; Ox, oxacillin; Sm, streptomycin; Su, sulphonamide; Tc, tetracycline.

R1. Sequence variation within the β-lactamase gene may confer a change in substrate profile (see Table 5). The second subdivision is typified by Tn*21*, a much larger transposon of approximately 20 kb that encodes resistance to streptomycin, sulphonamides and mercuric ions (Grinsted, de la Cruz & Schmitt, 1990). This element was originally discovered on the resistance plasmid R100. Between them, the Tn*21* subset of elements carry many different resistance genes and, arguably, are the single, most successful type of transposable element in terms of the spread of drug resistance genes, being responsible for much of the drug resistance seen in gram-negative bacteria.

The Tn*3*-like family of transposons comprises a set of elements related through their transposition functions. All such elements have terminal inverted repeats, usually of 35–38 bp. For a particular element these may be precise repeats, e.g. Tn*3*, or may have one or two mismatches, e.g. Tn*21*. Each element has a transposase gene of 3 kb encoding an enzyme of about 1100 amino acids specific for the elements IRs. Some, but not all of these enzymes are sufficiently similar as to be able to complement null mutations in the transposase genes of other elements (Grinsted *et al.*, 1990). In addition, most elements of this type encode also an enzyme called a resolvase, which is a site-specific recombinase that requires a particular site on the transposon termed *res*. Resolvase releases the final transposition product from an intermediate called a cointegrate. These two enzymes, transposase and resolvase, mediate a 2-stage transposition mechanism (Sherratt, 1989). The first stage replicates the transposon sequence specifically, in the process generating a joint molecule, the cointegrate. This structure has the entire donor DNA molecule, i.e. that carrying the original copy of the transposon, inserted at the target site on the receiving DNA molecule, with a copy of the transposon sequence at each junction of the insertion forming a direct duplication. This intermediate is then resolved by recombination across the direct repeats of the transposon (using either *res*/ resolvase or homologous recombination) to yield the final transposition product, simultaneously regenerating the donor molecule. Hence each transposition event results in an increase in the number of copies of the element, clearly an advantage with respect to lateral spread of resistance genes carried by the element.

The Tn*3*-related drug-resistance transposons comprise, as stated, two main sub-groups, within which the transposition functions of elements in one group are much more closely related to each other than to those elements in the other sub-group. The Tn*3* sub-group comprises essentially three elements, Tn*3*, Tn*1331* and Tn*1000* (also called *γδ*), the last being a cryptic element found on the *E. coli* fertility factor, F. Other members of the group are simply variants of Tn*3*. In contrast, the Tn*21* sub-group contains many elements that differ with respect to the number and variety of drug resistance genes carried and the degree of relationship between their transposition functions and size (Grinsted *et al.*, 1990). Although in general, somewhat larger than the elements in the Tn*3* sub-set, the basic transposon, namely, the combination of IRs, transposase and resolvase genes, is the same size in each element. A transposable element comprising just these components, with a size of 3.6k and clearly related to the group, Tn*5403*, has been described (Rinkel *et al.*, 1994; M.-C. Lett, personal communication). It differs sufficiently from both Tn*3* and Tn*21* to be considered to be representative of a third sub-set of Tn*3*-like elements.

Tn*21*-like elements between them code for many different types of drug resistance, including resistance to *β*-lactams, aminoglycosides, chloramphenicol, tetracycline, erythromycin, trimethoprim, sulphonamides, as well as to mercuric ions (Grinsted *et al.*, 1990). The majority of these elements have the same basic structure, with the cluster of genes encoding resistance to mercuric ions at one end of the element beside an IR sequence, conventionally the left end, and the transposition genes at the other, with the transposase gene terminating within the second IR. The different resistance genes on the various elements are located next to the transposition genes (Grinsted *et al.*, 1990). Many of these elements carry the *aadA* gene encoding an adenylyl transferase that mediates resistance to streptomycin and spectinomycin, and *sul*, encoding a sulphonamide-resistant dihydropteroate synthase. Additional resistance genes are normally located on one side or the other of *aadA*, if this is present, or replace it, if it is not. These resistance genes are small genetic casettes that are added or deleted from the basic structure by a site-specific recombination enzyme termed an integrase, encoded by the *int* gene on the transposon. The insertion of a drug resistance casette is formally equivalent to the integration of bacteriophage lambda into the *attP* site on the *E. coli* chromosome (Campbell, 1962) and the mechanism is basically the same. The resistance gene casette is first released as a small circle of DNA, accommodating one or two resistance genes, which is then integrated into the transposon structure via one of a small number of specific sites on the transposon. The integrase belongs to the family of site-specific recombinases of which lambda integrase is the typical enzyme (Argos *et al.*, 1986).

Sequence analysis of the DNA surrounding a number of such resistance genes has identified a 59 bp, loosely conserved palindromic sequence associated with the ends of the casettes (Hall, Brookes & Stokes, 1991). A copy of the sequence is present at one end of the *aadA* gene on Tn*21*. These palindromic sequences provide the sites at which the integrase acts and are equivalent to the *attP* site used by lambda integrase. The general arrangement of drug resistance genes (DRGs) and integration sites (is59) found on Tn*21*-like elements is:

.P..is58a.DRG1.is59b.aadA.is59c.DRG2.is59d.....

where DRG indicates a drug resistance determinant and is59 indicates a copy of the integration sequence (Hall *et al.*, 1991), and all resistance genes form an operon expressed from a promoter(s), P, to the left of DRG1 (see Davies, 1994).

Resistance genes commonly found as components of Tn*21*-like transposons, such as the *aadA1* and *sul* genes, are also found on plasmids in the absence of other Tn*21* features. An examination of some of these genes and the sequences that surround them has revealed than the homology extends for 1–2 kb on either side of the resistance gene cluster. Analysis of the ends of these structures, where homology between them disappears, revealed no similarities between the ends of the common structure, e.g. the presence of IR sequences, which might have indicated the existence of another class of transposable element. The DNA common to all these structures, disregarding the resistance casettes, is 3.4–4 kb and accommodates an integration site with a copy of the integrase gene, *int*, on one side and the *sul* gene, together with a gene, *qacEΔ1*, encoding resistance to quaternary ammonium compounds and an unidentified ORF on the other (see Davies, 1994). This structure has been termed an integron (Stokes & Hall, 1989), and it has been suggested that integrons are genetically mobile, because copies have been found at several, unrelated sites; but genetic mobility has not yet been demonstrated in laboratory experiments. Integrons are undoubtedly responsible, in part, for the spread of drug resistance genes among clinically important bacteria, not least because these genes are carried on Tn*21* and like elements.

The Tn*21* family of transposons are responsible for much of the drug resistance seen in gram-negative bacteria, particularly in multiply resistant bacteria, and the elements are almost always found on bacterial plasmids. A recent survey of approximately 1000 gram-negative clinical isolates from a multicentre European study, using a set of gene probes for different parts of Tn21, identified Tn21-like elements in almost 10% of the isolates, while another 10% displayed features that indicated they probably had part of a Tn21-like element (Zuhlsdorf & Wiedemann, 1992). A similar study of gram-negative bacteria isolated from several environment water systems identified Tn21-like elements in approximately 5% of the strains tested, although the distribution was not uniform (Roux et al., 1993). These studies serve to show that Tn3-like elements, particularly those of the Tn21 subgroup, are now widely distributed in nature and are responsible for disseminating many of the drug resistance genes found in clinical isolates.

In gram-negative bacteria drug resistance genes are primarily associated with large plasmids, most of which are conjugative. This probably reflects the dominant mode of transmission in this division, namely conjugation, in that plasmids have to have a certain minimum size to accommodate the genetic information for a conjugation system. This appears to be about 15–20 kb, as seen for the broad host range IncW resistance plasmid R388 (Valentine & Kado, 1989), although larger Tra systems are known, such as those of the E. coli fertility factor, F, and resistance plasmids R1 and R100 which occupy approximately 30 kb (Ippen-Ihler & Skurray, 1993). In contrast, in gram-positive bacteria such as Staph. aureus many drug resistance genes are located on small plasmids, often as single determinants, perhaps reflecting the fact that they are often transmitted by transduction (Lacey, 1975). Further, conjugation systems in both Staphylococcus and Enterococcus spp. appear to require somewhat less plasmid encoded information (Macrina & Archer, 1993).

CONJUGATIVE TRANSPOSONS

In gram-positive bacteria, such as Enterococcus faecalis, another type of transposable element has been described, called a conjugative transposon (Clewell & Flannagan, 1993). These are discrete elements that not only can transpose into both plasmids and the bacterial chromosome, but can also transfer from one cell to another. The mechanism of transposition involves a stage where the transposon is formally released from one site as a free, circular intermediate which can then be reintegrated into another site on the same DNA molecule or into a site on another DNA molecule. Alternatively, the free intermediate can transfer to another cell where it then transposes into a suitable replicon. Both transposition and transfer functions are encoded by the transposon.

This family of transposons comprises a set of related elements, sizes ranging from 16 kb, e.g. Tn918 conferring resistance to tetracycline, to more than 65 kb, such as Tn3951 (67 kb), most of which encode resistance to tetracycline. Resistance to erythromycin, chloramphenicol and kanamycin may also be specified, in addition to tetracycline resistance (Table 4) These

Table 4. *Some conjugative bacterial drug resistance transposons*

Transposon	Size (kb)	Markers	Species of origin
Tn916	16.4	Tc	*Enterococcus faecalis*
Tn919	16	Tc	*Streptococcus sanguis*
Tn920	20	Tc	*Enterococcus faecalis*
Tn1545	25.3	TcEmKm	*Streptococcus pneumoniae*
Tn3951	67	CmEmTc	*Streptococcus agalactiae*
Tn5253	65.5	CmTc	*Streptococcus pneumoniae*

Data from Murphy (1989) and Clewell & Flannagan (1993).
Resistance genes: Cm, chloramphenicol; Em, erythromycin; Km, kanamycin; Tc, tetracycline.

elements have also been found in *Strep. pneumoniae*, *Strep. pyogenes*, *Strep. agalactiae* and *Lactococcus lactis* (Clewell & Flannagan, 1993).

TRANSFER *IN VIVO*

Many of us are used to thinking of transfer events as they happen in laboratory experiments. In these, transfer systems can be manipulated to operate very efficiently, to maximise the number of transfer events (Willetts, 1988). Conjugation systems that routinely achieve transfer frequencies of 1% or more (number of transconjugants/donor, when donor/recipient <1) are readily established, providing the physical conditions of the mating are optimized and sufficient time is allowed for the cross. Transformation efficiencies of $10^8–10^9$ transformants/ug DNA can be obtained in some systems using chemically induced competence or electroporation (Sambrook *et al.*, 1989). In nature, such high transfer frequencies are unlikely, and indeed, are unnecessary. Transfer of genetic information in natural populations can be thought of as a strategic event, rather than as a tactical device, in the sense that the majority of cells within a natural population will rarely, if ever, be genetically altered. Rather, a few individuals will be changed; then, if the conditions subsequently favour the propagation of these new variants they will outgrow the rest of the population, in due course, to replace it.

DEVELOPMENT OF RESISTANT POPULATIONS

The use of an antibiotic is an immediate challenge to the population of cells exposed to the agent. Accordingly, susceptible cells must adapt or die and populations will change to reflect the antibiotic in the cells' environment. Clinical isolates collected before widespread use of antibiotics, in general, display uniform susceptibility to most antibiotics (Datta & Hughes, 1983), in contrast to present day isolates, many of which display multiple drug resistance. The continuous acquisition of new resistance genes and the inexorable rise in the percentage of isolates resistant to a particular anti-biotic while that compound is in general clinical use is an elegant testimony to the adaptability of bacteria and the power of natural selection. The rapid emergence of sulphanilamide resistant and penicillin resistant strains as a consequence of the use of sulphanilamide and benzylpenicillin in the 1940s and 1950s bears witness to this and these resistance genes persist widely in clinical isolates today, e.g. Neu (1985) showed that 95% of nosocomial staphylococci isolated in the developed world produced beta-lactamase. These cases serve as paradigms for the development of resistance in response to clinical use of an antibiotic. Resistance has been reported to virtually all antibiotics currently in clinical use, although for some the incidence of isolation of resistant strains is, thankfully, still rare and spasmodic. How long this will remain so is questionable.

As resistance to benzylpenicillin developed in the *Staphylococcal* spp. so a search for other β-lactams, natural and semisynthetic, which were not inactivated by staphylococcal β-lactamase was undertaken, resulting in the production of the penicillin derivatives, methicillin and the oxypenicillins, oxacillin and cloxacillin, all with good anti-*Staphylococcus* activities. It was not too long, however, before the isolation of methicillin resistant *Staph. aureus* was reported, organisms now commonly referred to as MRSA. The mechanism of resistance is not by the production of another β-lactamase, but rather by the production of a novel enzyme that replaces those inhibited by methicillin. The targets for β-lactams are the enzymes that catalyse the final stages of cell wall synthesis, the so-called penicillin-binding proteins, or PBPs. Most cells have multiple enzymes of this type and *Staph. aureus* is no exception. It possesses four such enzymes, three of which are essential and all of which are inhibited by methicillin (Song *et al.*, 1987; Spratt, 1994). MRSA strains have acquired an additional enzyme, PBP2' (also PBP2a), with a low affinity for methicillin. This new enzyme, when necessary, assumes the tasks of those essential PBPs inhibited by methicillin (but see de Lencastre *et al.*, 1994). PBP2' may be plasmid or chromosome encoded. The gene *mec*, is part of a transposon. Expression of *mec* is often inducible. Analysis of the genetic structure of this system has revealed that the *mec* gene has been fused to the first few codons of the inducible staphylococcal β-lactamase gene (*bla*), effectively replacing *bla* and placing *mec* under the same control system as that for the staphylococcal β-lactamase (Bennett & Chopra, 1993). Both systems are readily induced by methicillin. Where *mec* originated, or how the rearrangement occurred is not known, but it is likely that it was necessary to provide a suitable promoter for the gene. The gene appears to have been acquired only once and then spread among *Staphylococcus* species (Spratt, 1994). This development has provided *Staph. aureus* not only with resistance to methicillin, but also to most other β-lactams, because PBP2' has generally reduced affinities for all β-lactams. With the acquisition of resistance to a variety of other antibiotics (Cookson, 1993;

Lyon & Skurray, 1987; Witte & Hummel, 1986), some strains of *Staph. aureus* are almost untreatable. Virtually the only antibiotic left for these strains is vancomycin, and this is under threat, in that it has been demonstrated in laboratory experiments that the *vanA* determinant of *Ent. faecalis* can be transferred to and expressed in *Staph. aureus*.

The work that led to the development of methicillin also gave the broad spectrum β-lactam, ampicillin, with good activity against many gram-negative, as well as gram-positive bacteria. Again, however, within a few years of its introduction into clinical use, the isolation of gram-negative strains previously uniformly sensitive but now resistant to ampicillin was reported (Anderson & Datta, 1965; Datta & Kontomichalou, 1965). In these cases resistance was again due to the production of β-lactamase, an enzyme subsequently called the TEM β-lactamase. By 1970 this enzyme had been found in ampicillin resistant strains of several enterobacteriaceae, including *E. coli*, *Shigella* spp. and *Salmonella typhimurium*. It was next reported in strains of *Pseudomonas aeruginosa* where it conferred resistance to carbenicillin (Datta *et al.*, 1971), the first of the effective anti-pseudomonas β-lactams. The successful spread of the TEM β-lactamase gene is, without doubt, due to it being part of a transposon (Hedges & Jacob, 1974) (Table 4). Initially two versions were identified, Tn*1* and Tn*3*, producing slightly different β-lactamase, apparent because the two enzymes have different pI values. The difference is due to a single amino acid change (see Table 5), which doesn't alter the substrate profile significantly; but these two enzymes were to become the progenitors of an extended family of β-lactamases.

With more extensive and widespread use of β-lactam antibiotics such as ampicillin and carbenicillin throughout the 1970s, reports of ampicillin resistant gram-negative bacteria became more common. As these strains were analysed it became clear that that resistance was not always due to production of one of the TEM β-lactamases but to one of several other β-lactamases, such as the SHV and PSE enzymes (Bush, 1989), although many of these other β-lactamases have been identified on only one or two occasions (Amyes, Payne & du Bois, 1992). This increase in variety led to the requirement for some form of classification scheme. Several have been proposed, the most widely used of which has been that of Richmond and Sykes (1973). The documentation of new beta-lactamases has barely slackened in the last 20 years, and in the most recent classification scheme (Bush, 1989) 85 β-lactamases are listed, and this does not include a number of variants of some of these enzymes.

In 1972 penicillinase-producing strains of *Haemophilus influenzae* were identified and by 1976 penicillinase-producing *Neisseria gonorrhoeae* (PPNG) (Sanders, 1989; Saunders, 1984). Until then, these bacteria had been uniformly sensitive to β-lactams and had not produced penicillinase. In both cases, the development of these new abilities reflected the acquisition of the gene encoding the TEM β-lactamase, carried on bacterial plasmids. Unexpectedly, the plasmids carrying the TEM β-lactamase gene from the different species were found to be closely related, although not identical. As the result of further analysis it became clear that the PPNG had acquired their plasmids, in the first instance, from *H. influenzae*, by either conjugation, as a result of mobilization because the plasmids identified are not themselves conjugative plasmids, or, possibly by transformation, given that *N. gonorrhoeae* displays natural competence. From where *H. influenzae* acquired the TEM β-lactamase is not known, but it is likely that it was acquired by transposition to an indigenous plasmid from a plasmid probably acquired by conjugation, perhaps one that survived only transiently in *H. influenzae*. By the time the TEM β-lactamase gene spread into *H. influenzae*, it was already well established within the enterobacteriaceae, including *E. coli*, *Shigella* spp. *Salm. typhimurium*, *Klebsiella pneumoniae*, *Serratia marcescens*, and in *P. aeruginosa*. In the majority of cases, if not all, the gene for the enzyme was carried on a plasmid, and in the majority of these it was likely that the gene was on a transposon such as Tn*1/3*. As the gene spread, so the clinical efficacy of ampicillin and carbenicillin declined. So long as the strain to be irradicated was sensitive, then ampicillin was effective, but the probability that a particular strain would prove to be a TEM β-lactamase producer became unacceptably high, so that in many cases empirical therapy with ampicillin, which initially had been very effective, became inappropriate, if not self-defeating.

The development of widespread resistance to ampicillin and carbenicillin, due largely to the spread of the TEM β-lactamase genes, once again spurred development of new classes of β-lactams. These were based, not on the penicillin nucleus, but on a related compound, cephalosporin C. This compound, also produced by a mould, has little antibiotic activity, but, by altering the side chains, very effective antibiotics, with antibacterial activities 1000 times or more greater than that of cephalosporin C, were developed and, in particular, ones that were stable to many of the plasmid encoded β-lactamases then known. Compounds such as ceftazidime and cefotaxime have proved to be valuable antibiotics, but inevitably resistance to them has emerged and spread, and in more than one way.

Although not susceptible to hydrolysis by most plasmid encoded β-lactamases, ceftazidime and cefotaxime are hydrolysed by a number of chromosome encoded β-lactamases, such as those found in *Citrobacter freundii* and *Enterobacter cloacae*. However, resistance is only seen if cells produce significant quantities of one of these enzymes and in the majority of strains enzyme production is inducible with basal production being negligible. Although β-lactamase production is induced by some β-lactams, e.g. cefoxitin (see Bennett & Chopra, 1993), neither ceftazidime nor cefotaxime is an inducer and so most natural isolates of *C. feundii* and *E. cloacae* are sensitive to these antibiotics. However, clinical use of these antibiotics has led to a significant increase in the number of isolates in which production of the chromosome encoded β-lactamase is derepressed (Reeves, 1994; Sanders, 1987; Sanders & Sanders, 1987; Tzelepi *et al.*, 1992). Because this resistance results from mutation of chromosomal genes, spread of resistance in these cases involves clonal spread rather than gene transfer or the emergence of new mutant clones. This is likely to be more of a nosocomial problem than a community one.

It is said that nature abhors a vacuum and will take steps to fill it. The development of resistance to ceftazidime and cefotaxime is a case in point. Development of resistance by derepression of expression of a chromosome encoded β-lactamase gives protection to those species fortunate to possess an appropriate enzyme, but is of little help to those that do not. Hence, there is certain to be selective pressure for the development of some form of transmissible resistance, if this is possible. In the case of ceftazidime and cefotaxime, one development was mutation of the SHV and TEM β-lactamase genes (Kliebe *et al.*, 1985; Payne, Marriot & Amyes, 1990; Petit *et al.*, 1990; Rice *et al.*, 1990). Relatively minor changes were needed to the primary amino acid sequences of both proteins to alter the substrate spectra to accommodate one or other or both β-lactams, to produce sets of what are known as extended spectrum β-lactamases (Table 5). Many variants of both the TEM and SHV β-lactamases are now known, yet again elegant testimony to the flexibility of biological systems and the power of natural selection, and the possibilities may not yet be exhausted. These enzyme variants are not yet particularly common (MacGowan *et al.*, 1993; Reeves, 1994); it remains to be seen if they become so in the future. If the progenitor systems, TEM1, TEM2 and SHV are harbingers, the auspices are not very good, given continued, perhaps increased use of ceftazidime and cefotaxime, as seems likely.

A third development with respect to resistance to β-lactams such as ceftazidime and cefotaxime has recently been reported, namely, the migration of a chromosome β-lactamase gene to a plasmid (Payne, Woodford & Amyes, 1992; Fosberry *et al.*, 1994). The BIL1 β-lactamase, discovered in *E. coli*, is encoded by what is clearly a gene originating from the chromosome of *C. freundii*. The rearrangement has provided a promoter that renders expression independent of the normal induction mechanism. How the rearrangement occurred is not yet known, but the involvement of IS elements is a distinct possibility.

Hence the threat to bacteria of new β-lactams such as ceftazidime and cefotaxime has been/is being met in several ways. The response of the pharmaceutical companies has been to introduce yet more variations on the β-lactam theme. One of the latest is a class of compounds called carbapenems, in which the sulphur atom of the thiazolidine ring is replaced by a carbon atom. Antibiotics of this type, such as imipenem and meropenem, are poorly hydrolysed by the majority of plasmid and chromosome encoded

Table 5. *Extended spectrum-β-lactamases derived from TEM-1, TEM-2 and SHV-1*

	Amino acid at position							
β-lactamase	39	104	164	205	237	238	240	265
TEM-1	Gln	Glu	Arg	Gln	Ala	Gly	Glu	Thr
TEM-4	–	Lys				Ser		Met
TEM-5			Ser		Thr		Lys	
TEM-6		Lys	His					
TEM-9		Lys	Ser					Met
TEM-10			Ser				Lys	
TEM-12			Ser					
TEM-15		Lys				Ser		
TEM-17		Lys						
TEM-19						Ser		
TEM-26		Lys	Ser					
TEM-2	Lys							
TEM-3	Lys	Lys				Ser		
TEM-7	Lys		Ser					
TEM-8	Lys	Lys	Ser			Ser		
TEM-11	Lys		His			?		
TEM-16	Lys	Lys	His					
TEM-18	Lys	Lys						
TEM-24	Lys	Lys	Ser		Thr			Lys
SHV-1	Gln	Asp	Arg	Arg	Ala	Gly	Glu	Leu
SHV-2						Ser		
SHV-3				Leu		Ser		
SHV-4				Leu		Ser	Lys	
SHV-5						Ser	Lys	

Data from Bennett & Hawkey (1991), Collatz, Labia & Gutmann (1990), Davies (1994) and Jacoby & Medeiros (1991). TEM-1, TEM-2 and SHV-1 are parental types without extended spectrum activities.

β-lactamases, if at all. None the less, resistance has emerged in one of two ways. The first is exemplified by the development of imipenem resistance in *P. aeruguinosa*. The mechanism of resistance is dependent on loss of the outer membrane protein OprD2 (Nikaido, 1989, 1994; Quinn *et al.*, 1991). OprD2 is a porin that provides an outer membrane channel through which imipen passes to the periplasm at pharmacologically relevant concentrations. Loss of OprD2 eliminates this means of uptake, conferring moderate resistance to imipenem specifically. Resistance to β-lactams in general does not result, because most do not cross the outer membrane via OprD2. The second mechanism of resistance is the production of β-lactamases. Bacteria such as *Aeromonas hydrophila*, *Aeromonas sobria*, *Bacteroides fragalis* and *Xanthomonas maltophilia* possess what are called metallo-β-lactamases, because the activities are dependent on Zn^{2+} ions, rather than on an active site serine group (Payne, 1993). These enzymes, which are predominantly chromosome encoded, hydrolyse imipenem quite efficiently. Hence, use of imipenem will, given the appropriate situation, tend to select for these organisms. Possession of a chromosome encoded metallo-β-lactamase doesn't preclude carriage of genes encoding serine active site enzymes. Indeed, *A. sobria* can produce three chromosome encoded β-lactamases, one a metallo-enzyme, and expression of all three is coordinately regulated (Walsh *et al.*, 1994). Again, this form of resistance is not readily transmitted to other gram-negative bacteria. However, transmissible imipenem resistance, by virtue of production of a serine active site β-lactamase, has recently been reported in *S. marcescens*.

The development of clinical resistance to penicillin, due to production of a penicillinase, was first seen in *Staph. aureus* in the 1940s. Recently, plasmids carrying the *bla* gene encloding the β-lactamase in *Staph. aureus*, together with a gene for gentamicin resistance have spread to *Enterococcus* species in North and South America (Murray, 1992; Murray *et al.*, 1991, 1992). The plasmids in *Staphylococcus* spp. and those in *Enterococcus faecalis*, however, are not the same (Wanger & Murray, 1990).

Penicillin resistance has also now been found in *Strep. pneumoniae*, but in this case it is due not to β-lactamase production but to changes to the genes encoding the target enzymes (Spratt, 1994). As recorded elsewhere in this volume (Spratt), several PBP genes in penicillin-resistant strains of *Strep. pneumoniae* have been subject to extensive rearrangements by recombination with chromosomal genes from other, inherently more resistant enterococcal species, such as *Ent. mitis* (Spratt, 1994). This has been possible because *Strep. pneumoniae* is naturally transformable and so has access, in principle, to a wide range of chromosomal DNA, some of which will be sufficiently homologous to the cell's own DNA to permit recombination, and hence, gene remodelling, as has clearly occurred to create the penicillin-resistant *Strep. pneumoniae* strains. Penicillin-resistant strains of *Strep. pneumoniae* are now distributed world-wide, due in part, at least, to the ease of mass travel. This type of gene remodelling giving penicillin resistance is also known to have occurred in *H. influenzae* and in *N. gonorrhoeae*, two other naturally transformable species (Clairoux *et al.*, 1992; Spratt, 1994).

The development of multiple drug resistance among enterococcal species has narrowed the range of antibacterial agents available for therapy, so that reserve antibiotics such as vancomycin are now more commonly used. This has had the almost predictable effect of selecting strains resistant to vancomycin, and several resistance determinants are now known (Arthur & Courvalin, 1993). This was something of a surprise when it was first detected (Reynolds, 1992), given the mechanism of action of vancomycin. The drug's target is the terminal D-Ala-D-Ala dipeptide of the pentapeptide chain of the cell wall precursor, to which it binds to inhibit formation of peptide cross-links. The binding depends on several H-bonds between the drug and its target, including the amino group of the ultimate alanine group. Resistance is achieved by replacing D-Ala with lactate. This change requires a fairly major remodelling of the terminal stage of cell wall biosynthesis, and this is precisely what happens (Arthur & Courvalin, 1993). The necessary new enzymes are provided on a transposable element, Tn*1546*. Given the sophistication and complexity of this change, it clearly has not evolved in response to recent clinical use of vancomycin. Rather, the relevant genes must have evolved elsewhere, for what reason(s) we don't know. It is also unlikely that the generation of the transposon is of recent origin, whereas the transfer to and establishment in *Enterococcus* species is, almost certainly, a recent phenomenon.

The variety of resistance determinants for β-lactams is, in part, matched by the variety of resistance determinants for other types of antibiotics. Resistance to aminoglycoside-aminocyclitol (AGAC) antibiotics can be conferred by target site change, usually by mutation (e.g. streptomycin resistance in *E. coli* as the result of mutations in *rspL* which encodes the S12 protein of the small ribosomal subunit), or by acquisition of one of three types of enzyme that modify the drug to inactivate it. These enzymes phosphorylate, acetylate or adenylate the AGAC. Many different enzymes of each type are now known and among one type many are clearly related (Shaw *et al.*, 1993). The genes for these enzymes are found predominantly on plasmids and transposons. Similar variety has been reported among resistance determinants for chloramphenicol (Shaw & Leslie, 1989), tetracycline (Salyers, Speer & Shoemaker, 1990; Sheridan & Chopra, 1991), trimethoprim (Amyes *et al.*, 1992) and others (Leclercq & Courvalin, 1991; Neu, 1991).

THE ORIGINS OF RESISTANCE GENES

Given the prevalence of antibiotic resistance genes, it is pertinent to enquire as to their origins. In most instances we do not know. Certainly resistance to particular antibiotics is inherent to some species, either because the target enzyme is not susceptible to inhibition by the drug, or the drug cannot gain access to the cell because of permeability problems or because the species naturally elaborates a destruction mechanism. Hence, *P. aeruginosa* is clinically resistant to many beta-lactams either because they cannot penetrate the outer membrane or penetrate only slowly and are destroyed by the chromosome encoded β-lactamase (Nikaido, 1994).

Many bacterial species possess β-lactamase genes as part of their normal complement, and the spectrum of activity of those we know about is impressive (Bush, 1989). Given the relative ease with which genes can be mobilized, it is no real surprise that there are many β-lactamase genes on plasmids and transposons, although the primary source in particular cases has rarely been established. The BIL-1 enzyme is an exception (Fosberry *et al.*, 1994). Some β-lactamases have almost certainly evolved from the same ancestral genes that gave the present day PBPs.

Inherent resistance to aminoglycosides, because of species-specific chromosome encoded enzymes is well established (Shaw *et al.*, 1993), although the genes may not be expressed under laboratory conditions unless activated by mutation (Champion *et al.*, 1988). It is unlikely that these particular enzymes have evolved to confer protection against a subset of aminoglycosides, but what their natural substrates are is not known. Given that these aminoglycoside modifying enzymes essentially phosphorylate, acetylate or nucleotidylate amino sugars, it seems likely that their natural substrates are also amino sugars, perhaps those involved in LPS construction.

Most clinically useful antibiotics are natural compounds, or semi-synthetic derivatives, that are secondary metabolites of bacteria such as *Bacillus* spp. or *Streptomyces* spp. The producing organism itself requires protection (Cundliffe, 1989) and in some cases resistance genes have been found linked to the genes encoding the enzymes of antibiotic synthesis (see Davies, 1994). It is likely that some, at least, of the resistance genes found in clinical isolates, particularly those on plasmids, had their origin in bacteria which produce the antibiotic, or a closely related one (Cundliffe, 1989; Saunders, 1984). The spread of these genes may have been inadvertently facilitated by antibiotic production systems, because it was recently demonstrated that antibiotic preparations can be contaminated with DNA from the producing strain (Webb & Davies, 1993). This, of course, would then be administered together with the antibiotic, time and time again. Given appropriate conditions, this DNA may have been taken up and incorporated into a bacterium with a natural transformation system. From there it could, in principle, spread to other bacteria in the same ecological niche, widening the constituency of the gene. In due course, the gene would spread to bacteria of clinical importance, which would likely indicate the beginning of the end for therapy with that particular compound. This scenario is highly speculative, but not impossible.

It is important to appreciate that the genetic changes needed to generate a resistance phenotype are not events that occur in response to antibiotic use. These genetic events occur all the time, irrespective. The use of a drug does, however, establish the necessary conditions for outgrowth of resistant cells from the basic drug-sensitive bacterial flora. So given that there is no inherent biochemical or biological reason why the change(s) necessary for drug resistance should not occur, it is likely that somewhere, sometime resistance will evolve (or, indeed, may have already evolved). Exposure to the drug will amplify the numbers of the particular drug-resistant variant(s), providing an expanded genetic pool for any subsequent developments, e.g. more widespread dissemination as a consequence of association with plasmids and transposons. Again, these genetic changes would occur as the consequence of the normal behaviour and biological properties of DNA and not through any active participation on the part of the antibiotic. The antibiotic simply enhances the chance of survival of particular rearrangements. Antibiotic use may accelerate the process by causing resistant bacteria to proliferate at the expense of sensitive organisms, to the point where resistant ones become the dominant component of the bacterial population, providing an amplified pool of genes for further development.

REFERENCES

Amyes, S.G.B. & Gemmell, C.G. (1992). Antibiotic resistance in bacteria. *Journal of Medical Microbiology,* **36**, 4–29.

Amyes, S.G.B., Payne, D.J. & duBois, S.K. (1992). Plasmid-mediated β-lactamases responsible for penicillin and cephalosporin resistance. *Journal of Medical Microbiology,* **36**, 6–9.

Amyes, S.G.B., Towner, K.J. & Young, H.-K. (1992). Classification of plasmid-encoded dihydrofolate-reductases conferring trimethoprim resistance. *Journal of Medical Microbiology,* **36**, 1–3.

Anderson, E.S. & Datta, N. (1965). Resistance to penicillins and its transfer in Enterobacteriaceae. *Lancet,* **i**, 407–9.

Argos, P., Landy, A., Abremski, K., Egan, J.B., Haggard-Ljungquist, E., Hoess, R.H., Kahn, M.L., Kalionis, B., Narayana, S.V.L., Pierson III, P., Sternberg, N. & Leong, J.M. (1986). The integrase family of site-specific recombinbases: regional similarities and global diversity. *EMBO Journal,* **5**, 433–440.

Arthur, M. & Courvalin, P. (1993). Genetics and mechanisms of glycopeptide resistance in enterococci. *Antimicrobial Agents and Chemotherapy,* **37**, 1563–71.

Bennett, P.M. (1991). Transposable elements and transposition in bacteria. In *Modern Microbial Genetics*, Streips, U.L. & Yasbin, R.E. eds, pp. 323–364. Wiley-Liss, New York.

Bennett, P.M. & Chopra, I. (1993). Molecular basis of β-lactamase induction in bacteria. *Antimicrobial Agents and Chemotherapy,* **37**, 153–8.

Bennett, P.M. & Hawkey, P.M. (1991). The future contribution of transposition to antimicrobial resistance. *Journal of Hospital Infection,* **18** (Suppl A), 211–21.

Berg, D.E. (1989). Transposon Tn5, In *Mobile DNA*, Berg, D.E. & Howe, M.M., eds, pp. 185–210. American Society for Microbiology, Washington, DC.

Bush, K. (1989). Characterization of β-lactamases. *Antimicrobial Agents and Chemotherapy,* **33**, 259–76.

Campbell, A.M. (1962). Episomes. *Advances in Genetics,* **11**, 101–45.

Champion, H.M., Bennett, P.M., Lewis, D.A. & Reeves, D.S. (1988). Cloning and characterization of an AAC(6′) gene from *Serratia marcescens*. *Journal of Antimicrobial Agents and Chemotherapy,* **22**, 587–96.

Chandler, M., Clerget, M. & Galas, D.J. (1982). The transposition frequency of IS*1*-flanked transposons is a function of their size. *Journal of Molecular Biology,* **154**, 229–43.

Charlebois, R.L. & Doolittle, W.F. (1989). Transposable elements and genome structure in Halobacteria. In *Mobile DNA*, Berg, D.E. & Howe, M.M. eds, pp. 297–307. American Society for Microbiology, Washington, DC.

Chopra, I. (1988). Mechanisms of resistance to antibiotics and other chemotherapeutic agents. *Journal of Applied Bacteriology Symposium Supplement*, pp. 149S–66S.

Clairoux, N., Picard, M., Brochu, A., Rosseau, N., Gourde, P., Beauchamp, D., Parr, Jr., T.R., Bergeron, M.G. & Malouin, F. (1992). Molecular basis of the non-β-lactamase-mediated resistance to β-lactam antibiotics in strains of *Haemophilus influenzae* isolated in Canada. *Antimicrobial Agents and Chemotherapy*, **36**, 1504–13.

Clewell, D.B. (Ed). (1993*a*). *Bacterial Conjugation*. Plenum Press, New York.

Clewell, D.B. (1993*b*). Sex pheromones and the plasmid-encoded mating response in *Enterococcus faecalis*. In *Bacterial Conjugation*, Clewell, D.B., ed, pp. 349–367. Plenum Press, New York.

Clewell, D.B. & Flannagan, S.E. (1993). The conjufative transposons of Gram-negative bacteria. In *Bacterial Conjugation*, Clewell, D.B., ed, pp. 369–393. Plenum Press, New York.

Cohen, S.N., Chang, A.C.Y. & Hsu, L. (1972). Nonchromosomal antibiotic resistance in bacteria: genetic transformation of *Escherichia coli* by R-factor DNA. *Proceedings of the National Academy of Sciences, USA*, **69**, 2110–14.

Collatz, E., Labia, R. & Gutmann, L. (1990). Molecular evolution of ubiquitous β-lactamases towards extended-spectrum enzymes active against newer β-lactam antibiotics. *Molecular Microbiology*, **4**, 1615–20.

Cookson, B.D. (1993). MRSA: major problem or minor threat? *Journal of Medical Microbiology*, **38**, 309–10.

Craig, N.L. (1989). Transposon Tn7 In *Mobile DNA*, Berg, D.E. & Howe, M.M., eds, pp. 211–225. American Society for Microbiology, Washington, DC.

Cundliffe, E. (1989). How antibiotic-producing organisms avoid suicide. *Annual Reviews of Microbiology*, **43**, 207–33.

Datta, N., Hedges, R.W., Shaw, E.J., Sykes, R.B. & Richmond, M.H. (1971). Properties of an R-factor from *Pseudomonas aeruginosa*. *Journal of Bacteriology*, **108**, 1244–9.

Datta, N. & Hughes, M.V. (1983). Plasmids of the same Inc groups in Enterobacteriaceae before and after the medical use of antibiotics. *Nature*, **306**, 616–17.

Datta, N. & Kontomichalou, P. (1965). Penicillinase synthesis controlled by infectious R factors in enterobacteriaceae. *Nature*, **208**, 239–41.

Davies, J.E. (1991). Aminoglycoside-aminocyclitol antibiotics and their modifying enzymes. In *Antibiotics in Laboratory Medicine*, 3rd edn., Lorian, V., ed, pp. 691–713. Williams & Wilkins, Baltimore.

Davies, J. (1994). Inactivation of antibiotics and the dissemination of resistance genes. *Science*, **264**, 375–82.

de Lencastre, H., de Jonge, B.L.M., Matthews, P.R. & Tomasz, A. (1994). Molecular aspects of methicillin resistance in *Staphylococcus aureus*. *Journal of Antimicrobial Chemotherapy*, **33**, 7–24.

Fosberry, A.P., Payne, D.J., Lawlor, E.J. & Hodgson, J.E. (1994). Cloning and sequence analysis of blaBIL-1, a plasmid-mediated class C β-lactamase gene in *Escherichia coli* BS. *Antimicrobial Agents and Chemotherapy*, **38**, 1182–5.

Foster, T.J. (1983). Plasmid-determined resistance to antimicrobial drugs and toxic metal ions in bacteria. *Microbiological Reviews*, **47**, 361–409.

Galas, D.J. & Chandler, M. (1989). Bacterial insertion sequences. In *Mobile DNA*, Berg, D.E. & Howe, M.M., eds, pp. 109–162. American Society for Microbiology, Washington, DC.

Grindley, N.D. & Joyce, C.M. (1980). Genetic and DNA sequence analysis of the kanamycin resistance transposon Tn903. *Proceedings of the National Academy of Sciences, USA*, **77**, 7176–80.

Grinsted, J., de la Cruz, F. & Schmitt, R. (1990). The Tn21 subgroup of bacterial transposable elements. *Plasmid*, **24**, 163–89.

Gruneberg, R.N. (1992). The clinical implications of acquired bacterial resistance. *Journal of Medical Microbiology*, **36**, 23–5.

Hall, R.M., Brookes, D.E. & Stokes, H.W. (1991). Site-specific insertion of genes into integrons: role of the 59-base element and determination of the recombination cross-over point. *Molecular Microbiology*, 5, 1941–59.

Hedges, R.W. & Jacob, A.E. (1974). Transposition of ampicillin resistance from RP4 to other replicons. *Molecular and General Genetics*, 132, 31–40.

Heffron, F., McCarthy, B.J., Ohtsubo, H. & Ohtsubo, E. (1979). DNA sequence analysis of the transposon Tn3: three genes and three sites involved in transposition of Tn3. *Cell*, 18, 1153–64.

Heffron, F., Sublett, R., Hedges, R.W. Jacob, A. & Falkow, S. (1975). Origin of the TEM β-lactamase gene found on plasmids. *Journal of Bacteriology*, 122, 250–6.

Iida, S. & Arber, W. (1980). On the role of IS1 in the formation of hybrids between the bacteriophage P1 and the R plasmid NR1. *Molecular and General Genetics*, 177, 261–70.

Ippen-Ihler, K. (1989). Bacterial conjugation. In *Gene Transfer in the Environment*, Levy, S.B. & Miller, R.V., eds, pp. 33–72. McGraw Hill, New York.

Ippen-Ihler, K. & Skurray, R.A. (1993). Genetic organization of transfer-related determinants on the sex factor F and related plasmids. In *Bacterial Conjugation*, Clewell, D.B., ed, pp. 23–52. Plenum Press, New York.

Jacoby, G.A. & Medeiros, A.A. (1991). More extended-spectrum β-lactamases. *Antimicrobial Agents and Chemotherapy*, 35, 1697–704.

Kleckner, N. (1989). Transposon Tn10. In *Mobile DNA*, Berg, D.E. & Howe, M.M., eds, pp. 227–268. American Society for Microbiology, Washington, DC.

Kliebe, C., Nies, B.A., Tolxdorff-Neutzling, R.M., Meyer, J.F. & Wiedemann, B. (1985). Evolution of plasmid-coded resistance to broad-spectrum cephalosporins. *Antimicrobial Agents and Chemotherapy*, 28, 302–7.

Kokjohn, T.A. (1989). Transduction: Mechanism and potential for gene transfer in the environment. In *Gene Transfer in the Environment*, Levy, S.B. & Miller, R.V., eds, pp. 73–97. McGraw Hill, New York.

Lacey, R. (1975). Antibiotic resistance plasmids of *Staphylococcus aureus* and their clinical importance. *Bacteriological Reviews*, 39, 1–32.

Leclercq, R. & Courvalin, P. (1991). Bacterial resistance to macrolide, lincosamide, and streptogramin antibiotics by target modification. *Antimicrobial Agents and Chemotherapy*, 35, 1267–72.

Leelaporn, A., Paulsen, I.T., Tennent, J.M., Littlejohn, T.G. & Skurray, R.A. (1994). Multidrug resistance to antiseptics and disinfectants in coagulase-negative staphylococci. *Journal of Medical Microbiology*, 40, 214–20.

Levy, S.B. & Novick, R.P. (Eds.) (1986). *Banbury Report 24: Antibiotic Resistance Genes: Ecology, Transfer, and Expression*. Cold Spring Harbor Laboratory.

Lyon, B.R. & Skurray, R. (1987). Antimicrobial resistance of *Staphylococcus aureus*: genetic basis. *Microbiological Reviews*, 51, 88–134.

MacGowan, A.P., Brown, N.M., Holt, H.A., Lovering, A.M., McCulloch, S.Y. & Reeves, D.S. (1993). An eight year survey of the antimicrobial susceptibility patterns of 85,971 bacteria isolated from patients in a district general hospital and the local community. *Journal of Antimicrobial Chemotherapy*, 31, 543–57.

Macrina, F.L. & Archer, G.L. (1993). Conjugation and broad host range plasmids in *Streptococci* and *Staphylococci*. In *Bacterial Conjugation*, Clewell, D.B., ed, pp. 313–329. Plenum Press, New York.

Morrison, W.D., Miller, R.V. & Sayler, G.S. (1978). Frequency of F116 mediated transduction of *Pseudomonas aeruginosa* in a freshwater environment. *Applied and Environmental Microbiology*, 36, 724–30.

Murphy, E. (1989). Transposable elements in gram-positive bacteria. In *Mobile DNA*, Berg, D.E. & Howe, M.M., eds, pp. 269–288. American Society for Microbiology, Washington, DC.

Murray, B.E. (1992). β-lactamase-producing enterococci. *Antimicrobial Agents and Chemotherapy*, **36**, 2355–9.

Murray, B.E., Lopardo, H.A., Rubeglio, E.A., Frosolono, M. & Singh, K.V. (1992). Intrahospital spread of a single gentamycin-resistant, β-lactamase-producing strain of *Enterococcus faecalis* in Argentina. *Antimicrobial Agents and Chemotherapy*, **36**, 230–2.

Murray, B.E., Singh, K.V., Markowitz, S.M., Lopardo, H.A., Patterson, J.E., Zervos, M.J., Rubeglio, E., Eliopoulos, G.M., Rice, L.B., Goldstein, F.W., Jenkins, S.G., Caputo, G.M., Nasnas, R., Moore, L.S., Wong, E.S. & Weinstock, G. (1991). Evidence for clonal spread of a single strain of β-lactamase-producing *Enterococcus* (*Streptococcus*) *faecalis* to six hospitals in five states. *The Journal of Infectious Diseases*, **163**, 780–5.

Neu, H.C. (1991). Bacterial resistance to other agents. In *Antibiotics in Laboratory Medicine*, 3rd edn., Lorian, V., ed, pp. 714–722. Williams & Wilkins, Baltimore.

Neu, H.C. (1985). Contribution of β-lactamases to bacterial resistance and mechanisms to inhibit β-lactamases. *American Journal of Medicine*, **79**, Suppl 5B, 2–12.

Nikaido, H. (1989). Outer membrane barrier as a mechanism of antimicrobial resistance. *Antimicrobial Agents and Chemotherapy*, **33**, 1831–6.

Nikaido, H. (1994). Prevention of drug access to bacterial targets: permeability barriers and active efflux. *Science*, **264**, 382–8.

Oka, A., Sugisaki, H. & Takanami, M. (1981). Nucleotide sequence of the kanamycin resistance transposon Tn*903*. *Journal of Molecular Biology*, **147**, 217–26.

Pato, M.L. (1989). Bacteriophage Mu. In *Mobile DNA*, Berg, D.E. & Howe, M.M., eds, pp. 23–52. American Society for Microbiology, Washington, DC.

Payne, D.J. (1993). Metallo-β-lactamases – a new therapeutic challenge. *Journal of Medical Microbiology*, **39**, 93–9.

Payne, D.J., Marriot, M.S. & Amyes, S.G.B. (1990). Characterisation of a unique ceftazidime-hydrolysing β-lactamase, TEM-E2. *Journal of Medical Microbiology*, **32**, 131–4.

Payne, D.J., Woodford, N. & Amyes, S.G.B. (1992). Characterization of the plasmid-mediated β-lactamase BIL-1. *Antimicrobial Agents and Chemotherapy*, **30**, 119–27.

Petit, A., Gerbaud, G., Sirot, D., Courvalin, P. & Sirot, J. (1990). Molecular epidemiology of TEM-3 (CTX-1) β-lactamase. *Antimicrobial Agents and Chemotherapy*, **34**, 219–24.

Quinn, J.P., Darzins, A., Miyashiro, D., Ripp, S. & Miller, R.V. (1991). Imipenem resistance in *Pseudomonas aeruginosa* PAO: mapping of the OprD2 gene. *Antimicrobial Agents and Chemotherapy*, **35**, 753–5.

Reeves, D.S. (1994). Trends in resistance to antibacterials. In *Rapid Methods and Automation in Microbiology and Immunology*, Spencer, R.C., Wright, E.P. & Newsom, S.W.B., eds, pp. 489–502. Intercept, Andover.

Reynolds, P.E. (1992). Glycopeptide resistance in gram-positive bacteria. *Journal of Medical Microbiology*, **36**, 14–17.

Rice, L.B., Willey, S.H. & Papanicolaou, G.A., Medeiros, A.A., Eliopoulos, G.M., Moellering, R.C. & Jacoby, G.A. (1990). Outbreak of ceftazidime resistance caused by extended-spectrum β-lactamase at a Massachusetts chronic-care facility. *Antimicrobial Agents and Chemotherapy*, **34**, 2193–9.

Richmond, M.H. & Sykes, R.B. (1973). The β-lactamases of gram-negative bacteria and their possible physiological role. *Advances in Microbial Physiology*, **9**, 31–88.

Rinkel, M., Hubert, J.-C., Roux, B. & Lett, M.-C. (1994). Identification of a new transposon Tn*5403* in a *Klebsiella pneumoniae* strain isolated from a polluted aquatic environment. *Current Microbiology*, in press.

Roux, B., Lebaron, P., Hubert, J.-C. & Lett, M.-C. (1993). Occurrence of transposable elements in aquatic bacterial strains: involvement in the mobilization of pBR-type plasmids. *Microbial Releases*, 1, 223–8.

Salyers, A.A., Speer, B.S. & Shoemaker, N.B. (1990). New perspectives in tetracycline resistance. *Molecular Microbiology*, 4, 151–6.

Sambrook, J., Fritsch, E.F. & Maniatis, T. (1989). *Molecular Cloning: A Laboratory Manual*, 2nd edn., pl. 75. Cold Spring Harbor Laboratory Press, Cold Spring Harbor.

Sanders, C.C. (1987). Chromosomal cephalosporinases responsible for multiple resistance to newer β-lactam antibiotics. *Annual Reviews of Microbiology*, 41, 573–93.

Sanders, C.C. (1989). Bacterial proteins involved in antimicrobial drug resistance. In *Perspectives in Anti-infective Therapy: Current Topics in Infectious Diseases and Clinical Microbiology*, Jackson, G.G., Schlumberger, H.D. & Zeiler, H.J., eds, pp. 115–121. Friedrich Vieiveg and Sohn Braunschweig, Wiesbaden.

Sanders, C.C. & Sanders Jr, W.E. (1987). Clinical importance of inducible β-lactamases in gram-negative bacteria. *European Journal of Clinical Microbiology*, 6, 435–7.

Saunders, J.R. (1984). Genetics and evolution of antibiotic resistance. *British Medical Bulletin*, 40, 54–60.

Saunders, J.R. & Saunders, V.A. (1988). Bacterial transformation with plasmid DNA. *Methods in Microbiology*, 21, 79–128.

Saye, D.J., Ogunseitan, O., Sayler, G.S. & Miller, R.V. (1987). Potential for transduction of plasmids in a natural freshwater environment: effect of plasmid donor concentration and a natural microbial community on transduction in *Pseudomonas aeruginosa*. *Applied and Environmental Microbiology*, 53, 987–95.

Saye, D.J. & Miller, R.V. (1989). The aquatic environment: consideration of horizontal gene transmission in a diversified habitat. In *Gene Transfer in the Environment*, Levy, S.B. & Miller, R. V., eds, pp. 223–259. McGraw Hill, New York.

Shaw, K.J., Rather, P.N., Hare, R.S. & Miller, G.H. (1993). Molecular genetics of aminoglycoside resistance genes and familial relationships of the aminoglycoside-modifying enzymes. *Microbiological Reviews*, 57, 138–63.

Shaw, W.V. & Leslie, A.G.W. (1989). Chloramphenicol acetyltransferases. In *Microbial Resistance to Drugs*, Bryan, L.E., ed, pp. 313–324. Springer-Verlag KG, Berlin.

Sheridan, R.P. & Chopra, I. (1991). Origin of tetracycline efflux proteins: conclusions from nucleotide sequence analysis. *Molecular Microbiology*, 5, 895–900.

Sherratt, D. (1989). Tn*3* and related transposable elements: site-specific recombination and transposition. In *Mobile DNA*, Berg, D.E. & Howe, M.M., eds, pp. 163–184. American Society for Microbiology, Washington, DC.

Song, M.D., Wachi, M., Doi, M., Ishino, F. & Matsuhashi, M. (1987). Evolution of an inducible penicillin-target protein in methicillin-resistant *Staphylococcus aureus* by gene fusion. *FEBS Letters*, 221, 167–71.

Spratt, B.G. (1994). Resistance to antibiotics mediated by target alterations. *Science*, 264, 388–93.

Starlinger, P. (1980). IS elements and transposons. *Plasmid*, 3, 241–59.

Stewart, G.J. (1989). The mechanism of natural transformation. In *Gene Transfer in the Environment*, Levy, S.B. & Miller, R.V., eds, pp. 139–164. McGraw Hill, New York.

Stokes, H.W. & Hall, R.M. (1989). A novel family of potentially mobile DNA elements encoding site-specific gene-integration functions: integrons. *Molecular Microbiology*, **3**, 1669–83.

Thomas, C.M. (Ed). (1989). *Promiscuous Plasmids of Gram-negative Bacteria*. Academic Press, London.

Tzelepi, E., Tzouvelekis, L.S., Vatopoulos, A.C., Mentis, A.F., Tsakris, A. & Legakis, N.J. (1992). High prevalence of stably derepressed class-1 β-lactamase expression in multiresistant clinical isolates of Enterobacter cloacae from Greek hospitals. *Journal of Medical Microbiology*, **37**, 91–5.

Valentine, C.R.I. & Kado, C.I. (1989). Molecular genetics of IncW plasmids. In *Promiscuous Plasmids of Gram-Negative Bacteria*, Thomas, C.M., ed, pp. 125–163. Academic Press Limited, London.

Walsh, T.R., Payne, D.J., McGowan, A.P. & Bennett, P.M. (1994). A clinical isolate of *A. sobria*, strain 163a, possesses three chromosomally mediated, inducible β-lactamases: a cephalosporinase, a penicillinase and a metallo-β-lactamase. *Journal of Antimicrobial Chemotherapy*, in press.

Wanger, A.R. & Murray, B.E. (1990). Comparison of enterococcal and staphylococcal β-lactamase plasmids. *The Journal of Infectious Diseases*, **161**, 54–8.

Watanabe, T. (1963). Infective heredity of multiple drug resistance in bacteria. *Bacteriological Reviews*, **27**, 87–115.

Webb, V. & Davies, J. (1993). Antibiotic preparations contain DNA – a source of drug resistance genes? *Antimicrobial Agents and Chemotherapy*, **37**, 2379–84.

Willetts, N. (1988). Conjugation. *Methods in Microbiology*, **21**, 49–77.

Witte, W. & Hummel, R. (1986). Antibiotic resistance in *Staphylococcus aureus* isolated from man and animals. In *Banbury Report 24: Antibiotic Resistance Genes: Ecology, Transfer and Expression*, Levy, S.B. & Novick, R.P., eds, pp. 95–103. Cold Spring Harbor Laboratory.

Young, H-K. (1993). Antimicrobial resistance spread in aquatic environments. *Journal of Antimicrobial Chemotherapy*, **31**, 627–35.

Zuhlsdorf, M.T. & Wiedemann, B. (1992). TN21-specific structures in Gram-negative bacteria from clinical isolates. *Antimicrobial Agents and Chemotherapy*, **36**, 1915–21.

INDEX